Star Laws
Foundations for Convergence Buoyancy

Book I: Physics
&
Book II: Metaphysics

by

Ross Lee Graham, Ph.D.

Interrogation directed to the discovery of hidden presuppositions helps us sustain the attitude that a system of analysis is a tool for us to use, to improve, to revise, and even to replace.
                    ---the author

Time-Module Books
Braşov, Transylvania

Star Laws: Foundations for Convergence Buoyancy
Book I and Book II in one volume

All Rights Reserved
© 2009 and updates © 2011 by Ross Lee Graham, PhD

No part of this book may be reproduced or transmitted in any form or by any means, graphic, electronic, or mechanical, including photocopying, recording, taping, or by any information storage retrieval system, without the permission in writing from the copyright holder.

Softbound: ISBN **978-0-558-00229-3**
Hardbound: ISBN **978-0-557-11154-1**

Time-Module Books
Published in Braşov, Transylvania

Printed in the USA by lulu.com
Distributed by Barnes & Noble

to
My first Science Mentor

Dr. Ralph Baxter Larkin
of
Claremont, California

Acknowledgments

Deep thanks to:

Encyclopedia Britannica, Inc., for the quotations cited herein from their 1952 edition of the *Great Books of the Western World*.

Princeton University Press, for their open cooperation with scholars for quoting from their publications.

...and to the many authors of the influential books listed in the Bibliography.

Star Laws

An Introduction to Convergence Buoyancy

Book I
A Treatise on Physics

by

Ross Lee Graham, Ph.D.

Doubtless all exclusive pursuits tend to produce partial views, and it may be, that a mind long and deeply immersed in the contemplation of scenes over which the dominion of a physical necessity is unquestioned and supreme, may admit with difficulty the possibility of another order of things.

--- George Boole, *Laws of Thought*, p441

Time-Module Books
Brașov, Transylvania

Star Laws: An Introduction to Convergence Buoyancy

All Rights Reserved © 2004, 2009 by Ross Lee Graham, PhD

No part of this book may be reproduced or transmitted in any form or by any means, graphic, electronic, or mechanical, including photocopying, recording, taping, or by any information storage retrieval system, without the permission in writing from the copyright holder.

Time-Module Books
Published in Brașov, Transylvania

Printed in the USA by lulu.com

Two Volumes in One ISBN: on page 2

Distributed by Barnes & Noble

TABLE OF CONTENTS

BOOK I: Convergence Buoyancy Theory

Apology 9
Prologue: Book Outline 13

PART ONE: Science-Systems Analysis

Chapter 1: Convergence and the Illusion of Free-Fall 17

PART TWO: Science-Systems Design

Chapter 2: CB Inversion Spheres and the Illusion of Mass ... 79

Chapter 3: Impetus Weight and the Illusion of Inertia 145

Chapter 4: Radiation and Heat 197

Chapter 5: The Third Perpendicular 235

PART THREE: Science-Systems Implementation

Chapter 6: A Different Interface for Our Universe 291

Chapter 7: Dialogue 347

Epilogue 369

Bibliography (in Book II: *Star Laws Addenda: Duality Theory*)

...The unexamined life is not worth living.
--- Socrates

TABLE OF CONTENTS

BOOK II: Duality Theory 371

    Abstract 375

    Foreword 377

ADDENDA: Science-Systems Analysis

    Addendum 1: Science-Systems 385

    Addendum 2: Status of Logic in Science-Systems 419

    Addendum 3: Formal Duality in Finite Logics 449

    Addendum 4: Data-Field Duality 483

    Addendum 5: Reference-Frame Duality 507

    Addendum 6: Ptolemy, Copernicus, Kepler 543

    Addendum 7: Fermat's Last Theorem 569

Appendix: Number References for Calculations 587
Bibliography 589

    ...everything that comes into being comes into being from its contrary...
                                         --- Aristotle, *On the Heavens*

APOLOGY

Star Laws: an Introduction to Convergence Buoyancy is written with an intended rigor, but in the spirit of having a wider audience to include students, teachers, historians, philosophers, and scientists that have a special interest in the foundations of physical science. The subject matter centers on the foundations of gravitation theory and its associated fields of study.

Mathematical Space vs Physical Fields. Mathematical physicists that use mathematics in a manner that is functionally independent of the physics like to speak of 'curved space', a mathematical concept, rather than 'curved field', a physical concept. They usually do not admit the need of a Euclidean space to render the curved fields intelligible. For me it is more intelligent to presume the field matrix for conceptualization is a Euclidean space in which we may formulate many field shapes. It is the field shapes rather than space that influence the shapes and movements of our existence. In this text 'space' is used in two senses, one is mathematical and herein restricted to Euclidean space that underpins the conceptual matrix for various shaped fields, the other is to reference interplanetary or interstellar space with the same vagueness as normal usage.

Parametric Variables. There are many examples for the use of parametric variables in determining physical magnitudes. In Physics, one use for reference-frames is for positioning (locating) points and from this to determine physical distances and shapes. Thus a reference-frame dimension set provides the parametric variables, e.g., (x, y, z, t), that can define points that may be used for determining physical distances and shapes, etc. Once the determination is made there are many cases where the result can be regarded as functionally independent of the reference-frame used for that determination. The effectiveness of parametric variables is well exhibited in the determination of surface gravity accelerations of celestial objects. It is a matter of simple arithmetic to estimate the surface gravity acceleration of the Moon once you have a few facts about its mean orbit. It is also a matter of arithmetic to estimate its relative 'mass'. In effect, I argue in the main text that the reason Newton's system gives such good estimates for relative surface gravities is that his use of mass and centripetal force are parametric and the results determined by his use of parametric variables may be regarded as functionally independent of the mode of determination. In a sense they are like catalysts that can be used for doing a job, getting a result, and remaining independent of the result. It is also true that these Newtonian estimates have limited accuracy and remain impressive only so long as greater accuracy is not required.

Esoteric vs Exoteric Presentations. Many scientists today have come to recognize the esoteric aspect of their work and how it renders their work invisible to an intelligent public. The concerned response has been an increase in the number of more exoteric presentations, which though expressed with a rigor requiring intelligence, do not require the usual years of extreme specialized initiation.

The presentation of a new reference-frame begins with the qualitative distinctions required for its definition. As with any new interface for observations, some adjustments to usage and some new vocabulary are helpful for the presentation. In the main text, any required specialized terms are defined in the context where they first appear.

The advanced mathematical model for the gravitational field structure required for Convergence Buoyancy is not presented in this volume. Most of the arguments herein are qualitative and only require a good sense of logic. For the most part, when formulas are used for apodeictic reductions, the only requisite is a common knowledge of algebra and some geometry.

Galilean Attitude. Galileo was among the first to offer scientific discussion couched in terms that a non-specialized intelligent audience could follow. Though Galileo wrote his mathematical proofs in the Latin customary for intellectuals in his time, he wrote the discussions in his Italian vernacular. Galileo knew that in general the incumbent holders of the intellectual catechism of his time were hostile to his work. By writing in Italian he opened his work to a possible wider local support. As it turned out he spent his latter days in house arrest condemned by the Pope. It was during this period that he completed his most influential work, *Dialogues Concerning Two New Sciences*.

This book has already taken too long to get to press and yet the editing still seems unfinished. The author strives for clarity and appreciates any suggestions for improving the presentation.

Ross Lee Graham
Mid Sweden University
Sundsvall, Sweden
Email: drrosslg@gmail.com

The Ides of March 2004

PROLOGUE
Book Outline

Most of the effort in this work focuses on qualitative analysis. The quantitative aspects depend on this qualitative framework developed for the intelligibility of Convergence Buoyancy (CB).

Chapter 1 contains a critique of Galileo and Newton in their analyses of 'free-fall'. Defects in their arguments are pointed out and the discussion ends with a strong rejection of their respective interpretations. 'Free-fall' is a pseudo-concept that yields engineer approximations. Boyle's Tube, sometimes miscalled Newton's Tube, fails in its purpose. A dual use of Newton's Law of Gravitation leads to the denial of the Equivalence Principle in his own terms. This discussion is followed by the foundational definitions for *Convergence Buoyancy*.

Chapters 2-4 continue the development of CB as an anti-Newtonian system. These chapters contain convergence buoyancy interpretations and implementations for laws that correspond to Newton's system. We are then left with the task of determining the coherence and applicability of the anti-system, i.e., we are ready to test it.

Chapter 5 introduces speculations about possible convergence field interpretations on the microparticle level, a physics theory interpretation. In particular, polarity for a third perpendicular in the line of light propagation is suggested for synthesis with the electromagnetic field. Electricity and magnetism form perpendicular polarities about each other. The photon-graviton polarity is suggested for the third perpendicular. With respect to material bodies this polarity propagates to and from the body center. The microparticle analysis also leads to new considerations for the interpretation of temperature. This leads to a novel critique for entropy.

Chapter 6 is a general overview of the implications of CB in describing the Universe from our locale. In effect, it represents a foundation statement describing a different design for a systemic interface to our Universe. However, galactic dynamics are not discussed in any detail. This Chapter is highly speculative in the old sense of the term ('something to look at').

Chapter 7 is a dialogue with a variety of questions that suggest possible interpretations that favor the Convergence Buoyancy reference-frame. Circumstantial evidence, critical experiments, and previously unexplained phenomena are brought under the scope of this theory.

Addenda are placed in a second volume entitled *Star Laws Addenda: Duality Theory*. Each addendum focuses on aspects that underpin Convergence Buoyancy theory or its duality with the Newtonian system.

I regard this work as a compendious logical outline for Convergence Buoyancy reference-frames. Even so, some future mind who can read this work with more detachment than I can may find a better way to communicate what I wanted to say.

End Prologue

PART ONE

Science-Systems Analysis

Chapter 1: CONVERGENCE and the ILLUSION OF FREE-FALL

INTRODUCTION. *The notion of **convergence** is conceived as having two modes that direct movement toward a center on different principles: Convergence Attraction (CA), Convergence Buoyancy (CB). Here the notion of **attraction** is restricted to use only for indicating the interactions of opposite polarities, such as {N, S} for magnetism and {+, -} for electric charges, etc. A simplistic bifurcation into poles such as {+, -} and its neutral form {0} may be misleading. A strong positive charge relating to a weak positive charge can appear to be a relation between a positive and a negative charge, etc., as Maxwell pointed out. The underpinning logic here leads to a new basis for convergence relationships. The CB model uses the concept of **buoyancy** as a kind of measure for this convergence of like polarities.*

For Sir Isaac Newton (1642—1727) gravitation does not fit in CA or CB. It leaves out the possibility of convergence buoyancy therefore is not CB; his 'attraction' is a monopole and therefore is not CA.

Free-Fall. A careful search for an examination of the genesis of ideas can lead us back to arbitrary or even wrong turns and may help render other possible turns visible to us. This new visibility can put vim into the search for and design of new intelligibility frameworks. This in turn may enhance or may challenge current beliefs.

Galileo's conclusion for dropping two objects of different weights, barring the influence of air resistance, is that they fall to the Earth at the same rate of acceleration. Boyle's tube, designed and implemented to demonstrate Galileo's conclusion, contained a piece of solid gold and a bit of down in a long vacuum tube. Tip the tube to vertical and the two objects appear to 'free-fall' at the same rate. In the Apollo 15 mission (1971), the view on the Moon of Commander David R. Scott dropping a geologic hammer and a falcon feather that strike the ground at the "same" time was also intended to demonstrate this conclusion (unauthorized by NASA). Galileo's thought experiment was designed to dethrone Aristotle's view that heavier objects fall proportionately faster than lighter objects. Galileo's conclusion is consistent with his own presuppositions, but this can be shown as based on a false assumption and that it is inconsistent with the presuppositions for the Newtonian formula, **W = mg**.

Galileo Experiments. Due to the rapidity of falling objects in small-scale observations, Galileo had few choices that could yield significantly accurate results in his day. He could experiment on a scale such as dropping objects from 'high' structures. From a distance, objects falling from high places look as if they are moving in slow motion. If Aristotle had observed two stones falling off a cliff (a scene readily available to Aristotle), the larger heavier one might clearly have been seen falling at a significantly faster rate than a smaller lighter one. Under certain conditions, falling snow reveals the same 'confirmation'. The two difficulties, which were first solved by Galileo, are how to eliminate the effects of the air resistance on *terminal speed* (when air speed and any resistance are in equilibrium, the measured speed appears constant) and how to render the 'fall' rates easier to measure. In current belief this air resistance terminal speed is a constant for a given air pressure and for a given spherical bulk (quantity of matter). In air; 'heavy' augments terminal speed, 'bigger' diminishes terminal speed, (shape also affects results) and terminal speed is inversely proportional to air pressure. Galileo chose a more remarkable approach in terms of mathematical and practical ingenuity, namely, rolling balls down an inclined plane. This smaller scale and slower speed reduce the effect of air resistance and renders the speed changes (acceleration) easier to observe. Then he reasons by limits to eliminate the effects of air resistance altogether. (Today, more exact measurements require accounting for the rolling that also reduces the 'falling' speed.)

Aristotle could not solve this problem because his reference-frame for judgment created an obstacle to the consideration. He believed all objects that fall or move must do so in a medium and Galileo was trying to find out what the descent acceleration would be without a medium.

Galileo Method. For Galileo the task is restricted to finding the properties of naturally accelerated motion, deducing the mathematical consequences, and testing by reputable experiments. There is no causal nexus applied to weight in this argument or in this method. Within the accuracy limits of his experiments the descent acceleration appears functionally independent of weight.

Put in another way, this precludes the dual analysis that considers the weight of the Earth with respect to the object. Just as the object has weight relative to the Earth, the Earth has weight relative to the object. The objects gravitating to the Earth are presumed by Galileo not to contribute an influence to the total gravitational influence.

Thought Experiment: Let us keep tying more weights to the weight in 'free-fall' until we have a body the size of the Moon. According to Galileo's argument structure, this Moon-size object would still fall toward the Earth at the same uniform acceleration as any other 'falling' body, and he would not say that the Earth also 'falls' toward this object. Today our basis for judgment has changed and we can regard 'free-fall' as a one-sided (*ex parte*) view of an interaction. Though a one-sided ***view*** of an interaction may exist, a one-sided interaction is self-contradictory and does not exist. The Earth also 'falls' toward the Moon.

AN EQUIVALENCE PRINCIPLE CONTRADICTION

Introduction. *This section addresses the problem of 'falling' objects as an interaction. Newton's Second Law of motion is used in a new way to criticize Galileo's ingenious thought experiment on how all falling objects fall at the same acceleration, regardless of weight. I show that Newton's Law and Galileo's thought experiment are incompatible; they cannot both be true. The rationale for Newtonian gravitational attraction supports the concepts of inertial mass and gravitational mass. This support tacitly imposes an equivalence for these designations of mass so differently defined. Using a dual application of Newton's Universal Law of Gravitation leads to a denial of 'free-fall'. This in turn leads to the denial of the Equivalence Principle in Newton's own terms. At the foundation of the Einstein relativistic theory of gravitation, the Equivalence Principle is stated as a postulate. Therefore, if the argument holds, then the Einstein theory of gravitation is based on a false principle.*

> **Note:** One of the most mythic figures in 20th Century science, Albert Einstein, in letters to his first wife, Mileva, on relativity writes of it as "our theory."

When published he completely obliterates her role in founding this theory and presented it only as his own. Her name was removed from the final draft.

Galileo's 'Free Fall' Thought Experiment. Galileo's thought experiment governing 'free-fall' is found in his *Dialogues Concerning Two New Sciences*. This work is a set of dialogues cast with three persons. These three characters, Salviati, Sagredo, and Simplicio, meet and talk on four consecutive days. Their subject of conversation centers on the work of someone they refer to as the "Author" (context implies Galileo). It is around the middle of the First Day that Galileo puts into Salviati's role the recitation of this famous thought experiment for falling objects.

Let us consider this thought experiment:

"SALVIATI: But, even without further experiment, it is possible to prove clearly, by means of a short and conclusive argument, that a heavier body does not move more rapidly than a lighter one provided both objects are of the same material and in short such as those mentioned by Aristotle. But tell me, Simplicio, whether you admit that each falling body acquires a definite speed fixed by nature, a velocity which cannot be increased or diminished except by the use of force or violence.
SIMPLICIO: There can be no doubt but that one and the same body moving in a single medium has a fixed velocity which is determined by nature and which cannot be increased accept by the addition of momentum or diminished except by some resistance which retards it.
SALVIATI: If then we take two objects whose natural speeds are different, it is clear that on uniting the two, the more rapid one will be partly retarded by the slower, and the slower will be somewhat hastened by the swifter. Do

you not agree with me in this opinion?" (*Great Books of the Western World*, ed. 1952, vol. 28)

Comment: Notice here that no consideration is given to the influence of the gravitating bulk. The only consideration is speed in a medium. For Galileo, the gravitating quantity adds no effect to the falling. Galileo believes that all gravitating objects 'fall' to the Earth, this belief prevents him from considering the dual case that the Earth also falls toward these objects. We might reason that in laboratory scale experiments the Earth fall is minuscule, a tiny tidal effect. **For Galileo, the Earth does not fall toward other objects.**

"SIMPLICIO: You are unquestionably right.
SALVIATI: But if this is true, and if a larger stone moves with a speed of, say, eight while a smaller moves with a speed of four, then when they are united, the system will move with a speed less than eight; but the two stones when tied together make a stone larger than that which before moved with a speed of eight. Hence the heavier body moves with less speed than the lighter; an effect which is contrary to your supposition. Thus you see how, from your assumption that the heavier body moves more rapidly than the lighter one, I infer that the heavier body moves more slowly." (*op. cit.*)

Comment: He shows that, under the given restraints, Aristotle's assumption disallows Eudoxian weight. 'Eudoxian' indicates a species of magnitude that varies linearly in its measure. It is a well-defined characteristic and in current literature is often referenced as 'Archimedean' (see Addendum 1).

Consistent with the presuppositions of the argument, Salviati turns Aristotle's logic to contradiction and puts Simplicio in a state of confusion. Note that Galileo's speed measure seems to be a scalar quantity with no units designated as we do. There is no evidence in any of his work that he regarded speed as distance per unit time, nor did he ever use mixed unit ratios for any of his units (the magnitude dimensions of like quantities reduce to a dimensionless unity). He treats speed as if it had its own simple units of measure to which he gives no name other than 'speed' (see Addendum 5).

> "SIMPLICIO: I am all at sea because it appears to me that the smaller stone when added to the larger increases its weight and by adding weight I do not see how it can fail to increase its speed or, at least, not to diminish it." (op. cit.)

Comment: Salviati was waiting for this one. Simplicio presumes that adding the smaller stone to the larger actually increased the total weight in the fall.

> "SALVIATI: Here again you are in error, Simplicio, because it is not true that the smaller stone adds weight to the larger.
> SIMPLICIO: This is, indeed, quite beyond my comprehension.
> SALVIATI: It will not be beyond you when I have shown you the mistake under which you are laboring. Note that it is necessary to distinguish between heavy objects in motion and the same objects at rest. A large stone placed in a balance not only requires additional weight by having another stone placed upon it, but even by the addition of a handful of hemp its weight is augmented six to ten ounces according to the quantity of hemp. But if you tie the hemp to the stone and allow them to fall freely from

some height, do you believe that the hemp will press down upon the stone and thus accelerate its motion or do you think the motion will be retarded by a partial upward pressure? One always feels the pressure upon his shoulders when he prevents the motion of a load resting upon him; but if one descends just as rapidly as the load would fall how can it gravitate or press upon him? Do you not see that this would be the same as trying to strike a man with a lance when he is running away from you with a speed which is equal to, or even greater, than that which you are following him? You must therefore conclude that, during free or natural fall, the small stone does not press upon the larger and consequently does not increase its weight as it does when at rest." (op. cit.)

Comment: In effect, in 'free-fall' an object is reasoned to be weightless, therefore adding bulk makes no change in the free-fall since weightless plus weightless equals weightless. Galileo presents an analysis that forms a powerful argument consistent with the presupposed conditions. The alternate case of placing the larger free-falling body on the smaller free-falling body is handled with similar rigor and Simplicio finally admits:

"SIMPLICIO: Your discussion is really admirable; yet I do not find it easy to believe that a birdshot falls as swiftly as a cannon ball." (op. cit.)

Counter-Argument: Salviati continues the argument but it is here that we can step in and argue for Simplicio to show that Salviati's praised reasoning may also be turned to contradiction. This turning is not simplistic. Within Galileo's universe of discourse his argument is consistent. This implies that we have to revise the conditions for judgment in order to turn the argument. To do so we have to show that Galileo has 'proven' his case by using an invalid assumption. I propose that there is such an assumption and that it is tacit and therefore hidden from Galileo's direct consideration. This hidden assumption is independent of the reasoning that concerns the weight of objects in 'free-fall'. The sense of this hidden assumption is difficult to convey. The difficulty arises because there are two distinct modes of adding weight that need be considered. Galileo cannot consider both of them because he does not use a reference-frame that allows direct access to the second mode. In fact, his point of departure creates an obstacle that disallows considering the second mode. Here these modes are treated as distinct cases.

Case One: One accredited conclusion for which Galileo's thought experiment is useful is expressed in current terms as: We never increase the gravitational intensity of the Earth by increasing the material quantity of the falling object (until it lands). This is equivalent to saying that increasing the material quantity of a body falling to Earth does not increase the gravitational acceleration due to the Earth that is acting on this object. At first glance this seems to be Galileo's line of argument. However, this is not what Galileo was saying. Galileo states that he does not want to seek causes of this falling. He considers all causal explanations of it as "fantasies" and their examination "not worthwhile". He observes that weighted objects fall to the Earth. There is weight, but no attempt is made to find its cause. He never speaks of its causes and so he never speaks of free-fall acceleration *due* to the Earth or any other body.

In 'free-fall', objects are weightless. This leads to the Galilean conclusion that, barring medium resistance, objects of different weights cannot free-fall at different rates. Note again that this conclusion is consistent with Galileo's assumptions, but one of these assumptions needs further examination. Galileo would deny that the bulk of the falling object has its own gravitational intensity and that its own gravitational intensity participates in the total gravitational intensity. The tacit supposition for Galileo is that falling objects do not participate in the generation of gravitational intensity, do not participate in the determination of the total falling rate. In effect he implies that the object in 'free-fall' does not attract the Earth.

Note that in *Gravitation and Inertia* by Ciufolini and Wheeler, they refer to this case of "equal" falling rates as the Galilei equivalence principle and say that it is at the base of most viable theories of gravitation (p91). This is the current mainstream belief. As long as we never investigate the conceptual presuppositions involved, this belief can be sustained. The following arguments undermine this belief.

Case Two: This is the case that was not considered by Galileo. It answers the question: Does the Earth fall toward a body that is falling toward the Earth? We could pose the question this way: Does a falling body participate in the gravitational intensity of its falling? Going further, if we put this question in terms closer to the way Galileo talks, then it becomes apparent why he did not pose the question. For him the quantity of matter in a body was represented by its weight. He might have asked: Does the weight of the falling body influence the falling rate? But this is precisely the question that Galileo addresses. For Galileo this is the question that is answered in the thought experiment, in the negative. This follows from his reference-frame obstacle that was mentioned above.

Today we know that the Moon has its own gravitational intensity. We correlate this intensity to its size and compactness (concentration). It seems reasonable to presume that the Moon's gravitational intensity influences the falling rate we measure between the Moon and the Earth. The question opened here is how to characterize this influence. The description of the Earth-Moon relation has the two objects revolving around each other about a common center. The ratio of Earth material to Moon material is about 80:1 in Newtonian calculation. If we treated the gravitational connection between the Earth and Moon as a lever, then dividing the mean distance from the centers (384,403 km) by 81 gives 4,745.72 km. The lever principle (wrongly used) puts the fulcrum of the Earth-Moon movement at about 4,745.72 km from the Earth center. That is 1,661.28 km below the surface of the Earth. For the gravitational line between the Earth and Moon to be used as a lever in this manner assumes that the measure along the line is everywhere alike (Eudoxian). This is not true.

Assuming the Inverse Square Law, for the dynamic analysis we have two inverse square functions in opposition to each other. Granting these functions and assuming there is no physical fading or other changes than that deriving from geometric spread over distance, we can estimate the Earth-Moon Convergence Reversal Node (CRN) at about 37,638.3 km from the Moon center. Both the Moon and the Earth have an equal and opposed convergence (reversal of gravitational effects) at this node. Note that the Convergence Reversal Node (CRN) and the 'fulcrum' are not in the same location for the Earth-Moon relation. The fulcrum about which they revolve has a mean distance of about 31,231.3 km (37,638.3 minus Earth radius) above the Earth surface. This fulcrum distance between the Earth and Moon relative to the Sun continually varies with their revolution about each other, and in a manner that dampens the varieties of Earth distances from the Sun. Even so, despite inaccuracies, the secret of almanac long-term weather predictions is the 19-year Moon cycle (more later).

If, as Newton assumes, gravitational intensity increases directly with the quantity of matter, then it follows that increasing this quantity of matter increases the gravitational intensity we find associated with it. This argument is further developed in the following section.

USING NEWTON FORMULAS. Newton agreed with Galileo's argument. The Robert Boyle (1627-1691) vacuum tube experiment described by Newton could not contradict the conclusion. Newton proposed that it is true that all falling objects fall at the same acceleration but that heavier objects fall with more force. We can implement a direct measure of this force when we try to stop the falling. The Newton formula for this is derived from his Second Law of motion; **F** (force) is directly proportional to **m** (mass) times **a** (acceleration). The derived weight formula is expressed as: Weight (force) = (mass) x (acceleration due to gravity), or **W = mg**.

The unit for this mode of weight measure is named the 'newton' and abbreviated 'N'. The use of mass in this formula is a use invented by Newton. It has great convenience in his system of analysis, though it does not seem entirely consistent with "Hypothesis non fingo." It appears to be a handy mathematical device for quantifying matter independently of any particular value for **g**.

It is popular in many of the industrial nations to use kilograms as weight, though Newtonian physicists currently use 'kilograms' for mass and 'newtons' for weight. Mass is a parametric variable and its scale is defined by convention for the reference-frame that supports it.

The mass scale used by Newton falls out of his use of Earth-bound weight measures as the basis for mass units. This generates a misleading scale, as shown below. According to current practice, when a person weighs 90 kilograms (on-Earth) this person is using the Newton mass as weight (rather than the weight units now named 'newtons'). This same person weighs in at 15 kilograms on the Moon, and this weight does not represent the Newton mass. According to Newton the mass remains unchanged. If one has a mass of 90 kilograms on Earth, one has a mass of 90 kilograms on the Moon as well, even though one weighs less there.

Currently, Earth **g** is given a Standard Station value of 9.80665 m/s^2, where **m** is in meters and **s** is in seconds. The **g** value is minimum at the equator and maximum at the poles (differs by c0.5 percent). Therefore one weighs less at the equator than at the poles. On Earth, **g** changes every day. Consistent with Newton, it would increase as Earth material increases through daily accretion of space debris. **g** also varies according to latitude and altitude (inverse square, etc.). Also the Sun and Moon both have measurable influences on **g**, which is to say that the Earth neighborhood influences the effects of **g** (tidal effects). In elementary physics problems, **g** is treated as a constant, which it is not.

Closer to observations, it seems that the effective surface **g** measure is a function of what the bulk (quantity of matter), **b**, appears to be, and restraints from the neighborhood (tidal effects), **n**, at the location, **L**, where **g** is measured (location defined on a particular celestial body as latitude, longitude, altitude, and time/date):

$$g = f(b, n)@L$$

I say "what the bulk appears to be" because we have no direct means of measuring bulk as quantity of matter. This is another way of saying that we have no proof that gravity is proportional to the mass of material in the direct manner that Newton presumed.

Precise Russian measurements made at sea level on sufficiently small objects show that:

> North Pole f(g) ≈ 9.83210 m/s^2
> Standard Station:
>   45 degrees f(g) ≈ 9.80620 m/s^2
>   Equator f(g) ≈ 9.78038 m/s^2

The accuracy of any **g** number is put into question by the variation influence of the neighborhood tidal effects. Geometrically, the Sun **g** influences the fourth decimal place; the Moon **g** influences the fifth decimal place, etc. So there seems some affectation in exactitude presenting **g** beyond the second decimal place, unless accompanied by an analysis of the tidal effects at the time/date and place of the measure.

Engineer Calculations. Admitting the problems of accuracy, using the **g** value appropriate to location, we can calculate the mass of a range of known weights on the Earth. One of the determinants of the range is that the weight must not be too large. This range restraint is the same problem that is found hidden in Galileo's analysis. The greater the weight, the greater is its participation in the measurement of **g**. How to characterize this participation becomes paramount to understanding gravitational relations. But, in a sense, this is where our Lilliputian laboratory experiments can have an advantage for engineering estimates. For engineering uses we can assume that for these small scale experimental results that the gravitational influence of these small objects, in terms of percent error, is virtually zero. Multiplying results many times over (as if Eudoxian) still does not bring the enlarging error into the measurable range (shown below).

When the weights are large enough for measurable gravitational participation, then **g** becomes a perceptible mixture problem. This relates to why Kepler's Third Law shows significant errors for the larger planets. The CB account for why they are positioned further out than his Third Law predicts is presented in a later context. That is why I say that **g** as a 'constant' is *about equal* to the values indicated above for a range of weights.

There is a further possible source of error for Newtonian calculations of **g** for various celestial objects. The Newtonian inverse square calculations require that the orbits be stable. For a celestial object that is not yet in a stable orbit, its surface g-power cannot readily be determined by Newton's system.

Quantification of Newtonian Mass. In some respects, in Newton, the quantification of mass is like the quantification of planetary distances in Copernicus. For Copernicus the ratios of planetary distances were approximate, and they are fairly consistent with current values, but they are only ratios. It was not until we could determine the distance from the Earth to the Sun that we could assign distances to these ratios.

Some physicists believe that by using mass values assigned from 'on-Earth' values that we have an extensional magnitude for mass that will satisfy the proposed mass ratios. In the case of mass for these planets we have Newton determined ratios, but the arbitrariness of the current on-Earth values is not fully recognized. There are two sorts of arbitrariness, the easy one is in regard to the chosen units such as pounds or kilograms, and the other is in regard to how we determine the characteristic magnitude that we want to measure. With respect to chosen units the problems are related to standards and translations between them. With respect to the characteristic magnitude the problem is more difficult to describe. With planetary distances it was a matter of finding an 'actual' distance of one case that can be used for defining all the distances. 'Distance' is readily granted the status of an extensional magnitude. It is not so easy to grant extensionality to the current modes for determining Newtonian mass because Newton's formulas only lead us to an *intensional* quantification such as **m = F/a** or **m = a(r+H)^2/G** (the inertial and gravitational mass formulas, respectively).

Newton's contextual use of the quantification of mass in the inertial formula, **F=ma**, from which we derive **W=mg**, etc., is a peculiarity of his system and has an implicit circularity. Since we cannot know mass unless we can know acceleration-due-to-gravity, and we cannot know acceleration-due-to-gravity unless we know the gravitational influence due to the mass, we get to swing in full circular definition, a closed loop. This is another way of saying that the formula presents no direct means for calculating mass. This is the source of a major complaint from Ernst Mach (1838-1916) concerning Newton's mass concept, and this is why I said "*appears* to be a handy mathematical device," etc.) To date, no one has found a clear way out of the circle because the problem tends to be ignored. Furthermore, for Newton, the **g** in the formula **W=mg** has no variation due to **m** even though **g** must be a composite magnitude (see below).

Newton Free-Fall. By Newton's analysis, the relation between the Earth and the Moon is that of falling. The Moon falls toward the Earth. Assuming gravitation is a $1/r^2$ force; this falling is calculated with Hooke's suggested use of the inverse square law. It is Newton's originality to prove that the inverse square law works for conic sections, though he ignores the possibility of a physical fading or other changes than that derived from the geometric spread over distance. Using the radius of the Earth as 1, the distance to the Moon is about 60 of these radii. The inverse square law gives $1/60.27^2 = 1/3,632$. This gives us, relative to the Earth, that the Moon falls at 1/3,632 the rate of objects at the surface of the Earth. Calculating the Moon's falling acceleration we get 1/3,632 times 9.78 (Earth equatorial **g**) equals 0.002,679 ms^{-2}. This is the *dynamic acceleration* of the Moon toward the Earth. This shows that Earth **g** can affect the third decimal place of Moon **g**. Note that the Moon surface **g** is greater than the Earth effect at the Moon surface by a factor of about 605.5.

In like manner we can calculate the *dynamic acceleration* of the Earth toward the Moon. This time we set the radius of the Moon as 1. Then similarly we calculate that 1/48,951.88 times 1.622 (**g** at Moon surface). This equals 0.000,033,14 ms^{-2}, the dynamic acceleration of the Earth toward the Moon. This shows that the overhead Moon **g** affects the fifth decimal place of Earth **g**. Note that the Earth surface **g** is greater than the Moon effect on Earth **g** by a factor of a little over 289,949.6.

The ratio of these two dynamics accelerations, Earth to Moon, is about 80:1. This is also taken as the ratio of the respective Newtonian masses of the Earth and the Moon. For Newton the measurement of gravitational acceleration rates and mass are intertwined.

It is true that we can reason mathematically *ex parte* that all gravitational objects descending to the Earth are equally influenced at equal distances from the Earth. This is consistent with: *The gravitational intensity of the Earth is not in-*

creased by increasing the bulk of these gravitating objects. This is not the same thing as saying that objects of different weights that are placed above a fixed point on the Earth descend to the Earth under the influence of the local Earth **g** only. Every different weight that falls to the Earth from a given height and location, etc., falls to the Earth at a rate composed of the Earth **g** and the 'falling' object **g**, etc. We have yet to show the nature of this composition.

The attraction (at a given location) due to the Earth is equal for all gravitational objects ('masses') equidistant from the Earth. Gravitational objects of different 'masses' equidistant from the Earth at a given location are subject to this equal attraction from the Earth. In this sense we accredit Newton's phrase "descend with equal velocity," if we are only regarding the descent due to the Earth. However, there is also a descent of the Earth toward the gravitating objects. Small objects would contribute little to the attraction between them and the Earth. The Moon, however, is not a small object relative to the Earth. The Earth measurably falls toward the Moon. These two objects are 'falling' toward each other. Using a language consistent with Newton, the Moon falls toward the Earth at 0.002,692,7 meters per second per second, the Earth falls toward the Moon at 0.000,033,14 meters per second per second. Now what do we do? It seems that these two accelerations should relate to some composite 'falling' rate between the two objects. If gravitation is an attractive force, as supposed by Newton, then it seems logical that the acceleration should be an additive composite of the effects from both sources.

Universal Law of Gravitation. Newton gives a formula for the gravitational interaction of two masses; though one of my teachers, I. Bernard Cohen, notes that its derivation in the *Principia* is highly abbreviated:

$$F = G\frac{M}{(R+h)^2}m$$

Interpreting this for objects 'falling' to the Earth, we have **R** as the Earth radius, **h** is the height to the center of mass that is above the Earth (using **R** as unit). **M** is mass for the Earth; **m** is mass for the 'falling' object. big-G functions both as a 'constant of proportionality' and as the factor that makes the equation appear dimensionally correct. These dimension corrections require big-G to have the dimensions of $[F][L^2][M^{-2}]$ for which in one mode of implementation we write, Nm^2/kg^2, where **N** is newtons [F], **m** is meters [L] and **kg** is kilograms [M], big-G ≈ 6.67 x 10^{-11} Nm^2/kg^2. These required dimension corrections have led physicists to regard big-G as the Constant of Universal Gravitation, as if it were an entity of some kind, rather than a result of fudging engendered from mistaken initial suppositions. To date no one has been able to prove that this 'constant' is the same for every gravitational interaction. Furthermore, none of the results from some of the best laboratories on Earth, which have tried in recent years to measure big-G, agree on its value. Furthermore, one is hard pressed to find any statements concerning tidal effects in these precision results. However, when we assume that **G** is actually the same for every gravitational interaction, and we assume that Kepler's Third Law ($T_1^2/T_2^2 = R_1^3/R_2^3$) is correct (there are significant aberrations to this law for the Jovian planets), then we have an apparent means for estimating (calculating) the masses of other celestial objects.

In laboratory cases on Earth the 'falling' experiments in a manner of speaking look relatively microscopic, **h** expressed in terms of **R** as unit renders differences that are invisible to current modes of measure. If we only wanted estimates we could set **h** equal to zero, keeping in thought that the formula for quantitative analysis is fractionally falsified, and the formula without **h**, for qualitative analysis, is no longer valid.

By **F = mA** we have:

$$F = G\frac{Mm}{(R+h)^2} = mA \quad \text{or} \quad A = G\frac{M}{(R+h)^2}$$

In the case where **M** is the mass of the Earth, **m** is the mass of the 'falling' body, and **R** is the radius of the Earth, we see that the acceleration, **A**, is the same for each body 'falling' to the Earth from a height, **h**. This can be taken as Newton's first rationale for "descend with equal velocity." **F = ma** uses mass from an inertial reference-frame and the Universal Law of Gravitation uses masses in a gravitational reference-frame. Equating these forms of Force without proof tacitly assumes the Equivalence Principle. It is this assumption that the following arguments put in question.

From Newton's text it seems that this argument is consistent in that he can determine from his formula the acceleration due to the Earth, and this determination is functionally independent of the acceleration partition that is due to the "falling" object. At this point the argument is logically consistent. But then he uses Boyle's experiment (feather and gold coin in a vacuum tube) as if it shows that the "descend with equal velocity" is demonstrated by it. The assumption that Boyle's tube shows this "equal" descent introduces an inconsistency that can be demonstrated by using a dual argument to determine the gravitational influence partition due to the object that is in the so-called "free-fall" acceleration.

To account for the Earth falling toward that 'falling' body, we must use the dual approach and reverse the roles of the Earth and the 'falling' body and use **f = Ma**, where **M** is the mass of the Earth and the acceleration, **a**, is due to the 'falling' body. We set up the new equation and then divide out the big **M**'s, as follows:

$$f = G\frac{Mm}{(r+H)^2} = Ma \quad \text{or} \quad a = G\frac{m}{(r+H)^2}$$

The contribution of mass, **m**, to the fall towards the Earth is the acceleration, **a**, when the Earth is at height, **H** (for their geometric magnitude, **r+H = h+R**, but they differ in number because **H** counts how many **r**'s and **h** counts how many **R**'s.). This, of course, leaves out the contribution of the Earth to the total 'falling' force. Newton did not consider this dual case and therefore by default showed he did not recognize the problem.

EQUIVALENCE PRINCIPLE. In 'falling' objects, following Newton principles, the acceleration *increases* proportionally to the gravitational mass and *decreases* proportionally to the inertial mass. If we assume that all objects 'fall' to the Earth with the same constant acceleration then it follows that gravitational mass and inertial mass are equivalent even though so differently defined. This is taken as implying that the velocity of free-fall does not depend on the mass of the falling object.

From the dual argument for the 'falling' object we may conclude that since the magnitude of **a** can vary according to the object while **A** is fixed, that any composition of these accelerations will vary as **a** varies. If this is true, then the so-called Equivalence Principle fails to be true. Catechism physicists have so much faith in this "Principle" that a longer argument may be required to convince them.

***Ex Parte* Acceleration and Mass**: Using Newton terms, the ratio of my mass to the Earth mass is 90 kilograms to 5.976×10^{24} kilograms. This reduces to the scalar quantity 1.506×10^{-23}. Since, in terms of their scalar values, **m/M = a/A**, to calculate the acceleration of gravity due to me we use **a/A**, where **a** is due to me and **A** is due to the Earth. If we use $\mathbf{A} = 9.81$ ms^{-2}, then solving for **a** yields:

$$\mathbf{a} = 1.477 \times 10^{-22} =$$

$$0.000,000,000,000,000,000,000,147,7 \text{ ms}^{-2}$$

General practice in countries using the metric system is to use kilograms for measured weight. My measured weight in this popular sense is 90 kilograms. Weight in terms of newtons is usually confined to science literature and often avoided through preference to using 'mass' rather than 'weight' in physics analysis. Science convention interprets this Earth measured 'weight' of 90 kilograms as my mass and calculates my Newtonian weight to be mass times acceleration due to gravity or 90 kg x 9.81 ms^{-2} = 882.9 newtons.

For the following argument I abandon the use of 'newtons' and use the popular sense of Earth-measured weight as mass and use this to estimate my participation in my measured weight of 90 kilograms. That this value is not functionally independent of the Earth reference-frame does not vitiate the argument.

The idea here is to partition my Earth-measured weight into what I contribute to it and what the Earth contributes to it. Again I use the ratio of **m/M** and separate 90 kilograms into two quantities that match this ratio. My contribution to my measured 90 kilograms weight ('Newtonian mass') is some 22 orders of magnitude less than 90, viz.,

$$m = 1.355 \times 10^{-21} =$$

0.000,000,000,000,000,000,001,355 kg

This small **m** (to date immeasurably small) corresponds to the mass that I contribute to the total 90 kg as derived on Earth from Newton formulas. I call this result my ***ex parte* mass**. The rest of the 90 kg is what I call the ***residuum***. The residuum is the Earth effect on the *ex parte* mass. If the total 90 kg mass were a Eudoxian magnitude then this Newtonian calculation would work and the residuum would be a multiple of the *ex parte* mass. This Earth residuum is many thousands of billions of times greater than my *ex parte* mass. The subtraction of my *ex parte* mass from the 90 kilograms is a change that cannot be measured by current technology. The overhead Moon affects the fifth decimal place of this total mass (in kilograms). On Earth, if the Moon were above or below me (other side of Earth), its participation in my Newton mass is many orders greater than my *ex parte* mass. However, the Moon's maximum participation in my *ex parte* mass on Earth is five orders less than the influence of Earth. This implies a relatively powerful physical lunar influence on humans, but it does not change the magnitude of the *ex parte* mass. The *ex parte* mass remains invariant with respect to any neighborhood.

Using the *ex parte* mass to indicate the quantity of mass is a semi-arbitrary choice. The actual scales and numbers used for representing magnitudes are contingent on usefulness. There is no necessary relationship between the chosen units and what they represent. In this respect the measuring system is functionally independent of the measured weight.

If the *ex parte* mass remains invariant with respect to these neighborhood changes then this allows us to use it as a basis for comparing various quantities of matter independently of any neighborhood induced changes. This sustains Newton's underlying intention for his concept of 'mass'. I say that this choice, though somewhat arbitrary, is useful for further comparisons with Newton measures.

If my *ex parte* mass does not change from planet to planet, etc., then, under the assumption that weight is Eudoxian, if I weighed myself on the Moon (in the popular mode using spring scales calibrated to Earth weight) I would be able to calculate the Newton mass of the Moon from this weight measure. I take my *ex parte* mass and divide it into my measured weight and then multiply the measured weight by this quotient. The result corresponds to the Newton mass of that celestial object. Even if weight is non-Eudoxian (non-Archimedean, nonlinear variation), the Lilliputian scale on which my measured weight is taken is sufficiently small such that the error remains small over a very long range even when calculating as if it were Eudoxian. In the **Convergence** section below I argue that weight is non-Eudoxian.

At present I can only know my *ex parte* participation in my Earth weight through plausible reasoning and calculation. It is beyond all known means of direct measurement. However, astronauts in 'free-fall' trajectories in space report that standing in the shuttle bay they can push satellites with one finger, satellites that weigh several metric tons on the Earth surface. It was the ease of this pushing that was a surprise. The present argument suggests (using terms consistent with Newton) that the 'multi-ton' Earth objects cannot have the 'inertia' magnitude in their space trajectories that is equivalent to their 'inertia' on the Earth surface. Newtonian physics predicts that it is just as difficult to overcome the inertia of such quantities of material in space as on the Earth surface on a low friction concentric sphere surface (approximate inertial reference-frame). This may not be the case in fact.

Catechism physicists still claim that these observations in 'free-fall' trajectories are consistent with Newton. Some believe that the consideration of friction on the Earth surface provides a sufficient argument. On Earth even in Galileo's time there have been demonstrations of one person on a dock pulling a ship. This may suggest that we have a case in point for the astronaut experience. However, the Earth-bound determined magnitudes led to an anticipation that had far overestimated the actual required effort in the space-shuttle.

Big-G. If we had defined the newton (**N**) or some of its factors in some other way then we could make big-G equal to 1. If we had made big-G equal to dimensionless 1, then we would be in the position of seeking a new unit for the measure of gravitation. In a manner of speaking, big-G is an artifact of Newton's reasoning. It is prior commitments in the units that make big-G a constant (with dimensions) to search for. However, I remark again that no two laboratories have matching results for this 'constant'. It would be interesting to measure differences that relate to the position of the Sun or Moon at the time of measure, or other tidal effects, etc. Inconsistencies may also generate from the actual 'masses' used for the experiments. It may be that all that is being measured is a kind of Bernoulli effect of gravitational convergence passing between the two objects.

The torsion balance used for this experiment was originally designed by English geologist John Michell (1724-1793) and after his death the apparatus was taken over by the British chemist and physicist Henry Cavendish (1731-1810) who used it to "weigh the Earth." He expressed his result as a specific gravity of 5.48 g/cm^3, though his actual data yield 5.448. Current accepted value is 5.518. The torsion balance was similar in conception to the torsion balance invented independently at a later date by Charles Augustin Coulomb (1784) to determine the formula for electrostatic charge, $\mathbf{F_{el}} = \mathbf{Q_1Q_2/R^2}$, a formula almost isomorphic to Newton's Universal Law of Gravitation. I say "almost" because Coulomb's *original* version does not have an encumbering and peculiar constant of proportionality that Newton's Law requires. I regard the so-called universal gravitational constant, big-G, in much the same manner as I regard phlogiston. It is an artifact generated from a mode of reasoning.

Visible and Invisible g-Accelerations. As already argued, when we say that I weigh 90 kilograms on Earth, it is also true that the Earth weighs 90 kilograms on me. Reasoning with these dual forms we can say that if my acceleration participation in this measured weight were doubled that we would have:

$$2a = 2 \times 1.477 \times 10^{-22} \text{ ms}^{-2}$$

or

$$0.000,000,000,000,000,000,000,147,7 \text{ ms}^{-2}$$

$$2a = 2.954 \times 10^{-22} \text{ ms}^{-2}$$

which shows no change in the order of magnitude. It remains beyond current measurement technology to detect this difference in acceleration, yet I measure twice the weight.

What is supposed here, consistent with Newton, is that by increasing the mass of a body we increase the g-power (acceleration due to gravity) of that body. This g-power is the effect this body has on the Earth. By doubling the g-power of the smaller body, the measured weight on Earth is twice what it was before. The question we ask is what effect this has on the 'falling' acceleration. As shown above, we cannot measure the difference in the falling rates of such objects so small relative to the Earth. It will appear as if the two weights, one double the other, fall at the same acceleration. Furthermore, it is clear from the order of magnitude of these weights that even an object multiplied to a hundred million times heavier would appear to our measuring instruments as 'falling' at the same rate as the original object.

Consistent with Newton's notion that g-power augments with mass, we can grant that if we could keep increasing the mass of the smaller object that finally its surface g-power would come into the range of our measuring instruments. Of course the surface g-power is a function of both mass and size. Euler is the first to use the notion of density as mass per unit volume. The expressed surface g-power of a material object depends on the type of material, its compactness, its quantity within that compactness. The question we now address is how to determine the total 'falling' acceleration.

Newton states that all bodies descend to the Earth at the same acceleration. In the two body problem, this statement is only consistent if we are *not* talking about the total acceleration between the two bodies. With respect to the total acceleration between the two bodies gravitationally approaching each other, we cannot have a variety of bodies with different g-powers descending at the same acceleration toward a body with a fixed g-power. This is because the

Earth does not descend to them with the same acceleration and therefore the total acceleration between the Earth and these other bodies must vary. The compounding of the Earth participation and the 'other object' participation leads to different totals for the acceleration between the two bodies. Assuming mutual attraction, for spherical bodies (with adequate distance for mass point calculations) we have:

$$A_t = a_1/(r_1 + h_1)^2 + a_2/(r_2 + h_2)^2$$

For purposes of dual interpretations, A_t is confined to the scalar value of the total acceleration of the two bodies (body$_1$, body$_2$), a_1 and a_2 are the respective surface g accelerations expressed as scalar quantities, r_1 and r_2 are the respective radii, h_1 and h_2 are the distances in terms of the respective radii as units. The total acceleration, A_t, has no assignable direction unless, e.g., one of the two bodies, body$_1$ or body$_2$, is chosen as the reference-frame. Then it takes the direction toward the chosen reference-frame, etc.

Example: For the Earth, $a_e = 9.78$ ms^{-2} (equator) and the Moon, $a_m = 1.62$ ms^{-2}:

$$A_t = a_e/(r_e + h_e)^2 + a_m/(r_m + h_m)^2$$

or

$$A_t = 9.78/3{,}632 + 1.62/48{,}952$$

or

$$A_t = 0.002{,}692{,}7 + 0.000{,}033{,}1 = 0.002{,}726 \text{ ms}^{-2}$$

Thus if the Earth and the Moon were actually falling toward each other their *total* acceleration, A_t, relative to the Earth has the Moon descending at 0.002,726 ms^{-2}, and relative to the Moon has the Earth descending at the same acceleration. The ratio between 0.002,692,7 and 0.000,033,1 is close to 80 to 1, which is also taken as the ratio of their masses. It is their revolution about each other that sustains the composition of their accelerations without collision.

Acceleration differences fall into two general classes. Either acceleration differences are too small to measure or they are large enough to measure. For our Lilliputian experiments on Earth it is evident that most differences in the *ex parte* acceleration of laboratory objects are too small to measure. These small differences are *invisible* to us. (In general, the Lilliputian arguments also ignore the stipulation for the Newtonian condition that requires the Earth be regarded as a mass point even though the Lilliputian distances may invalidate regarding it as one.)

Newton expressed the force between two objects as directly proportional to their masses multiplied together. Thus if you have two spherical objects approaching each other (gravitationally) that are the same size and kind (yet distant enough to be treated as mass points), then force is proportional to the multiplication of these masses. If one of the two masses is reduced to half of what it was, then you cut the force between them in half as well. If you cut both of the equal masses in half, then the original force is cut to one-fourth of what it was, etc. Furthermore the force diminishes inversely with the square of the distance **r + H**, where **H** is the distance from the surface in terms of a count of the radius **r** as unit.

Regarding Newton's Universal Law of Gravitation relative to me:

$$F = G\frac{Mm}{(r+H)^2} = Ma_{me}$$

from which we get an expression for a_{me}:

$$a_{me} = G\frac{m}{(r + H)^2}$$

It would follow that if my mass were doubled that the acceleration due to me would be doubled. In the relationship between the Earth and me, the Earth would descend toward me at twice the acceleration, i.e., $2a_{me}$. Since at the Earth surface, $r + H = 1$, the acceleration rate at the surface would be approximately $A_t = 2a_{me} + a_e$. The small variations from this would relate to the mass-point distance requirement. Neither a_{me} nor $2a_{me}$ are large enough for our current measuring instruments to detect. Both values are invisible to us. My new acceleration increments the total acceleration between me and the Earth, but this resultant is many orders below doubling the total acceleration. Though this increases my measured weight (popular mode) to 180 kilograms, my new 'free-fall' rate cannot be distinguished from my old one because both magnitudes are too small for today's measurement technology.

However, this framework suggests that we can use this special case of measured weight to know whether the Lilliputian accelerations are similar or different. *Measured weight can be used as an indicator of these quantitative differences in acceleration that are otherwise invisible to us.*

Boyle's Vacuum Tube. Robert Boyle designed an experiment that Newton used for his own argument. Newton described it in the General Scholium, which ends the *Principia*, where he writes of air resistance and contrasts this with a vacuum. Newton says:

"Withdraw the air, as is done in Mr. Boyle's vacuum, and the resistance ceases; for in this void a bit of down and a piece of solid gold descend with equal velocity."

It is this description of Boyle's experiment that led some people to call this common Physics demonstrator 'Newton's Tube', a mistaken name. Newton should have written something like 'Mr. Boyle's vacuum experiment', etc., instead of 'Mr. Boyle's vacuum'.

This 'in the rough' experiment was intended to demonstrate Galileo's conclusion that all falling objects descend to the Earth with equal velocity. In the Apollo 15 mission (1971), the view on the Moon of Commander David R. Scott dropping a geologic hammer and a falcon feather that strike the ground at the "same" time was also intended to demonstrate this conclusion. In regard to their original purpose, these experiments were futile.

ANTI-EQUIVALENCE SUMMARY. Galileo proposed that all weights descend with equal velocity. This is another way of saying that the velocity of free-fall does not depend on the weight of the falling object. He tells us to notice that a grain of sand and a grindstone fall at very nearly the same rate. He devised experiments for rolling balls down an inclined plane and this slower 'fall' was easier to measure and it reduced the effects of air resistance. Using a rough notion of limits he concluded that small differences in velocity of bodies in 'free-fall' could be attributed mostly to air resistance (note that this addresses the influences on falling and not the causes of falling, i.e., Galileo remains consistent).

Newton also argues that the velocity of free-fall does not depend on the mass of the falling object. He took Boyle's 'in the rough' vacuum tube experiments as demonstrations of this assumption. Though this ostensibly shows his agreement with Galileo's argument, when Newton translated 'weight' to his notion of 'mass' and related gravitational strength as proportional to mass, the argument for 'free-fall' becomes self-contradictory in the Newtonian universe of discourse that requires the composition of accelerations between attracting bodies.

To this day the assumption that free-fall does not depend on the mass of the falling object has *not* been falsified by terrestrial laboratory experiments.

The Einsteins assume as confirmed fact that the velocity of free-fall does not depend on the mass of the falling object. Their belief in this equality was reinforced by the experimental results of the Hungarian physicist Roland von Eötvös (1848—1919), which had an accuracy of about 10^{-7} (cgs units). In Einstein Relativity this assumption is a foundational postulate for relativistic theory of gravity. By 1909 Eötvös had brought the accuracy to 10^{-8}, etc.

The later Robert H. Dicke (1916—1997) experiments were intended as more accurate confirmation of this equality. Dicke took the results to the eleventh decimal place (cgs) using Gold and Aluminum. It might also be interesting to test elements with less conductivity. Conductivity likely influences results. For Lilliputian terrestrial laboratory experiments, the 'falling' objects contribute so little to the action that Dicke's experiments (1960) are still many orders below the required accuracy for establishing the Physics principle here discussed. If you ignore what I am talking about, then Dicke's experiments contribute toward believing in the validity of Newton's phrase "descend with equal velocity," in the sense of Boyle's experiment.

The Soviet physicist V. Braginsky with Panov (1972) made attempts for more accuracy and they obtained accuracy to 10^{-13} (cgs units), still many orders below the accuracy required for the beginning of a significant measure difference. In terms of the laboratory scale these magnitudes are many billions of times below the required accuracy.

NASA plans to increase the accuracy by a factor of 1,000,000 using a satellite experiment called STEP (Satellite Test of the Equivalence Principle). But even this increase does not put the test within reach of the orders of precision required for a decision. If we take their estimate of 1,000,000 times more accuracy and apply it to the Braginsky and Panov value of 10^{-13} we get 10^{-19}. This is still over 1,000 times too small for a significant measure. The NASA experiment is ill conceived and a waste of money if this is not taken into consideration. They also do not seem to recognize the difference between microgravity (near parallel lines of gravitation, etc.) and neutralized gravity (what they call 'free-fall'). By their failure to recognize this difference they are tacitly assuming the Equivalence Principle even before they begin the experiment.

From the above arguments we conclude that the convergence acceleration partition that is due to the Earth at a given location is the same for any object at that location. However, the convergence acceleration partition that is due to the object differs according to the object. Otherwise said all objects with respect to a given Earth location descend with equal velocity to the Earth surface, but dually the Earth descends to these objects according to their differences in convergence power. In other words the Equivalence Principle fails.

Note: The arguments concerning this Equivalence Principle Contradiction were presented at the 2004 conference Physical Interpretations for Relativity Theory (PIRT-IX) and appear in the second volume of the proceedings under the title: *An Equivalence Principle Contradiction*.

CONVERGENCE

In Convergence Buoyancy (CB) it is supposed that in a Universe of two objects with like-pole surfaces that they can each be caught in the convergence field of the other. This interaction induces a convergence toward each other until they come to proximity where a sorting out of the repelling like-pole magnitudes takes place. This sorting varies in accordance with the convergence relationship between the objects and falls into three categories of buoyancy: float, neutral, and sink. In its application to gravitational convergence, these varied states of repelling generate a local model that is dual to the Newton notion of gravitational 'attraction'. In the text that follows, the phrase **'gravitational convergence' always implies the CB interpretation unless otherwise said**. *In the section below entitled* Material Categories for CB *these sorting states are identified in more detail.*

When I speak of *lines of convergence*, this is an apodeictic usage for indicating field shape and compactness (represented by shape and how many lines in the zone). It has use in CA and in CB modeling. At present I attribute no other physical dereference to these lines.

Implied in these foundational statements is a shift in the basis for analysis. In the Galileo, Kepler, Newton tradition the emphasis is on movement description. Though Newton introduced aspects of a viable dynamics, he recognized himself that his work was more centered on mathematical description of movement then on physical principles. This is also admitted by his chosen title: *The Mathematical Principles of Natural Philosophy*. The emphasis in CB is on physical principles, on how differences in power and/or polarity of the convergence account for the motions such differences generate. In the case of gravitational convergence it is the material distribution adjusted for distance that determines movement. Understanding the CB categories, the emphasis is placed on material distribution (exterior distances and interior compactness).

For two hand-size magnets not too different in power, the like-pole interaction has a visible repelling effect. William Gilbert (1544—1603) reasoned this out with great clarity in his *On the Loadstone and Magnetic Systems* (1600).

EXHIBIT 1.1: Faraday Fields

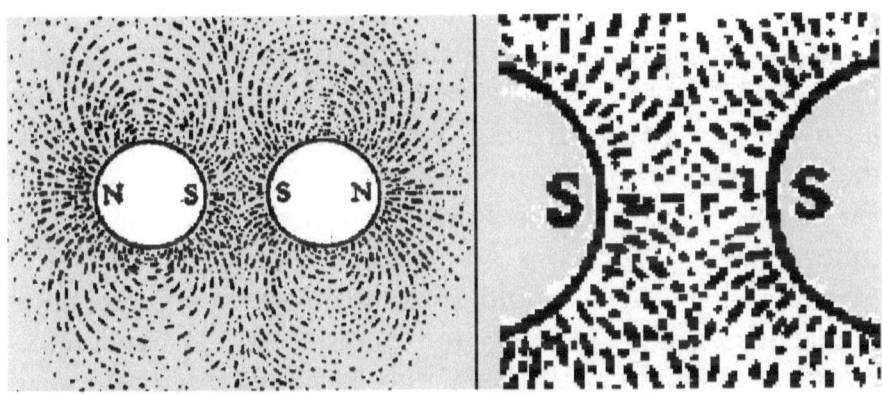

a. Dipolar magnetic fields repelling   b. Detail of like-pole interaction

The line shapes are like bent springs pushing the objects apart. Faraday assumes that the iron filings follow the general field pattern without considering their possible influence on the displayed pattern. Iron seems to conduct magnetism, which induces the filings to accrete in packets along apparent lines. There may be no packets or lines in the field itself. Once in place, the filings do not seem to move. This suggests that the fields may be static. Whether they are static or dynamic, their interaction has the dynamic effect of bending the shape of the lines. This change in shape shows a plausible structure for the repelling.

Michael Faraday. 1839-55, *Experimental Researches in Electricity*

Large magnets can overpower magnets of lesser power, regardless of polarity. Reasoning by analogy, a small gravitational object could be conceived of as repelled from the Earth yet be caught and overpowered by the neighborhood of the larger object because of the likely field structure they participate in. It is also possible to conceive of two similar gravitational objects with powers of sufficient magnitude to repel one another.

Though this may give a general notion about the reasoning, there are complications that make this book several times longer than it might have been.

Dereferencing the Term 'Object'. In this text, 'object' indicates 'body' or 'system of bodies' according to context. A body is regarded as well-defined when we know its boundaries well enough to decide whether any given point is on the body (surface), inside it, or outside it. A well-defined system of bodies also has such boundaries. The term 'body' remains difficult when pressed to an exactitude that exceeds ordinary usage. There is no consensus as yet for where the top of the Earth's atmosphere is but sometimes artificial decisions are helpful. I promote the belief that the term 'body' should sustain an openly agreed upon precision deemed sufficient for the argument at hand. Thus the more general term 'object' should be regarded as a variable that takes on the characteristics required for its contextual instantiations. The understood instantiations constitute the dereference for 'object'.

SUMMARY DEFINITIONS: *A summary definition is a mnemonic device for recalling the fuller discussion of that definition.*

**CB SUMMARY DEFINITIONS
(*SYSTEMIC* and *EX PARTE*)**
The following summary definitions reference the more detailed discussion in the Scholium and main text that follow.

SYSTEMIC DEFINITIONS:
  Definition I. Objects directed toward a center are in *convergence*.
  Definition II. Objects that have converged to a revolution around each other are in *float buoyancy (buoyance)*.
  Definition III. Objects that have converged to an object surface are in *sink buoyancy (accretion)*.
  Definition IV. Objects that are on the boundary between sink and float buoyancy are in neutral buoyancy (two forms; *boundary buoyancy* = natural, *neutralized buoyancy* = not-natural).
  Definition V. Insofar as objects are in float, boundary, or sink buoyancy they are i*sodynamic* (synonym: *natural movement*).

EX PARTE DEFINITIONS:
  Definition VI. A *neighborhood* convergence is the convergence influence exterior to the object under scrutiny (*ex parte* context).
  Definition VII. Neighborhood r*estraints* drive objects into isodynamic convergence and *constraints* drive objects away from isodynamic convergence.
  Definition VIII. An *impetus* is the total effect invested in an object from applied restraints and constraints; impetus = (restraint impetus) ∨ (constraint impetus).
  Definition IX. A *regression* is a diminishing of constraint impetus, an object with no constraint impetus is isodynamic with its neighborhood.
  Definition X. A sufficiently large object (*host*) that accretes small objects (*debris*) is said to *squelch* the buoyance of the debris it accretes.

SCHOLIUM FOR CB SUMMARY DEFINITIONS

We have a long history of separating events into constituent parts to aid our understanding. These parts are artificial separations from holistic events. Artificial separations impel us to treat the holistic event as a mixture problem. We attempt through suppositions and calculations to measure the contributions of each part to the whole event. What is imparted to or removed from these constituent parts is determined by and calculated according to systemic assumptions. This is another way of indicating that we cannot always directly measure the contributions of the supposed constituent parts of an event. It is for this reason that we must take care to be aware of the presuppositions of any systemic mode of analysis that we use. Interrogation directed to the discovery of hidden presuppositions helps us sustain the attitude that a system of analysis is a tool for us to use, to improve, to revise, and even to replace.

Heteronymic Nomenclature. There is a preference in the sciences to a reduction of terms, e.g., the heteronymic pair 'acceleration' and 'deceleration' are reduced to the homonymic usage of 'acceleration' with plus or minus values. In most cases this favoring of homonymic terms leads to concise statements. However, the concise statements can also be misleading (see end of Chapter 2 and Addendum 5 on Einstein Relativity). There are also contexts where homonymic terms complicate the discourse. I have opted to use heteronymic nomenclature to sustain a clear demarcation between the isodynamic and non-isodynamic drivers. Constraint/restraint is a heteronymic pair that distinctly references the non-isodynamic and the isodynamic drivers that invest impetus in an object, etc.

1. **Constraint and Restraint.** All actions are isodynamic or move toward an isodynamic state unless otherwise constrained. By CB Definition VII, non-isodynamic movement is constraint driven and isodynamic movement is restraint driven. In the case of vehicles used for going from the Earth to the Moon there is a mix of constraint impetus (invested by its engines, etc.) and restraint impetus (from its neighborhood). In CB any object invested with constraint impetus is a mixture of constraint impetus and restraint impetus as long as there is a neighborhood.

2. **Tidal Effects that Wax and Wane.** We characterize the changes invested in a particular object under restraint as *restraint impetus*. This restraint interpretation of impetus is context dependent. Since restraints may be characterized as interactions, this impetus has use as a term for the individual investment effects on any particular object subsumed under the restraint it is a part of at a given moment in time. Impetus is therefore an *ex parte* term. This *ex parte* term represents an artificial separation of the individual investment effects from restraints characterized as an interaction. Furthermore, this impetus in an individual object can wax and wane under the restraints. The waxing and waning of restraint impetus is conceived as synchronic with the effecting restraints of the neighborhood and these effects are also referenced as tidal effects. Any violation to this synchronicity is conceived as constraint driven and forms the basis for

defining the power of the constraint at a given instant.

3. **Impetus and Regression.** Since constraints and restraints may be characterized as interactions with an object, *impetus* may be used as a term for the individual investment effects on any particular object subsumed under these drivers. When we speak of the *regression* of impetus, it is an indirect way of asserting that there is a constraint impetus portion of the total impetus and that it is the constraint impetus that is diminishing. It is redundant to say that an object in motion continues its motion until otherwise acted upon, as there is always the universal presence of neighborhood restraints acting upon it. With no further constraint applied to the object, the constraint impetus must finally be dampened by the neighborhood restraints to no influence. At this point the object is isodynamic with its neighborhood. Any change from a non-isodynamic state toward an isodynamic state is called *regression*.

Comments: The term 'interaction' presupposes an individuality of participation. We must be wary of such terms as these (like 'sunset' though it is the Earth that turns) that may have a determining effect on how we think about a subject. Intelligent usage requires that we maintain awareness of the implied dereferences. 'Interaction' predispositions us to look at individual participations without regard to the appropriateness of such a framework (interface). Analysis into individual participations is an *ex parte* analysis. Rebuilding the event from this analysis may show a loss or distortion in the rebuilt representation for the event; emergent qualities may be lost, system artifacts may be gained.

Impetus may regress quantitatively and/or qualitatively. All cases of regression are toward an isodynamic state, whether or not they lead back to an original isodynamic state. We thus have two general classes of regression classed according to the mode that dominates; quantitative regression goes back to or near its original isodynamic state, qualitative regression goes to a different isodynamic state (qualitative differences in structure). Regression can be a function of both forms of influence.

In terms of movement, the neighborhood convergence functions as a restraint to the impelled motions as it 'interacts' with the convergence of the moving object. That it is neighborhood convergence implies a relativity such that more constraint impetus is required to move me on Earth than would be required to move me on the Moon. This difference in exertion is observed. This also implies that all self-induced movement of a human being may be regarded as constraint driven (counter to isodynamic).

An *impetus* (CB Definition VIII) in its original definition (pre-Galileo) denotes 'driving force'. In part, the term is rehabilitated as an anti-inertia term (for a non-inertial physics) to correspond to the typical *ex parte* analyses found in Newton's treatment of 'inertia'. It will be put to other uses as well, but even so, 'impetus' is not a fundamental term in this anti-Newtonian system. It is a derived term. It depends first on the dereferences for 'constraint' and 'restraint'.

The question that is never directly addressed in the Principia is, **"How is change in velocity physically specified to the object that is changing velocity?"** For Newton we could say that it is inertia that specifies to an object that it is undergoing a change in velocity. However, as Newton is first concerned with mathematical principles for description summaries, we are left with inertia as an axiom and no physics dereference. That he has named a resistance to change in motion as 'inertia' adds no comprehension to its physical nature. In other words, 'inertia' functions as an apodeictic fiction for Newton's physics. For CB, 'inertia' is filed in the same basket as 'phlogiston' and "Hypothesis non fingo" is contradicted by Newton (details in Chapter 3).

For CB the answer is in terms of like-pole fields in proximity. When the relationships between like-pole fields are understood, the answer to this question is understood. The thesis of Convergence Buoyancy, expressed in *ex parte* terms, is that neighborhood convergence field communicates with the object convergence field and this interaction specifies the relative status. Expressing this in a language independent of *ex parte* concepts would require a major overhaul of our language and our attitude in its usage.

The implications applied to the concept of entropy are discussed in a later context.

Gravitational Convergence Intensity. Gravitational convergence intensity has two main aspects that are open to measure. One aspect is the convergence power in a neighborhood locale; the other aspect is the neighborhood pressure at that locale. Because of our habituation to Newtonian thought, the power of this convergence is usually thought of in terms of the accelerations it can induce in objects and the pressure on these objects is ignored.

1. **Gravitational Convergence Power; [E]** $= E_n \vee E \vee e$. For CB arguments, the gravitational convergence power in a neighborhood locale is denoted by E_n.

This can be summarized as a vector value (a magnitude and a direction). Normally we can analyze this value into a function of values dependent on the objects in the neighborhood. In certain contexts it adds clarity to use the small **e** when we want to specify the gravitational convergence power at the surface of a debris body and capital **E** for hosts. In general, for debris-host relations we have $\mathbf{E_n = f(e,E,n)@L}$ where **n** is the neighborhood contribution (tidal effects). This relation emphasizes the composition elements of $\mathbf{E_n}$. In Newton physics this power ($\mathbf{E_n}$) at a specified location, L, on the surface of objects is expressed simply as **g**.

2. **Gravitational Convergence Pressure; [P].** For CB arguments the gravitational convergence pressure, **[P]**, is conceptually distinct from its power **[E]**. Whereas its power always has a resultant magnitude and direction, its pressure magnitude at any given point is omnidirectional. Given a point under water, or a point in the atmosphere, we can speak of the pressure at these points and these pressures are omnidirectional. At the same time as being omnidirectional we can speak of object buoyancy at these points. In the case of water in a given neighborhood, its compactness and molecular clustering mode are fairly uniform and this renders it highly isotonic even to the depths of ocean trenches. In the case of atmosphere, the compactness is not uniform and therefore it has a non-isotonic relation with the object buoyancy. Thus in water there is little variability in buoyancy for a given object (barring solubility and other possible interactions), therefore it rises to the top, sinks to the bot-

tom, or buoys in the water at non-fixed depths, and it is in one of these buoyancy states according to its relation with the water. In the atmosphere an object can also sink to the bottom, rise to the top (wherever we decide that is), but it cannot buoy to an unfixed depth. Each object has a well-defined position in the atmosphere in accordance with its isotonic relation (state of buoyancy) with the atmosphere.

Gravitational convergence compactness is more similar to the atmosphere in the way that it is non-isotonic. It therefore sustains a well-defined relation of convergence buoyancy with objects. Objects cannot buoy to an unfixed level in these non-isotonic fields.

With respect to the object under scrutiny a given convergence zone could be interpreted as under a positive or negative pressure. Until equilibrium is reached, the positive pressure would have a compression effect; the negative pressure would have an expansion effect. The negative pressure may also be called a *tension*. It is conceivable that (ignoring changes in state) an object the size of the Empire State Building could be placed in a pressure zone that would compact it to the size of a needle. And in reverse, an object from that zone the size of a needle that is put into our zone could expand to the size of the Empire State Building. This is the pressure that is referenced by the term 'convergence pressure'. Furthermore, this implies that convergence pressure influences isodynamic distances (more below).

Although it makes sense to regard the Earth convergence power as a pressure that holds debris to its surface, this mode of analysis is not adequately distinct for CB and therefore is avoided. This is a case where heteronymic terms reduced to a homonymic term can complicate the discourse.

For material objects, there are three natural isodynamic categories; floatation, accretion, and the boundary between them (a form of neutral buoyancy). All three categories apply to each of the three physical scales for material convergence: gravitational, surface tension, or atomic.

The demarcation between host and debris is context dependent. A discussion on how to quantify this demarcation is given in Chapter 2.

These CB categories should be regarded as the setting out of types. Within a given physics scale (barring the transition zone difficulties) we can class objects as hosts or debris. Hosts are regarded as 'large' and debris are regarded as 'small' within the latitude of the chosen physics scale.

1. **Category One:** flotation (Exhibit 1.2 A).
    Case One, host-host (large-large). It is conceived that two material objects of like scale {gravitational, or surface tension, or atomic}, of sufficient size within that scale, converge toward each other until their lines of convergence interfere with each other with sufficient strength to generate a repelling that holds them at a distance. Its isodynamic state is an equilibrium in **float buoyancy**.
    Case Two, debris-debris (small-small). This form of floatation is rendered possible when the ambient neighborhood convergence pressure is less than the convergence power of the debris. Its isodynamic state is an equilibrium in **float buoyancy**. The floatation level adapts to the overall ambient neighborhood convergence pressure. Under sufficient ambient pres-

sure the composite convergence power of the debris may be driven to an accretion state (Category Two, Case Two).
2. **Category Two:** accretion (Exhibits 1.2B, 1.2C).

Case One, debris-host (small-large). It is conceived that two like-material objects of like scale are different in size within that scale, such that the large object squelches the repelling effect between them. This squelching makes the small object sink to the larger object. Its isodynamic state is **sink buoyancy** (accretion). How quantities of matter on a human scale (Lilliputian) can appear Eudoxian is described in Chapter 2.

Case Two, debris-debris (small-small). This form of accretion is enhanced when the ambient neighborhood convergence pressure is larger than the convergence power of the debris. Its isodynamic state is **sink buoyancy** (accretion). Floatation is hampered when the ambient neighborhood convergence pressure is more than the convergence repelling power generated by proximate debris. From neighborhood pressure they merge into one object.

3. **Category Three**: neutral buoyancy (no exhibit).

Neutral buoyancy can be regarded as the upper limit of sink buoyancy and the lower limit of float buoyancy or as the boundary between sink and float. Its isodynamic state is **boundary buoyancy** (an unlikely exterior state for large objects due to stress on surface tension, etc. The interior case is considered in Chapter 2). Its constraint driven state is **neutralized buoyancy**. Neutralized buoyancy can be mechanically induced in a space-shuttle by matching its centripetal convergence acceleration, v^2/r, with the Earth **E** power adjusted for the altitude.

MATERIAL CATEGORIES FOR CB.

EXHIBIT 1.2; A, B, C: Material Categories for Buoyancy (not to scale and neutral buoyancy not shown.

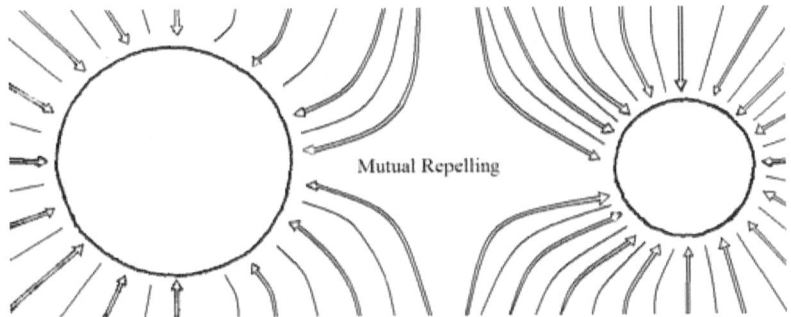

A) HOST-HOST (large-large), converges to buoyance: Like bent springs the facing convergence lines repel the objects apart with a power equal to the mutual convergence that draws them together.

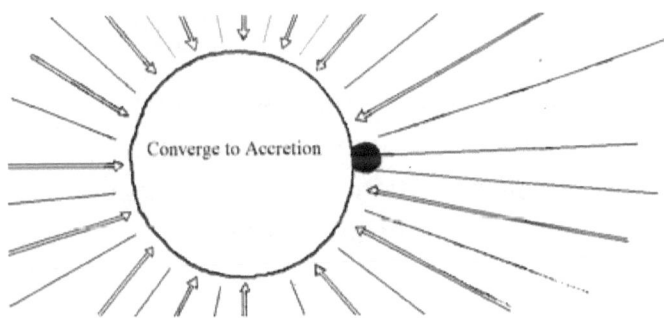

B) DEBRIS-HOST (small-large), converges to accretion: The host squelches the debris-host repelling and the debris accretes to the host object, augmenting the host object bulk.

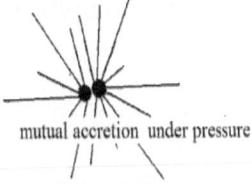

C) DEBRIS-DEBRIS (small-small) converges to a distance inversely proportional to the neighborhood pressure: Accretion occurs when two proximate convergent debris objects cannot sustain float buoyancy under that pressure. They can merge into one object.

Host-host accretion is improbable under a neighborhood pressure because the neighborhood pressure is dependent on the conglomeration of hosts in the neighborhood and this pressure is limited to the interlacing powers from these hosts that keep a distance proportional to their powers. If the unlikely neighborhood compression from initial conditions required for host-host accretion could be met, we can surmise that it would entail breaking the physical integrity limit of surface tension before accretion took place, etc.

Convergence Power Quantification Factors. Because it is toward a center, the convergence object resembles a monopole object. Three main factors influence the convergence power effect (magnitude of convergence effect):

1. The *cone angle* of the convergence subtended (geometric or directional polarity).
2. The *compactness* of the convergence (compactness is determined by isodynamic distancing).
3. The physics *polarity* of the convergence lines (matter-matter, matter-antimatter, antimatter-antimatter, etc.).

Chapter 5 gives a possible physical interpretation to this polarity and shows how CB polarities may be expressed as dipoles. For polarity, matter-antimatter relations are dominated by CA lines, not CB lines.

A gravitational convergence cone angle has its apex (vertex) at the center of the object and the cone base takes on the silhouette (along the lines of sight from the apex) of the subtending object. Without curvature, only parallel lines of sight could cover half the surface of the subtending object (as Euclid proves in his *Optics*). By definition, the subtending cone angle of gravitational convergence is directed toward its apex for this convergence and this angle influences the magnitude of the convergence effects on the object subtending it. For small objects such as us, it is logical that the direction of this convergence manifests as weak. What strength there is, is a combination of this angle of convergence and the physical compactness with respect to the convergence lines (more in Chapter 2).

We can compare the cone angle with its apex at the Earth center subtended by the Sun, and the cone angle with its apex at the Sun center subtended by the Earth. Within a given scale, the smaller the subtending object, the closer to parallel are the neighborhood lines affecting it. This explains why, for sufficiently small objects, such as gas corpuscles (noble or molecular gases), that large object gravitational convergence can have only a weak hold on them.

For gases, on the surface of host bodies the convergence lines toward the host are near parallel. The neighborhood convergence angle for a gas corpuscle on the Earth surface is submicroscopic, calculable but beyond any current means of direct measure, viz., it is the angle from the Earth center subtended by a surface gas corpuscle

This suggests the possibility that gases converge to large objects on a basis different from these convergence angles. In a later context, based on an interpretation of empirical observations made by Russian scientists in the MIR Space Lab, we can argue that these gases follow a radiation gradient. This helps describe how Venus, which is smaller than the Earth, can have an atmospheric pressure over 90 times the Earth's. Venus has an average surface temperature exceeding 480° C (hottest surface of solar planets). In part, surface tension may contribute to gas clinging to spherical surfaces. The pressure is generated by a physics still poorly understood and different from what is usually referred to as 'gravitational attraction', more about this later.

In the case of Mercury we observe a different result. The proximity of the Sun likely reverses the radiation gradient between them and this can conduce the leaching of the Mercury atmosphere (more in Chapter 4). Current belief credits the solar wind for stripping the Mercury atmosphere.

In the CA case of the Earth's magnetic field, the 'lines' of the field bend back on themselves and it is currently believed that they link with the opposite poles (a dipole view that follows from the Faraday field mappings) forming widening angles, angles much greater than the gravitational convergence lines distribution on the sphere surface. The widening adds magnitude to the directional polarity of the Earth's magnetic convergence.

The $1/r^2$ geometric dispersion rate for surface convergence is not sufficient for mapping this field form of magnetic power. This problem also adds to an argument that the $1/r^2$ dispersion rate for CB cannot be precise for two facing convergences in float buoyancy because the buoyancy is generated from opposed convergence lines.

Compression and Dilation of Objects and Distances. High pressure neighborhoods have a compaction effect on objects from a lower pressure neighborhood. Low pressure neighborhoods have a dilation effect on objects from a higher pressure neighborhood. These pressure differences also affect isodynamic distances between objects. Higher pressure shortens distances, etc. Thus barring other changes, if the Earth-Moon neighborhood pressure were higher, the Earth and Moon would move closer together. If the neighborhood pressure were lower, the Earth and Moon would move farther apart. This suggests that the further the Earth moves from the Sun, the farther apart the average distance for the Earth and the Moon would be.

We need to determine with more precision what these pressure effects have on the inverse square law as applied to gravitational convergence. Since the high-pressure neighborhoods tend to compress distances between objects we can ask how this compression influences $1/r^2$. With respect to revolutions, we expect a higher velocity when under higher pressure. This suggests a kind of compensation adjustment to the contractions of distances that could support a conservation law (more in Chapter 2).

Note: This notion of pressure does not imply that a space-ship in deep space will greatly expand from the release of gravitational pressure. Instead, there is a shift in dominance from gravitational compression to surface-tension compression. The Empire State building and needle under different pressures was used as an apodeictic fiction (see Appendum 2).

Sorting Formula. The restraint pressure for hosts can be expressed in terms of convergence power. If we only consider the geometric adjustment for restraint in a two-host universe we could write:

$$S_t = E_1/(r_1 + h_1)^2 - E_2/(r_2 + h_2)^2$$

where S_t is the total sorting acceleration and $E_1 \geq E_2$. The subtraction indicates the repelling power generated between E_1 and E_2 at distances in terms of their own radii, r_1 measures h_1, r_2 measures h_2.

This allows S_t to be zero even when E_1 and E_2 are different. When an isodynamic state renders $S_t = 0$ then for E_1 and E_2 the sorting is complete in accordance with their mutual buoyancy relation. When $S_t > 0$, then E_2 and E_1 have not reached an isodynamic state. When $S_t < 0$ then E_1 and E_2 are reversed. The interior neutral case is discussed in the Chapter 2 section on Franklin neutralization.

For host objects, the exterior neutral case has no easy determination because of the complexity of the induced changes in shape and the stress on surface tension that proximity generates. If the two objects are in a neighborhood of other objects, and when the total surface effect of the neighborhood pressure is sufficiently smaller than the total E_2 effect (so that the surfaces of the two objects under scrutiny are tangent (an unlikely state for insufficient rigidity)), we have the exterior form of boundary buoyancy. If the neighborhood pressure is greater, the objects are in accretion. If the neighborhood pressure is less, the objects are in float buoyancy.

> **Note:** 'Fall' is a common Newtonian term. 'Sink' is the corresponding CB term. CB action on an object, in this sense, could be characterized as a 'sinking' until it sorts into one of the three isodynamic states according to the buoyancy relationship with its neighborhood.

In the case of mechanically induced neutralized buoyancy, it is possible that a measurable directional polarity is sustained, possibly a strong one, in the 'weightless' environment. The power of the subtended convergence angle is neutralized by sinking at a calculable rate proportional to that angle. In other words, *neutralized buoyancy driven by mechanical constraints does not affect the directional polarity of convergence*. This is an important point to remember when interpreting experiments done in mechanically neutralized buoyancy.

In practice, the near-Earth neutralized buoyancy state is always an approximation for debris since it involves magnitudes too small to measure. When 'weightlessness' is experienced for a few seconds in a descending airplane (e.g., for astronaut training) we can reason that the **e:E** ratio is likely distinct for each person, but that any differences among the ratios are below the perception level. All the trainees seem 'equally' weightless.

Field Categories for Buoyancy. With similar considerations we can also categorize buoyancy *field relations* independently of the material objects that define their centers. For buoyancy fields of like convergence scale from {gravitational, surface tension, atomic}, we again have the three categories of natural buoyancy. Field characterizations of CB may be regarded as otherwise functionally independent of the material properties.

Microconvergence Buoyancy vs Neutralized Buoyancy. Microconvergence buoyancy and neutralized buoyancy are distinct in regard to their modes for quantification.

1. Microconvergence buoyancy is a state that is quantified in two ways, geometrically by near parallel lines of convergence or physically by weakened effects from the sources. Thus the quantification of convergence effects is dependent on distances and the gravitational power generated by the neighborhood bulks.

    a. Geometric microconvergence is qualified by how close to parallel the lines of host convergence are. For a state of geometric microconvergence with respect to the Earth, an object is either very small or it is far removed from the Earth convergence center, etc. The angles referred to are cone angles. The perception limit is defined when the convergence angle cannot be distinguished from parallel lines by our measuring instruments.

    b. Physical microconvergence is qualified by the source convergence power effects. Celestial bodies vary in their source convergence power. Physically, for a given distance, the Sun convergence angle has more power than the same angle of Earth convergence (cone angle measured from centers). The perception limit is defined when the physical convergence is invisible to our measuring instruments.

These fuzzy definitions are sufficient for the following arguments and a suggested physics for them is discussed in Chapter 5. There are already indications that accuracy in measurement has limits beyond which no decision can be made for the exact measurement. This implies a fuzzy limit to accuracy that under the current technology may not allow further improvement in the measure. Different scales of observation require a difference in kind for the demarcations of microconvergence according to the physics that dominates the chosen scale. Note also that we may not always be able to determine in any clear way between geometric and physical microconvergence in their possible combinations.

2. Neutralized Buoyancy simulates aspects of buoyancy from boundary to float. However, neutralized buoyancy is not isodynamic. It requires constraints. Boundary buoyancy is isodynamic. It does not require constraints. Neutralized buoyancy is sustained through constraints whereas microconvergence buoyancy dereferences only to a neighborhood condition. *Mechanically induced neutralized buoyancy sustains the neighborhood directional polarity*, whereas by definition, microconvergence has weak directional polarity (weakness from near parallel convergence lines or weakness corresponding to a weak source).

Near Parallel Lines of Sight. Related in its geometry to geometric microconvergence is the effect of near parallel lines of sight. When viewing distant celestial objects through near parallel lines of sight, these objects no longer measurably change in their apparent size. However, we observe variations in brightness. Since they are viewed with near parallel lines of sight (lines so parallel that we have no means for measuring an angle difference), if two very distant objects look alike in kind and size, then it seems logical to estimate that they *are* the same size. With same in kind, this implies that they generate about the same amount of light. Measuring relative apparent dimness, some astronomers believe that they can estimate the relative distances using the inverse square law. This may not be true. This contradicts the supposition that the lines are near parallel. That we have near parallel lines of sight implies that the *compactness* of light is no longer measurable in any meaningful way. That the lines of light are near parallel implies that any differences in brightness should be due to the effects of intervening space debris or other effects (independent of whether there is also a physical fading with distance). Under these conditions, to assume we know how the variation in brightness shows variation in distance introduces inferences that have not yet been adequately tested (2004 A.D.). A dim object could actually be closer than a brighter object, etc.

A side effect of near parallel lines of sight is the illusion of foreshortening of depth. This illusion is apparent in a low power telephoto picture where magnification makes the image appear flatter. It should therefore be expected to a greater extent in high-powered magnification set for a greater distance. Therefore, we should be more critical of announcements that observations of great distance reveal that the Universe is flat (Euclidean). The accurate interpretation is that the farther we look, the flatter our observations get and this should be expected from the predictable foreshortening of apparent depth with distance.

Related in its physics to gravitational microconvergence, any variation in brightness that is functionally independent of the geometric spread corresponds to a physical fading. There are at least three non-geometric forms for fading with distance:
1. intervening space debris.
2. neighborhood pressure:
   a. resistance
   b. distortion

These distant objects have only submicroscopic effects on our galaxy in regard to the gravitational (matter-matter) convergence. However, there are other possible modes of convergence that are more powerful than material convergence, e.g., the Convergence Attraction (CA) of matter and antimatter, which may influence larger scale convergences. Galaxies may form where antimatter objects have converged to sufficient size for attracting stellar size matter, etc. We may also conceive of anti-galaxies. M104 may be a candidate anti-galaxy with its bright central cluster belted with dark rings.

SUMMARY CONCLUSIONS

Star Laws: An Introduction to Convergence Buoyancy introduces a new model for gravitational convergence. This study addresses an analysis and design for an anti-Newtonian system, a non-inertial Physics. Special emphasis is given to the problem of 'falling' objects. Newton's Second Law of motion is used in a new way to criticize Galileo's ingenious thought experiment on how all falling objects fall at the same acceleration, regardless of weight. I show that Newton's Law and Galileo's thought experiment are incompatible; they cannot both be true. The rationale for Newtonian gravitational attraction supports the concepts of ***iner-***

tial mass and ***gravitational mass***. This support tacitly imposes an equivalence for these definitions of mass so differently obtained. Neither of these concepts implies a physics for the attraction described by Newton. Newton remains strictly with mathematical description for the observations. He did not talk about the physics of inertia or the physics of gravitation. Using a dual application of Newton's Universal Law of Gravitation leads to a denial of 'free-fall'. This in turn leads to the denial of the Equivalence Principle in Newton's own terms. The anti-Newtonian repelling is supported by the concept of ***convergence buoyancy*** with its three modes of expression, ***float buoyancy*** (buoyance), ***neutral buoyancy***, and ***sink buoyancy*** (accretion). The theoretical framework dereferences to the physics for these. Convergence Buoyancy forms a basis for deriving a different interface for viewing the Universe from our locale.

These categories were formulated for gravitational convergence, but may serve as a paradigm set for any CB scale {gravity, surface tension, atomic}. The fuller development of the physical principles forms the subject of the following chapters.

The restraint pressure for hosts can be expressed in terms of convergence power. If we only consider the geometric adjustment for restraint in a two-host universe we could write: $S_t = E_1/(r_1 + h_1)^2 - E_2/(r_2 + h_2)^2$, where S_t is the total acceleration. The analysis required for the CB quantification is in Chapter 4.

Since the Jovian planets have a stronger influence on the composition of repelling effects that position them, their distances from the Sun are proportionally augmented, which offsets Kepler's Third Law. This also implies that we have not correctly calculated their surface gravitational powers.

I would have preferred that the units for Newtonian mass be called 'newtons' to honor the inventor of this usage. In this interpretation the measured weight would be thought of as kilograms with mass redefined in terms of newtons. This new definition of 'newtons' is different from current science convention. Thus, if a measured weight on Earth is 90 kilograms, then its mass would be c90.0/9.81 = c9.17 newtons, etc. Even so, I regard this change as futile since CB leads to abandoning the Newtonian reference-frame altogether.

The laboratory objects used in the study of local gravitation are extremely small relative to the Earth. This smallness in size implies a corresponding smallness in gravitational intensity generated by the small objects. Quantity is not the only significance in science exposition. There are qualitative distinctions that have small quantitative representations of utmost importance for a generalization. Ignoring these small quantities creates obstacles in coming to terms with this problem in its fullest generalization. The object falls toward the Earth and the Earth falls toward the object. We could set ourselves up to misrepresent the relation between the Earth and the Moon. Understanding this error can help us adjust Kepler's Third Law to fit the Jovian planets and render it more accurate for the inner planets as well.

On the Lilliputian scale the 'free-fall' accelerations are dominated by the acceleration due to the Earth. On this Lilliputian scale the effects of the falling bodies on that total acceleration are invisible to us. 'Free-fall' is usually analyzed as if it were a one-sided action, and the acceleration of 'free-fall' is usually calculated as a function of the larger body. This is a misrepresentation that creates an obstacle to rigorous generalization. I would prefer to throw one-sided 'free-fall' out of science and into history.

If this analysis is correct, then we must assert that there is no such thing as 'free-fall'. If there is no such thing as 'free-fall' then we must abandon the Equivalence Principle. The velocity of falling objects is an interaction and therefore also depends on the mass of the 'falling' object. As the Equivalence Principle is a foundational postulate for the Einstein relativistic theory of gravity, the denial of this Principle implies that their theory of gravity is founded on an error. The replacement concept for 'free-fall' is the 'convergence rate'.

For Newton vs CB, if the movements of planets were based on Newtonian mechanical principles, our solar system would have been destroyed long ago, or for that matter, never put together. The structure of the solar system demands a self-correcting system, which the Newtonian system is not. Laplace and Lagrange worked hard to find some principle of stability for the Newtonian system. The smallest disturbance to its movements could introduce destructive chaos to the entire system. This fragility identifies the central defect in Newton's conceptualization and use of 'inertia', where everything rotates or coasts along until disturbed. Inertia has no dereference in CB. 'Impetus' is rehabilitated from pre-Galilean Physics for use as an anti-Newtonian term to correspond to the dual *ex parte* characteristics of Newton's 'inertia'. ***The Newtonian celestial mechanics is self-destructive. The Convergence Buoyancy celestial mechanics is self-constructive.*** This destructive and constructive duality underpins the systemic difference between Newton and CB.

With respect to CB, all restraints are foundationally anti-entropic with respect to the isodynamic state that they can construct. This is true even if they drive the material toward a homeostatic structure significantly different from the original state. This analysis gets more complex when we consider the evolution of star cycles. More details on entropy are given in the following Chapters.

I argue for another usefulness of this dual analysis because it brings to the fore some of the underlying presuppositions and inconsistencies hidden in the original Newton conceptualization. This aspect is of value even if the dual hypothesis fails to subsist.

END Chapter 1

PART TWO

Science-Systems Design

Chapter 2: CB INVERSION SPHERES and the ILLUSION OF MASS

INTRODUCTION. *In general we have a good estimate for the shape of the Earth. However, the opinions of geologists differ widely concerning the models of the Earth's interior. We know very little about the actual structure of this interior. Direct access has only penetrated to c12 kilometers out of c6407, that is, less than 0.27 percent of the depth. Many books and articles present cross-cuts of the Earth showing well-defined concentric spheres. They are often perceived with a misleading clarity. Every description of the interior of our globe is hypothetical. Developing improvements on Michell's torsion balance, Cavendish "weighed the Earth" and estimated from Newtonian calculations its mean density. The current value used for the Earth's mean density is c5.518 g/cm^3 and about 2.8 g/cm^3 for the mean density of its surface material. In Newtonian terms, this difference in density between the globe and its surface material requires that the center of the Earth be made of heavier material than the surface in order to compensate for this difference. One of the early suggestions was to assume that the Earth core consisted of iron. In 1935 the physicist Eugene Wigner (1902—1995), Nobel Laureate 1963, predicted that if Hydrogen were squeezed hard enough it will conduct electrici-*

ty just like a metal. Theoretically, gases in a metallic state could also satisfy the required higher weight.

Surface rocks are easier to drill than deep rocks of the same material. So far, the deeper we go into the Earth the more the rocks are resistant to drilling. This is true for the deepest drilling done in northwest Russia at the Kola Peninsula. In the completed c12 kilometers the first 7 kilometers consist of alternating layers of sediments and volcanic rocks. The next 5 kilometers consist of gneisses. This is direct evidence that the Earth interior is layered chemically.

MECHANICAL DISCONTINUITIES. It is regarded as certain that the Earth interior is also layered mechanically. This is exhibited by the echoes interpreted as mechanical discontinuities. Earthquakes and meteors send shock waves throughout the interior of the Earth. Seismographs around the globe register these disturbances. These waves are like messengers from the Earth's interior. All speculations about the deep structure of the Earth's interior are based on messages of this sort. What is not so well defined is how many discontinuities there are and what are their exact depths. We have to make estimates from the tracing of the observed shock patterns and make assumptions about the internal materials (chemical composition). No one yet has offered non-controversial reasons for these mechanical discontinuities to exist.

Newton's Proposition 70. In *Principia* Section XII (The Attractive Forces of Spherical Bodies) Newton considers the general problem concerning the gravitational action for a globe. The problem is divided into two parts. One is to determine the gravitational action on an object inside the globe (starting with the first proof in Section XII; Proposition 70, Theorem 30). The other is to determine the gravitational action on an object outside the globe (starting with Proposition 71, Theorem 31).

For an object outside the globe, he concludes that the globe forces sum up to a single force, and this single force is equivalent to the force of gravity as if the entire mass of the globe were concentrated at its center (*assuming an adequate distance*). For an object inside the globe, he concluded that the globe attractions toward the surface neutralize (sum to zero) so that these attractions have no gravity force on the interior object. He reasoned in the *Principia* in the following way:

BOOK I: Proposition 70. Theorem 30
If to every point of a spherical surface there tend equal centripetal forces decreasing as the square of the distance from those points, I say, that a corpuscle placed within that surface will not be attracted by those forces any way.

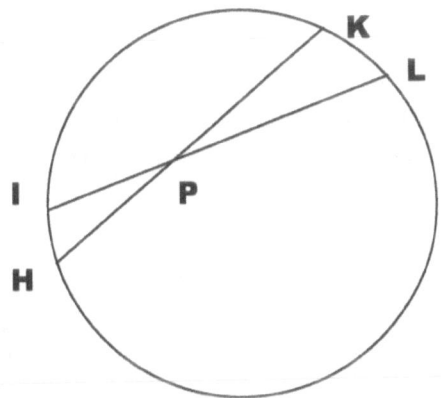

Let HIKL be that spherical surface, and P a corpuscle placed within. Through P let there be drawn to this surface two lines HK, IL, intercepting very small arcs HI, KL; and because (by Cor. III, Lem. 7) the triangles HPI, LPK are alike,

> **Note:** Lemma 7 states: The same things being supposed, I say that the ultimate ratio of the arc, chord, and tangent, any one to any other, is the ratio of equality. Cor. III, Lem. 7 states: And therefore in all our reasoning about ultimate ratios, we may freely use any one of those lines for any other.

those arcs will be proportional to the distances HP, LP; and any particles at HI and KL of the spherical surface, terminated by right lines passing through P, will be as the square of those distances.

> **Note:** Lemma 5 states: All homologous sides of similar figures, whether curvilinear or rectilinear, are proportional; and the areas are as the squares of the homologous sides.

Therefore the forces of these particles exerted upon the body P are equal between themselves. For the forces are directly as the particles, and inversely as the square of the distances. And these two ratios compose the ratio of equality, 1:1. The attractions therefore, being equal, but exerted in opposite directions, destroy each other. And by a like reasoning all the attractions through the whole spherical surface are destroyed by contrary attractions.

Therefore the body P will not be any way impelled by those attractions. Q.E.D. (*Great Books of the Western World*, ed. 1952, vol. 34)

Franklin Neutralization and the Franklin Boundary State. Benjamin Franklin (1706—1790) discovered independently that a neutralization occurs inside electrically charged 'empty' cans. When Franklin communicated this to Joseph Priestly (1733—1804), Priestly verified the results and told Franklin that Newton had shown that this was to be expected.

Note: We can credit Franklin with the first observation of this form of neutralization, but I shall argue that this is not the same form of neutralization that Newton wanted to show.

This was a disservice to Benjamin Franklin because we have no evidence that Newton applied anything like Proposition 70, Theorem 30 to *empty* cylinders or to electricity. Nor do we have any consideration or proof from Newton that this Proposition is functionally independent of whether the sphere is full or empty. Though Newton provided no proof for cylinders it is not difficult to adapt his reasoning to a proof of neutralization within *empty* cylinders (cans) that have a static surface charge of some kind. But neither did he treat gravitation or any other force as a static charge in a surface distribution.

Though Newton's reasoning starts from the spherical surface we can grant that his proof for Proposition 70 treats of attracting forces coming from every direction. He regarded it as undecidable whether this is a tension or a pressure, a push or a pull, etc. He simply refers to it as an attraction.

Note: French critics contemporary with Newton had a good snigger over Newton's theory of celestial bodies attracting one another.

These days we tend to think of surface gravitation only as a centripetal force (the power aspect), and ignore the pressure (or tension) aspects that are omnidirectional. CB references the convergence power (corresponding to Newton's centripetal force) at the surface gravitation with the name *E-power*. Therefore, I have four objections to Priestley's claim that Newton showed that Franklin's observation of neutralization was to be expected.

This phenomenon of neutralization is also observed in cylindrical magnetic fields. This empirical fact is used for magnetic shielding. Magnetic cylinders on the stem of cathode ray tubes use this fact. Coaxial cable uses this neutralization effect for electricity to avoid interaction of two parallel lines of electricity. Herein, any further reference to these forms of engineered neutralization is designated by **Franklin neutralization**. In the case of isodynamic occurrences in nature this state is referenced as the **Franklin boundary state**. These names shall be used for empty cylinders or empty spheres where inner surfaces may be regarded as holding or carrying a charge of some kind. Justification for using Franklin's name in this context can be drawn from above and from the following commentary where it is further argued that Newton did not intend to be treating hollow spheres (or empty cylinders), etc.

Commentary on Proposition 70 Theorem 30: For Newton, the material quantity (Newton mass) is a condition for the force due to gravity. This is implicit in his Universal Law of Gravitation. This is also another way of saying that if there is no mass, there is no gravity. In order for centripetal forces (forces converging toward a center) to be there, there has to be material that generates these forces. Otherwise we would have to say that these converging forces had an outside source, an option that Newton did not pursue.

From the context we know that Newton wanted to make sure that the gravitational force on an arbitrary corpuscle within a solid body is not directed in a direction different from the centripetal direction. This is how he used Proposition 70 for proving Proposition 73 (that a corpuscle placed within the sphere is "attracted by a force proportional to its distance from the center"), which could be called his *centripetal force theorem*.

The introductory sentence just before Section XII (The Attractive Forces of Spherical Bodies) states:

> "Let us see, then, with what forces spherical bodies consisting of particles endued with attractive powers in the manner above spoken of must act upon one another;

and what kind of motions will follow from them." (*op. cit.*)

The context implies that Newton intended his reasoning be applied to material spheres. He put his purely mathematical Propositions in a separate and previous section. All the other theorems in this section of the Principia are directed toward understanding forces within material spheres. If Proposition 70 is about a hollow sphere, it would be the only hollow sphere in the section. This in itself is not an argument against the hollow interpretation. However, we would be hard-pressed to find a use for a proof concerning an empty sphere being used for proving something about a non-empty sphere when the physical forces invoked require material to generate them.

Let us examine another argument to see if the objection can be made clearer

Take a solid homogeneous material sphere with a concentric spherical hollow core. The inner surface of this hollow core could be treated as a charged surface. Outside of it is overlaid by material that allows this treatment. We could conceive of the solid part of this sphere as thinning toward the outer surface. In other words, the sphere radius is held constant but the empty core is increasing in size by decreasing the material symmetrically about the center. Physically this implies that the surface gravity power, **E**, on the outer sphere is decreasing. The question we now ask is; what is happening to the gravitational effects on the inner surface? There can be no effective centripetal forces affecting the inner surface of this hollow sphere. In Newton terms the center has no mass and therefore no force of gravity originates there.

Whatever physical force is expressed from the physical sphere surface is thinned out inversely as the square of the distance. For the Franklin boundary this is not the case.

Whatever forces are expressed at the surface of the empty core, converge toward the common center of the spheres and this implies an augmentation through concentration of whatever forces apply. Furthermore, these forces have equal and contrary forces, which in Newton terms destroy one another.

To show in detail how this argument might go let us restructure Newton's Proposition 70 Theorem 30 to account for the Franklin Boundary in a sphere.

Franklin Boundary Theorem.
Given a homogeneous material sphere that contains a concentric empty sphere within it, if from every point of the inner sphere surface there tend equal centrifugal forces directed outward from the sphere center, I say, that a corpuscle placed within the hollow of the inner sphere will not be impelled by those forces any way.

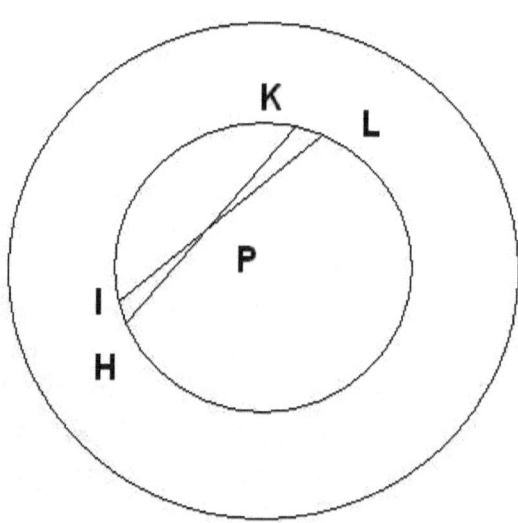

Let HIKL be that inner spherical surface, and P a corpuscle placed within this empty core. Through P let there be drawn to this inner surface two lines HK, IL, intercepting very small arcs HI, KL; and because (by Cor. III, Lem. 7) the triangles HPI, LPK are alike, those arcs will be proportional to the distances HP, LP; and any particles at HI and KL of the inner spherical surface, terminated by right lines passing through P, will be as the square of those distances (in the sense of Lem. 5).

Whatever forces are expressed at the surface of the empty core, converge at the common center of the spheres and this implies an augmentation through concentration of whatever forces apply.

But the forces of these particles exerted upon the body P are equal between themselves. For the forces vary inversely as the particles, and directly as the square of the distances. And these two ratios compose the ratio of equality, 1:1. The forces therefore, being equal, but exerted in opposite directions, destroy their attractions (impelling effects) on each other. And by a like reasoning all the attractions through the whole spherical inner surface are destroyed by contrary attractions.

Therefore the body P will not be any way impelled by those attractions.   Q.E.D.

Commentary on the Franklin Boundary Adaptation. The E-power on the inner surface is centrifugal. This E-power cannot generate from the center of the hollow sphere as there is nothing there. The inner sphere is like a gravitational lens that focuses at the center of the sphere. We therefore might think that the strongest gravitational force is at the sphere center. However, in Newton terms, due to the spherical symmetry all the forces are destroyed by contrary forces. The entire empty volume is neutralized, every position in it is isotropic (no preferred direction). Putting a corpuscle in there, its small effects would likely not render an observable change in this neutrality.

This adapted Theorem can treat the E-power loss with distance as functionally independent of any monotonic law for it, as long as it functions in symmetry from every direction within the hollow of the sphere. With a few adjustments this Theorem applies to empty cylinders, which directly applies to Franklin's original observations.

Using the inner surface as a charged surface was not what Newton intended to do. The outer surface force is centripetal. Though we could use a mathematical model with a trace to the common center (as in customary usage), this model functions as an apodeictic fiction (a fictional device used for argument) that though functionally independent of the physics, it has parametric validity as an effective simplification for using the inverse square law.

If we placed a solid gravitational globe in the center of the empty core we change the status of the corpuscles on the inner surface of the shell. The interior empty region is no longer neutral as there is now a material core that is producing a centripetal force effect on the inner surface of the outer shell. As this interaction is symmetric about the common center, the material inner core, if it has sufficient surface tension, would be centered by the interaction of forces.

Conceive of an *isolated* solid physical sphere of homogeneous material. Overlay a virtual sphere on its surface. Shrink the virtual sphere until there is just as much material effect inside as outside. This sphere now divides the material sphere into two equal sources of gravitational forces. If the force of gravity relates directly to material quantity and tidal effects, then we would have a division of material into two near equal quantities. What is the direction of gravity at this position of the virtual sphere? If a corpuscle is placed on this virtual sphere, which way does it get attracted?

If we believe Proposition 70 as applied to Proposition 73, the direction of gravity mathematically remains directed toward the center at any virtual sphere within the physical sphere. In order for this to be true we have to believe Proposition 70 actually has validity when applied to physical solid spheres. To show this in current terms, Subramanyan Chandrasekhar (1910—1995) used modern integration techniques and this, in effect, increases the dimensions of the variable by one dimension. Thus he also sustains the Newtonian development of a physical interpretation from the physical impossibility. In effect it begs the question. In a Newtonian universe, it is physically impossible for gravity to get expressed as a $1/r^2$ force within the very material that generates the gravity. Furthermore, in the hollow core example the forces from inner surface to center are expressed as r^2 forces, which is to say they are amplified as they concentrate toward the center.

Some readers suggest that his argument covers the case for an empty globe (cf Joseph Priestly), and if this were the case, then Newton would be in error for another reason. The empty globe is contra-indicated by the stipulation that the material is the condition for the attraction. Newton also uses Proposition 70 to prove other theorems concerning solid objects (e.g., Proposition 73). The propositions of the section clearly concern the interior or exterior of solid objects. However, as has been shown, the Proposition can be restructured to work for hollow spheres with inner surface forces that can be treated as charges.

CONVERGENCE INVERSION SPHERE (CIS).

Introduction. *Conceive again a virtual sphere enveloping the surface of the Earth. That the Earth is slightly oblate and bumpy does not vitiate the argument. Imagine that the* *sphere image matches and shrinks concentric to the Earth surface until there is just as much Earth material effect outside of the sphere as inside. With respect to the Earth this makes a division of its volume into two parts with equal material effects in opposition to each other. If quantity of gravitational convergence is associated with the quantity of material effects, then at this division there is an inversion of the gravitational direction, an inversion of the convergence polarity. Convergence and divergence have equal power here. This is the Convergence Inversion Sphere (CIS). It is a hypothesis here proposed to use for interpreting one of the major mechanical discontinuities in the Earth's mantle that divides the mantle into the upper and lower mantles. This division is the Transition Zone (Exhibit 2.1). The existence of this Transition Zone, characterized here as generated by convergence inversion, also suggests possible relations to other observed phenomena in astromechanics and variations in convergence.*

Outside the Earth's surface is the influence of the rest of the Universe. An isotropic convergence of this Universe with the Earth displaces the 'equal effects' division sphere slightly toward the Earth surface in what we call tidal effects. This follows from the assumption that its position is dependent on material distribution adjusted for distance, etc. There are stronger tidal distortions that give variable shapes to the CIS that derive from the Moon and the Sun, etc.

CIS Relative Location Problem. To find the relative position of the CIS.

A, B, and C are the surfaces of three concentric spheres. AB is the thickness of the material shell. BC is an empty kernel.

C is the surface of a material core. The distance CB is used for determining the effect of the E-power of C on the inner surface of B and vice versa. These calculations seem straightforward until we consider how the surface of the core material affects the inner surface of the shell material in the determination of their resultant interface effects. We have left to determine what the influence of a material core has on the neutralization that would occur without it.

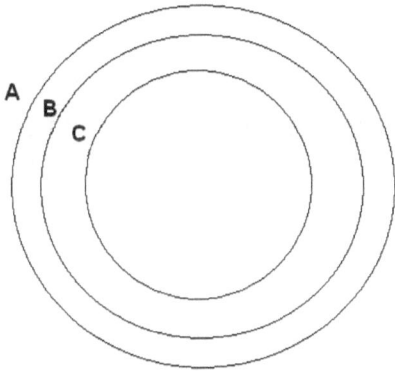

Whatever the E-power on surface B, we know that it will get amplified from B to C following the r-square rule, while the E-power of C to B is diminished following the inverse r-square rule. We now have to determine the position where these contrary forces are equal to each other. This equality marks the position of maximum mutual destruction (total destruction). The gravitation from B cannot pass this demarcation and neither can the gravitation from C. Therefore this sphere marks the locus of a gravitational inversion.

Note: If we ran a pipe from B to C and placed a marble at the inversion point then in a quasi-Newtonian interpretation there is a tension as the marble would be drawn equally in both directions. For the CB interpretation the marble, as debris, would be under a pressure and a tension. There is a slight repelling pressure that is equal with respect to both directions that is overpowered (squelched) by the opposing E-powers. The squelching powers in opposition produce a resultant tension on the debris bulk of the marble, a tension that derives from distinctly different principles from the Newtonian tension.

It is feasible that we can contrive a shell and a core model such that there is a virtual convergence inversion sphere (CIS) in the empty section. The CIS position is where there is just as much power effect calculable from the inner surface of B as there is from the outer surface of C. If we enlarge C by increments we want to know what effect this has on the location of the CIS. As this incrementing increases the E-power of C we expect that the CIS shrinks toward the center to keep the position of an equal power effect to that from B.

If we continue until the CIS is at the surface of C, we next consider two cases. The first case is that the empty space no longer exists. In this case the surface of C is the location of the CIS and we are done. The second case is that there is still an empty space. For this case we next increment the inner surface of the shell, B. As this incrementing increases the E-power at B we expect that the CIS expands toward the outer surface (moving toward a lesser effect from B) to keep the position of equal power with C. This implies that the first increment moves the CIS back into the empty space. Increments can be added until the CIS is located at B. Again we have two cases. The first case, if this takes us to where there is no empty space left then we have a full sphere with a CIS at B. If there is still some empty space left then we can again increment C, which moves the CIS toward the center putting it back into the empty space.

Incrementing the two surfaces back and forth in this manner, we have at the limit a virtual sphere that is interior to A and not at the center that has sufficient material effect above it and below it to render it into a CIS. Q.E.F.

For large celestial objects, from the CIS to the exterior surface is the *upper mantle*, here named the ***shell***. From the

CIS to the interior surface is the lower mantle, here named the *kernel*. If the center is hollow, this hollow volume is denoted the *core* to distinguish it from the shell and kernel.

Estimating the Depth of the CIS. For a material sphere to be divided by a concentric CIS implies that there is sufficient material effect above and below the CIS to mutually destroy the contrary powers. This is equivalent to having two equal quantities of matter juxtaposed. To make an estimate for the depth of the CIS we can start with the formula for the volume of a sphere, $V = (4/3)\pi r^3$ to find the geometric division for two equal parts of a homogeneous non-hollow sphere. From the surface, this division is on the radius at about 1 of 5 parts toward the center. Thus, in this virtual sphere, the natural position for a CIS independent of other neighborhoods and in a homogeneous distribution of corpuscles lies on the radius about four parts of five from the center. Barring tidal effects, this ratio likely holds closely for solid bodies even though the material corpuscles are actually under a pressure that varies with depth and material type. This is from the isodynamic symmetry of convergence power generated about the CIS, consistent with isostasy. The tension at this CIS is dependent on compactness, a function of content, size, and neighborhood pressure. We also have to account for the effects of the spherical shape on the two quantities. If the core is empty there is an increased flattening of the CIS (as it is nearer the surface). As we increment toward a flatter CIS this renders the physical depths of the shell and the kernel toward an equality (an improper limit that they can never reach).

If two objects could be of similar material and compactness that varies only with respect to size, a weaker CIS is associated with smaller objects, a stronger CIS is associated with larger objects. The CIS is subject to tidal effects (shape distortion) under the influence of the neighborhood. Two large objects buoyant about each other influence the shapes of their respective CISs. These tidal influences also exhibit isostasy. Thus if a tidal bulge generates on one side of the Earth, there is an equal and opposite bulge on the other side of the Earth.

CIS Heat Generation. The minimal number of dimensions required for CIS synchronic analysis is three. In certain modes of analysis more dimensions are required.

The CIS produces a tension driven agitation that has a heating effect with a radial symmetry distorted by tidal effects. The diachronic evolution of the CIS can also give rise to resonance effects (dampening or amplifying) that may account for cycles of heating and cooling in long term star evolution.

This inherent tension implies a material distribution in permanent perturbation at the CIS. Barring resonance effects, the degree of heating is directly proportional to the bulk of the celestial sphere and relates to the proportions of the various elements that it contains. The proportions of the various elements are determinate in how compact the material system becomes for a given quantity. This form of material disequilibrium guarantees amplification of material agitation, an agitation (characterized as corpuscular vibrations in a later context) that generates heat. All material celestial objects are subject to this agitation and therefore all such objects radiate heat. For now we ignore speculations concerning antimatter (more in Chapter 5).

As celestial objects accrete material over time, the energy level of the CIS disequilibrium in them rises. This is expressed in greater agitation. No 'fuel' is required for the heat that is produced from this amplified agitation. Nothing is used up to generate the agitation. It is rather a kind of effect. It is the CIS that renders possible and plausible the production of heat without the loss of material, without the need of mysterious unobservable carbon cycles, etc. Furthermore, if there is some primordial location in our Universe, first-generation stars would not even have any carbon.

The agitation level correlates to the object size, compactness, and its material content. The role of compactness is clarified by example. The interstellar gas or dust clouds do not generate the concentration of heat that the stars do. However, we could argue that the total heat in each of two objects, differing in compactness, could be the same with appropriate volumes. If a cloud of dust or gas is equivalent to a large enough multiple for any fixed average compactness, it could sustain just as much heat as a star. Thus the aftermath of a nova might contain just as much heat as the star that exploded, though the temperatures are far below those produced by the star. If heat is conserved, then temperature in homogeneous objects becomes an indicator of heat concentration.

> **Note:** The German physicist, Rudolf Clausius (1822—1888), conjoined the results of Joule with the theoretical work of Sadi Carnot in two Laws. His First Law of Thermodynamics states the conversions between heat energy and work energy. That heat can be 'converted' to work could be conceived as a translation of molecular motion, a micromotion to a macromotion, etc. He developed the concept of entropy (1865), which led to the formulation of his Second Law of Thermodynamics: entropy can never decrease in a physical process and can only remain constant in a reversible process. Formulated from a Lilliputian scale, these Laws do not vitiate the possibility of heat conservation on larger scales.

Since nova clouds may converge to form new stars, the explosions may seem as part of a reversible process. In this interpretation it is more accurate to speak of system entropy in a quasi-reversible process, though the post-nova convergence is not regarded as symmetric since we could have an increase in the number of heavier elements than that found in the original material, etc.

In a given convergence reference-frame such as the Earth, the heat level determined by the increasing bulk of the Earth produces the effect of increasing material agitation at the CIS. This agitation manifests as heat, but this agitation and its relation to surface tension may produce other effects not considered here. Though from certain states not all materials expand with heat (e.g., ice to water is a contraction, M. Fizeau discovered that silver iodide contracts with heat without changing state, etc.), in general, we associate corpuscular vibrations with a counter pressure to the convergence pressure generated in material objects.

Hollow Celestial Spheres. In 1962 Dr. Gordon McDonald, a NASA scientist, said "If the astronomical data are reduced, it is found that the data require that the interior of the Moon be less dense than its outer parts. Indeed, it would seem that the Moon is more like a hollow than a homogeneous sphere." He believed the data were faulty. However, other studies seem to confirm the original conclusion. On 20 November 1969 the Apollo 12 crew, after returning to the command ship, let the lunar module ascent-stage crash onto the Moon. It crashed about 60 kilometers from an ultra-sensitive seismograph. This created an artificial moonquake that reverberated like a bell for about an hour. Some following experiments yielded results up to four hours of reverberations. The conclusion, either the core is made of very light substances or it is hollow. Furthermore, its surface is heavy and difficult to drill, which likely renders the surface a better carrier of the reverberation.

The catechism scientists call people who believe the Earth is hollow, the 'Hollow Earthers'. There are certainly groups of these believers who seem more interested in a Jules Verne fantasy than in factual possibilities. Yet, granting the effects from convergence inversion within the Earth, there could be a hollow core (Exhibit 2.1). This is actually more likely to be the case in spite of the problem it introduces into Newtonian mass requirements for celestial objects. (One of the results of CB analysis is the elimination of the Newton mass concept.)

Hollow celestial spheres imply that these cores enclose a Franklin boundary state for gravitational convergence. Here note that the Franklin boundary state is functionally independent of the direction of gravitational convergence at a hollow core surface. These cores may contain gases, but gases are driven by the radiation gradient (from Russian experiments described in Chapter 4).

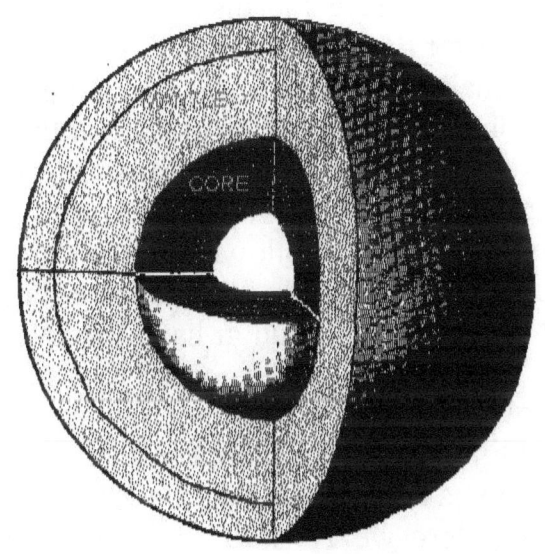

EXHIBIT 2.1: The CB Inversion Sphere divides the volume of material into two near equal parts. The material Earth is made up of the shell and the kernel. The Earth center may be hollow, which implies a Franklin boundary state that nullifies the interior gravitation. It likely contains gases, which complicates the interpretation of 'hollow'. The material shells are likely thinner than shown here.

In the case of hollow celestial spheres, *tidal effects applied to equatorial spin can generate a redistribution of material within the sphere*. The equatorial bulge could generate a mantle thinning from the rotation under tidal influences that would render the polar regions thicker than the equatorial regions. This could be confirmed by seismic studies in or near the polar regions, which should show greater depth for the CIS than in the equatorial regions. This redistribution of material could account for the difference between 9.78 m/s^2 (equator) and 9.83 m/s^2 (poles). *For CB, variations in acceleration due to the E-power for an isodynamic convergence reference-frame can only be accounted for in terms of material distribution*. If the Earth were not hollow then CB would only have the disposition of the chemical layers to account for the variations in **E**, etc.

The CIS is a major indicator of the driver power for shaping the large object form. Matter at the Earth surface and Earth center would condense in the direction of the CIS. If true, this suggests and allows for the existence of hollow celestial spheres. If stars are hollow, it follows that their current compactness ('density') estimates are falsely calculated. French scientists at Meudon have discovered that the Sun exhibits sphere surface harmonics that suggest that it could be hollow. However, gases may be trapped in the hollow in a Franklin boundary state but influenced by the radiation gradient on gas distribution (more..Chapter 4).

Tidal Effects. Here we consider land, water, and atmosphere tides. Tidal effects can in turn influence the shape of the CIS. On Earth the Sun tides are smaller in height than the Moon tides. This is due in part to the flattening effect from the more parallel solar angle of convergence. The tidal effects that derive from the Moon and Sun complicate the determination of the convergence power, **E**, at the Earth's surface. In particular, the tidal effects put isostatic bulges in the Earth CIS. These tidal effects that change the shape of the Earth produce physical handles that affect the Earth spin. The measured Earth spin varies according to these tidal restraints.

Rocks and dirt are less affected by tides than water, a fact related to the relative degrees of surface tension (hardness), elasticity, and compactness for these materials. Water tides rise higher than land tides. The molecular clustering in water that is required for maintaining the surface tension of its liquid state renders each cluster of molecules only several times larger than its vapor form. This suggests that for water on the Earth the influence of molecular clustering on solar tides may compound with the radiation gradient effects (Chapter 4, Radiation and Heat). We know that the quantity of clustering of water molecules associates with water temperature. What is wanted here is a clear determination of what effects gravity has on this clustering, independent of temperature. A critical experiment is wanted for this demarcation. If this demarcation exists, it should be even more evident in the water on board space flights under neutralized buoyancy.

This further suggests that the in-submarine experiments that were done to put down the fifth force hypothesis are rendered into rhetorical nonsense if no consideration is given to the radiation gradient effects on the molecular clusters.

Even so, the isodynamic requirements of CB account for the Stacey experiments (Australia) in a manner not requiring a fifth force.

The atmosphere also has tides but these are more strongly affected by the radiation gradient (more details in Chapter 4).

In Newton physics it is assumed that the Earth spin is sustained by inertia and that tidal effects slow the spin making the Earth turn slower and slower. In CB physics the dual assumption is that the Earth spin is generated by the interaction of its convergence with the neighborhood. The largest effect on Earth movement is from the Sun maelstrom. The isostatic *gc* tidal effects are like handles that become brakes of a special kind. If a car is going up a fixed incline at a given fuel mix and feed rate it will have a fixed speed. Any variation would correlate to changes in air pressure. If we steepen the incline then the car will slow to a lower fixed speed. Lower the incline and the car speeds up again. For the Earth to slow down to a lower fixed angular velocity the tidal restraints must proportionally increase. It seems that if the Earth could be released from them then its rotation would speed up to keep pace with the neighborhood maelstrom. However, this thought experiment is defective because release implies a diminishing of restraints, in effect, a disappearing neighborhood. If there is no neighborhood, there is no spin.

> **Note:** Bishop Berkeley attacked Newton's concept of absolute space in a work entitled *Motion* (1721). In this work he expressed a kinematics analysis of motion that follows a logic that corresponds to the CB dynamics arguments concerning the relativity required for movement intelligibility (visibility).

With respect to the Sun surface, the Sun subtends a much larger convergence angle from the Earth and this largeness thins the Earth convergence power expressed on the Sun surface over an area thousands of times larger than the Earth. The Sun convergence, **E**, also contributes to resistance to tidal effects due to the Earth, etc. We therefore should expect a highly reduced altitude for the Earth tidal effect on the Sun, an altitude that is many times smaller than the Sun tidal effect on the Earth.

Spin-Locking. The tidal effect that the Earth has on the Moon distorts the Moon CIS into an oval shape along the line to the Earth that is just extreme enough to physically restrain the Moon to librations of the same face (wobble effects). Relative to the Earth, the Moon is spin-locked; relative to the Earth, it cannot rotate.

The deepest CIS depth is likely on the obverse and the reverse side. This position is also the likely place for any magnetic dipole effects to concentrate. It is the extremity of this CIS shape that prevents the Moon from spin relative to the Earth.

This spin-lock is common among the moons in our solar system with respect to their host planets. We might wonder why the planets do not get spin-locked by the Sun. The answer lies in the shape of the Sun's tidal effects on the planets. With respect to the Earth, these tidal effects are very broad and flatter on the sphere then the Moon tidal effects. It is the broad spread of the Sun's tidal effect that allows the Earth to spin. Mercury and Venus both exhibit what seems a higher resistance to spin than all the other planets. Their proximity to the Sun generates higher, narrower tidal effects that conduce resistance to rotation. Venus will be discussed in more detail in a later context.

Conjecture: A molten CIS may allow slipping between the shell and kernel. This further suggests that the kernel could have a stabilization effect on Earth shell rotation, etc.

Conjecture: The duration of the Moon revolution around the Earth is very close to the axis rotation of the Sun. This suggests a near synchronization between the Sun spin and our Moon revolution. A more spurious conjecture concerns whether the size of the Moon orbit relates to the Sun. If the Sun pulses, then we should detect some effect of this pulsing in the Earth/Moon distance.

Harmonic Inversions in Celestial Rings. The division of a celestial sphere into two quantities at the CIS also suggests the possible further generation of weaker harmonic spheres that form other mechanical discontinuities. This exposition does not include a complete discussion of these possible harmonics. The CIS itself is the first main convergence harmonic.

Saturn's rings exhibit another form of these convergence harmonics. They are exhibited by the apparent gaps in the rings. The solar asteroid belt also exhibits distribution harmonics (Kirkwood gaps). All the Jovian planets have full or partial rings that exhibit divisions in harmonic distribution.

Saturn's rings have many divisions. The two main divisions consist of the Cassini Division 3,500 kilometers wide, and the Encke Gap, 325 kilometers wide, making three separate major rings. These separations make visible the accretion in circular lines and these align in a manner consistent with convergence inversions. The outer ring is in a ratio of about 1:4 to the middle ring (both are parts of ring A), as if the inner edge of this middle ring were something like a center of a solid object. This ring is likely a harmonic to the Cassini Division, the largest partition (dividing the rings into sets A and B). That the division is almost empty of material is consistent with supposing a convergence repelling effect of collected small objects in space into rings.

If the material in Saturn were evenly distributed, these rings would likely approach a spherical distribution and our analysis would require concentric spheres in a harmonic distribution enveloping Saturn. For CB, ring formations indicate that the polar regions have more material than the equatorial region and therefore repel debris toward the equatorial plane. Saturn's rings themselves are sparse of material, but from a distance we see three distinct rings with two markedly clear divisions. On closer view, many more divisions are visible; all of them are likely due to harmonic clustering from the convergence generated by the ring material (Exhibit 2.2).

The highest temperatures should be at the inversion rings, yet there seems to be little material to sustain the heat. It would be interesting to determine if the ice melts (or sublimates) at a faster rate when near the inversion rings.

It is likely that all spinning celestial objects with a well-defined CIS have rings, though for the smaller host objects the fineness of the ring material may not render easy visibility. The artificial satellites and pieces of space junk left by vehicles launched from the surface of the Earth are likely moving toward an equatorial torus ring position, as individual pieces sink toward the Earth. For spin-locked objects we do not expect a ring formation. A spherical distribution is expected that is distorted by tidal effects from the host and the rest of the neighborhood.

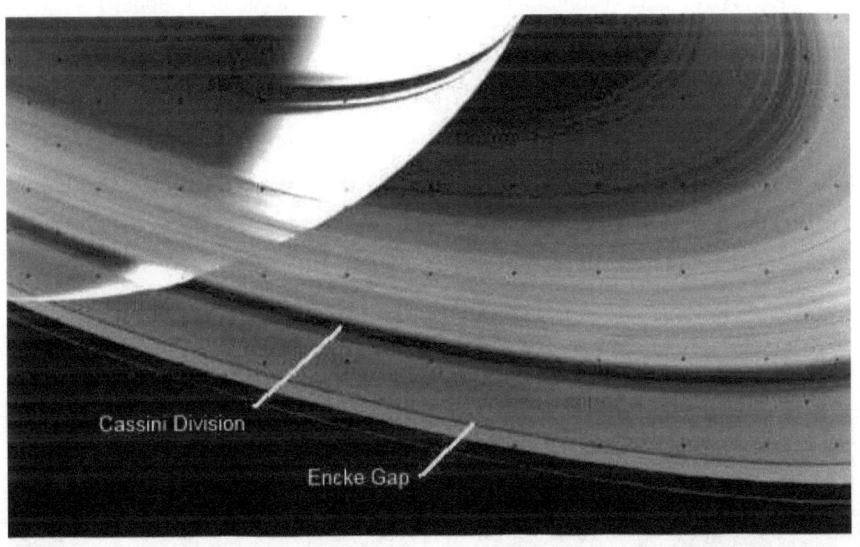

EXHIBIT 2.2 Saturn's rings: shows a clear view of the major Saturn ring partitions (NASA).

Shepherd Moons. All the Jovian planets have rings of some kind. There are a number of instances where it is evident that sectors of these ring structures are stabilized by what are now called shepherd moons. One example is the F-ring of Saturn where Prometheus and Pandora keep this ring narrow and well defined by the repelling generated between them (Exhibit 2.3). CB accounts for this phenomenon, Newtonian physics cannot account for it.

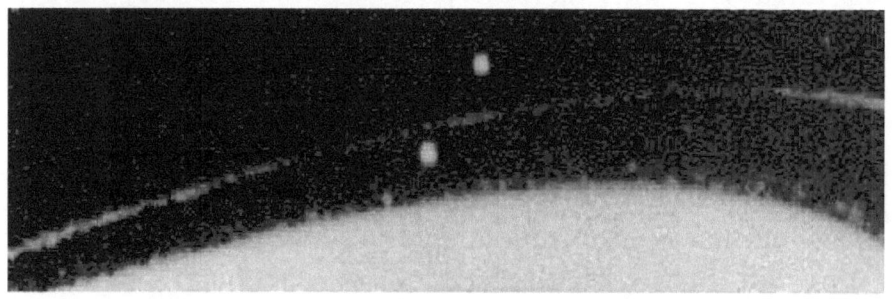

EXHIBIT 2.3a: Prometheus (inner) and Pandora (outer) shepherd the F-ring (NASA)

The high resolution of Exhibit 2.3b below renders visible a ribbon of faint material parallel to the inside of the F-ring showing it is repelled as Prometheus goes toward the right of this view. To the left we see the ribbon has been narrowed. The bright inner F-ring band has a slight bend in it that corresponds to a pushed tidal effect.

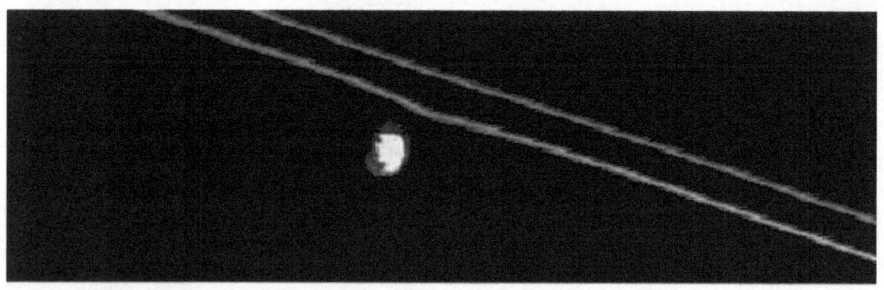

EXHIBIT 2.3b: Prometheus shepherding the inside of Saturn's F-ring (NASA).

Shepherd Planets. In a similar way Jupiter and Mars shepherd the Sun's rings, the asteroid belt. The distances to the asteroid belt rims correspond to the pressures exerted by Mars and Jupiter. Mars likely delays the number of asteroids that might otherwise cross the Earth's orbit. Jupiter also herds two groups of asteroids (called Trojans) at its orbital Lagrange points, one group ahead (L4), one group behind (L5).

When Worlds Collide. Janus and Epimetheus share closely parallel orbits (separated by c50 kilometers) around Saturn at about 151,472 kilometers from Saturn's center. As they approach each other they flip-flop around each other and trade positions, the inner satellite becomes the outer and the outer goes to the inner position. This flip-flop happens about once every four years. They do not collide because when they come near each other their CB is sufficiently strong to repel them around each other.

For Convergence Buoyancy it might seem that because two large objects in proximity repel each other that there would be a flattening of the CIS. This is not directly the case. It is the material quantity distribution and distances that affect the shape of the CIS. This is functionally independent of repelling effects. In plastic objects, the repelling effects would produce an increased pressure on the face toward the action and this has an effect on the material distribution but it is only the distribution that affects the CIS shape. For non-isodynamic encounters, this is also the mechanism for splooch craters (Chapter 6).

CIS Integrity Limit. In the accretion case of the Shoemaker-Levi 9 comet and Jupiter (1994), Jupiter's CIS likely had a slight tidal effect. The comet's CIS was likely at a low energy level, but it could still be drawn into a relatively extreme isostatic ovoid shape along the line of approach. On one of the passes around Jupiter (1992), the two ends of the shaping tension became too extreme for physical integrity and the comet broke into pieces that lined up in a platoon formation (one after the other). This breakup indicates that the sustainment of **CIS integrity** has a limit. This integrity limit associates a CIS mechanism with the Roche Limit; Edouard Roche (1820—1883). The Roche

Limit defines the driver of tidal breakdown in Newtonian terms.

ICONIC VECTORS. One of the difficulties in the arguments concerning convergence vectors below the surface of celestial spheres is in arguments using comparisons between different dimensions, line arguments for planes, plane arguments for solids, etc. These argument forms can lead to contradictions as Galileo has clearly shown (*Dialogues Concerning Two New Sciences*). Using calculus in physical analysis always entails a change in the number of dimensions. For functions of one variable, the first derivative diminishes the powers by one dimension; the first integral augments the powers by one dimension, etc.]

In billiard ball physics, vectors are well represented as lines with direction. They are lossy summaries. Sometimes a non-linear vector is more useful when we wish to keep the representation spatially similar to the phenomenon. Thus the shell and the kernel of a celestial sphere can be regarded as vectors, vectors with volume and direction. Here 'direction' has the larger scope of 'outward' or 'inward' rather than the usual linear notion of direction. The iconic vector of Earth through time could be characterized as a curving cylinder. This notion of iconic vector is synthetic, which renders it distinct from a tensor and its analytic representation.

> **Note:** A tensor is a generalization of a scalar (no index), a vector (one index), or a matrix (two indices) to an arbitrary number of indices.

An iconic vector is not thought of as made up of other vectors. It *is* the vector. The number of dimensions is contingent on the argument. Vectors that sustain a dimensional similarity to the drivers they represent are iconic, summary vectors are data reductions that can be lossy.

Care must be taken in mathematical representation of physical principles neither to analyze (differentiate) the phenomenon into a physical impossibility, nor to synthesize (integrate) a physical state from physically impossible elements. The thesis here is that using mathematics in a manner that is functionally independent of the physics that we want to describe may (but not necessarily) spawn inherent contradictions.

CB WEIGHT

Introduction. *According to current measurement theory, when we weigh an object it is a comparing of the masses under the influence of a gravitational convergence reference-frame. Using a balance we can use any host-object such as the Earth or Moon surface. Within definable but fuzzy limits the results are independent of the power of the host object convergence. A spring scale requires calibration to a standard reference-frame. If I weigh 90 kg on the Earth, it is also true that the Earth weighs 90 kg on me. Newton and CB interpret this in ways that are dual to each other. For CB the role that each object plays in this is a mixture problem. It is much like the Archimedes problem of calculating how much gold is in the king's crown, how much of the 90 kg is due to me, how much is due to the Earth. CB also uses a form of buoyancy for the solution.*

PHYSICAL COMPACTION. Physical states are modes of compaction. There are at least five general distinctions for material compaction states and they vary in their milieu requirements for sustainment:

       1. crystalline
       2. solid

3. liquid
4. vapor (gas)
5. ionic (plasma)

Another state, predicted as the Bose-Einstein Condensate, is thought to have been induced under laboratory conditions in atomic gases by lowering the temperature to near zero kelvins. This state is not entirely understood yet and may be just a more compact form of plasma. Currently it is supposed that the material at this temperature loses its particulate (corpuscular) nature, a sixth state. If this were true we might have a means to devise a critical experiment to determine the wave/corpuscle question in regard to light, etc.

Transitions between states (phase changes) may occur directly or mediately. For example, a solid may become a vapor directly, i.e., sublimate, or it may become a liquid first and then evaporate. Russians in a Siberian winter hang their clothes out to dry. The water freezes (solid) and in Sunlight the frozen water seems to sublimate (goes directly to vapor, no dripping). Our limited results in phase change studies are reflected in our lack of vocabulary for the twenty phase change possibilities. Each state has structural characteristics and neighborhood restraints required for each possibility. Phase changes that appear to skip steps may simply have a foreshortening of the steps, etc.

Changes in heat, temperature, and pressure are used for inducing a change in state. If you raise the temperature of paper to Celsius 233 (Fahrenheit 451) then, in Earth's atmosphere, certain of its constituents go directly from solid to ionic, which exhibits as flames. We do not know of a regression that can restore this paper and therefore this transition is regarded as irreversible. As this transition degrades the available thermal energy, it is also regarded as an example of entropy.

Squelched objects influence the effects that neighborhood convergence has on them in at least five ways, varying according to their physical characteristics:

1. opacity/transparency (flip-flop terms)
2. temperature
3. shape
4. bulk
5. compactness

These influences on the effects of neighborhood convergence have limited functional independence. Herein they are treated aspectually.

1. Opacity/Transparency. It is possible that various elements and conglomerations have different degrees of *transparency* to gravitational convergence effects. The more transparent a squelched object is to this convergence, the more diminished its weight relation with the host. This regard suggests the possibility that Aluminum (Al) is more transparent than Lead (Pb) for gravitational convergence effects. That Lead can be used as shielding against x-rays and other forms of radioactivity shows that it has a high opacity for a wide range of radiation effects and its relative 'heaviness' may indicate its opacity to gravitational convergence. This does not preclude the possibility that a neighborhood convergence could get into resonance with a small object and amplify or dampen the convergence relation. This consideration remains consistent with the hypothesis of transparency effects on convergence power.

The two crystal forms of corundum (Al_2O_3) used in lasers (light amplification by stimulated emission of radiation) are rubies and sapphires. Corundum crystals are in part transparent to visible light. The designed use of these crystals in lasers is currently explained as for light amplification through a resonance effect.

The range of ordinary glass transparency is very similar to the range of human vision. This glass scatters ultraviolet and resists the passage of infrared. Both extremes are invisible to humans. Even so, ultraviolet spectra sunburn human skin and destroy retinas. Furthermore, it is not sensed until the damage is done. That glass resists the passage of infrared explains in part why the infrared generated by the scattered ultraviolet that enters closed cars can render the interior hotter than the outside. This is the sector of the infrared spectrum that humans sense as a warming effect.

Note: the higher energy light, UV, scatters as it passes through the car windows strikes the interior furnishings of the car and generates the infrared, etc.

Transparency can be characterized as a kind of conduction. This suggests another interpretation for opacity, as resistance to conduction. What is conducted is likely related to a kind of polarization. I am talking about the sort of polarization that allows us to look at something and see it with clarity from any direction rather than as some vague thing that would be distorted by bumper-car photons. When I look at an object from one direction and then move to look at it from another direction, somehow the two directions do not interfere with each other. A form of polarization could allow for this. In effect, I am blind to the polarities that are not in line with my sight (more in Chapter 5).

With the possibility that objects are partially transparent to gravitational convergence disallows any simplistic use of compaction or the angle of subtending convergence in determining convergence neighborhood power. It also puts doubt into regarding gravity as directly proportional to mass when in fact *we are treating mass as directly proportional to gravitational interactions*. In terms of operational definitions in how we actually determine mass we must first assign a value to mass through observation of gravity accelerations, or more usually, the impedance of gravitational accelerations for material objects in a given reference-frame. The magnitude of CB bulk is determined through a mode of analysis of the impedance of gravitational accelerations for material objects in a given reference-frame.

Note also that as objects become large enough to develop a well-defined CIS, the convergence inversion (reversal of direction) suggests a possible dampening of their transparency and thus the neighborhood convergence may have a stronger effect on them. We presume that they in turn are gaining convergence power with their bulk. The CIS therefore also contributes to a non-linear augmenting of the repelling power, thus a compensatory heavy core may not be required, etc.

An opacity (transparency debit) to neighborhood convergence is likely an important factor in the measure of weight on host surfaces. Increasing opacity implies a proportional augmentation of weight, i.e., decreasing transparency increases the magnitude of the measured weight. If Aluminum is highly transparent to neighborhood convergence, its measured weight may be proportionally diminished. If Lead is highly opaque to neighborhood convergence, its measured weight may be proportionally augmented. This not only puts into question the determinations for material quantities, but also puts into question the determinations of atomic numbers and may explain why so many neutrons are postulated for heavier elements.

It is conceivable that gravitational convergence is also amenable to amplification and dampening through resonance. Furthermore, convergence transparency suggests the possibility of material gravitational lenses.

Conjecture: In a limited way, light amplification and spectrum shifting are used for the better night vision devices. For night vision, the Americans reverse engineered a light amplification system, which gives an unintentional spectrum shift toward the green (with a kind of filtering). Protracted use of these light amplification devices can damage the eyes permanently. This is a case where a better image imitation might be derived from virtual reality techniques. Virtual reality techniques could provide a shift result that is equivalent to a physical shift of the available spectrum toward the visible spectrum. We have a control over virtual reality that could render the effects of spectrum amplification and shifting without direct contact with physical changes. In the extreme this suggests the possibility of rendering images from apparently opaque walls, etc. If we could produce a virtual *spectrum shifter*, it is feasible that we could see through certain objects (that pass any unscattered radiation of any kind) by shifting the passed spectrum to the spectrum range of human vision.

2. Temperature. Resistance to transparency likely has direct effects on indigenous temperature inductions from radiation. Not every object exposed to a given quantity of radiation responds with the same temperature (specific temperature, as distinct from specific heat).

> **Note:** The *specific heat* of a material is expressed as a ratio: the heat magnitude required to raise a mass unit of the material one temperature unit to a reference material heat magnitude similarly obtained. The *specific temperature* of a material is expressed as the temperature magnitude of a bulk unit when the bulk attains a steady-state temperature relative to a reference level of radiation. Both definitions require recognition of physical integrity limits. Furthermore, the specific heat for a given substance can vary depending on the starting temperature as Maxwell points out.

If material objects have a degree of transparency for gravitational convergence, if transparency is a kind of conduction, and if temperature corresponds in some way to this conductivity, then this suggests the possibility that temperature itself can have an effect on the measured weight of the object. If a lowering of temperature of a material object could increase its conductivity of gravitational convergence then the material object would weigh less, etc. There is some evidence that metals can vary in weight with temperature. More experiments are required to determine the exact effect of the various differences in potential for radiation intensity. Observations also need a precise critical analysis regarding the effects of Archimedean buoyancy (based on author experiments done in Paris, 1983).

When raising the quantity of heat (corpuscular vibrations) in metals it is ambiguous whether weight changes are attributable to a change in convergence conductivity or induced from 'absorption-emission' activities (impetus radiation, etc.).

3. Shape. Given a spherical object with a fixed weight and adequate plasticity, what happens if we flatten it? Will it weigh less or more standing on its edge, etc. We might reason that flattening increases transparency through the thinner dimension and increases opacity through the longer dimension, barring new alignments or resonance effects. This possible shape effect interferes with having a simplistic determination of the demarcation under scrutiny. Also, on a Lilliputian scale (the human scale) there is already a measurement difficulty due to the extreme accuracy required. However, it is useful to know the *ex parte* influence of the measured object on the quantity measured. When shape influences transparency, it influences convergence power effects.

Thought Experiment. In Newtonian terms, whether or not the shape taken in a neighborhood can affect a weight measure can be settled in the affirmative without even doing the calculations. If we stretch a wire from the Earth to the Moon, how much does it weigh? If we coiled it on the Earth, how much does it weigh? We do not even have to bother with exact calculations to know that the same wire coiled on Earth has to weigh more on Earth. This follows from the fact that per unit volume it is subject to a higher average gravitational convergence power in its coiled form on the Earth surface. This is a logical conclusion derived from superficial facts (Moon convergence power is 1/6 of the Earth's, etc.). Showing the arithmetic is just busy work for detail. Here we note two influences on the outcome, the non-isotropic convergence fields and the shape of the object. This thought experiment shows an intrinsic difficulty in determining the isotonic aspect of the question that theoretically renders the object measurement independent of the neighborhood.

4. Bulk. Measured weight varies according to the neighborhood in which it is measured. What we want is an extensional magnitude for weight. What we get is a magnitude measure that can vary with the host. The quantification problem for a fixed quantity of matter has two main aspects to consider. First, we want a representation of this quantity that is functionally independent of the neighborhood it is measured in. Newton intended 'mass' for this independence as a distinction from 'weight' that varies according to **g**. Second, we want an extensional representation for it. In this, Newton's notion of 'mass' fails because he only provides intensional definitions for it, e.g., **m ~ F/a**. To date no one has found an extensional magnitude for weight that has the directness of measure that length has for distance. The usual magnitude representations are intensional Earth-bound standards derived from measured weights that have only a local parametric utility. We can separate a measured weight into two parts (each part is an *ex parte* magnitude), one part is the proportion due to the host, and one part is due to the debris (as argued in Chapter 1). If we can determine the *ex parte* debris weight on Earth and the *ex parte* debris weight on the Moon, these magnitudes should be equal from the fact that the object material quantity has not changed. This invariance allows its use as a weight measure independent of the neighborhood it is derived from. It also allows us to treat the *ex parte* debris weight as a Eudoxian magnitude.

> ***EX PARTE* WEIGHT QUANTIFICATION:** *Bulk* denotes the quantification of the *ex parte* weight of a material object and the dimension of its units is denoted by **[B]**.

We are left with the problem of how to solve for the *ex parte* weight.

In effect, to quantify the *ex parte* weight of debris is to find the measure of its bulk. The representation mode for a weight quantity can vary (e.g., kilogram, pound) whereas the quantity remains invariant. We can show that the direct measure of weight is not Eudoxian; therefore calculations with it have their attending limits. To develop transformational equations for translating an on-Earth measured weight into a measured weight that is local to a different surface convergence power we need a way of rendering the measure Eudoxian.

Our only access to bulk measure is through measured weight. In order to transform the debris measured weight into a Eudoxian bulk representation we have three things to do:

1. **Identify** the *ex parte* debris convergence power (includes elimination of constraint impetus, adjustments for Tidal Effects, and other effects of material distribution).
2. **Adjust** for Weight Decrement.
3. **Partition** this result into *ex parte* debris weight and host impetus

The first two steps are sufficient to render measured weight into a quantity that is independent of certain variations though not functionally independent of the locale. This is to say that it is rendered Eudoxian for the convergence reference-frame in which it is derived but it is not yet functionally independent of that reference-frame. If the first two adjustments can be accomplished then we have a Eudoxian magnitude, the *CB weight*. In the case of debris, if this result can then be partitioned into the ***ex parte* debris weight** and the ***ex parte* host impetus** (acceleration invested in the debris by host), then we have the bulk measure of the debris. By definition the bulk measure of the debris is its *ex parte* debris weight. To solve for it we must have an operational definition for each of the steps. *Weight Decrement* (discussed below) is the key concept for understanding the status of measured weight and why it is non-Eudoxian.

For now there is no reason to presume that my convergence, **e**, can be augmented or diminished by neighborhood convergence, **E** (barring possible resonance effects). If on the Earth my measured weight is 90 kilograms (Newton 'mass', but used as a common weight measure), then on the Moon my measured weight is about 15 kilograms (*not* Newton mass). For now, it seems reasonable to assume that my *ex parte* participation in these different measured weights is conserved. That is why I say that the *ex parte* debris weight is invariant by presumption. Measured weights do not directly reveal my participation. My participation in these measured weights is directly proportional to my own convergence power, and it is this convergence power that I claim remains constant (if physical integrity, including volume is conserved) in the various host environments. With it I can find the proportion of the measured weight that is due to me.

It is simplistic to assume that the chosen representation of the calculated bulk has an exact relationship with the quantity of matter. Bulk is a form of relative quantification and it is conceivable that a determination of bulk **A** is greater than a determination of bulk **B** and yet bulk **B** contains more matter. Such a state might be accounted for by one or more of the other four influences on bulk determination, such as transparency, etc. This is why I say that bulk is not a fundamental concept of CB. *Our bulk (*ex parte *weight quantification) only measures aspects of neighborhood CB interactions.* This is why for CB the Newtonian mass concept is an illusion derived from a misinterpretation.

Though much useful physics can be done that ignores these difficulties, we have come to a level of exactitude where these features begin to show their relevance. Until we can find a method for determining an intrinsic bulk that actually measures material quantity we are left with our faulty extrinsic methods. Even so, we find our crude methods useful as a relative quantification for matter of a certain kind in a certain state in terms of how it interacts with a neighborhood.

5. Compactness. If we can vary compactness, it seems logical that an increased compactness could yield an augmented opacity and hence of weight if the material alignments remain unchanged, etc. In the case of rubies and sapphires, their compactness in a crystal form is more transparent to light than the corundum non-crystal form (however, crystallographers traditionally partition all material states only into gas, liquid, or crystalline). This crystal form represents an overall re-alignment that may obtain a higher compactness (alignments can also obtain the reverse, such as water to ice). If the overall gravitational convergence opacity is decreased, then these crystalline forms should weigh less. By analogy this suggests we might find a material alignment for certain elements or compounds that are relatively more transparent to gravitational convergence than their other forms. Any directional influence implies the possibility of a gravitational convergence lens.

Using the interpretation of 'opacity' as a kind of resistance, it seems logical that measured weight per unit volume should increase though not necessarily linearly with opacity. The more resistance it offers to the passage of polarization, the more opaque an object appears to be, and conversely. This is likely why the measure of the 'speed' of light phenomena riding as photonic polarization yields slower speeds when passing through certain 'media'. It can also vary with the heat level. By analogy, it seems possible that measured weight magnitude should vary in proportion to the compactness (at least within the constraints of the CB weight concept explained below and barring change from resonance effects or change in transparency due to new alignments).

> **Note:** In 1999, Danish physicist, Lene Hau, at Harvard University managed to slow visible light to 60 kilometers per hour in a claimed Bose-Einstein Condensate. In 2001 she 'stopped' light in a vapor of Rubidium gas. In CB these light phenomena ride on the photon-graviton polarity (details in Chapter 5). It would be interesting to test different parts of the spectrum in these experiments to determine whether the rider velocities are equal.

The power of convergence is associated with a quantification of the *ex parte* weight, **bulk**. For CB, bulk is a mode of comparing convergence powers expressed as *ex parte* weight. This in turn can be associated with volume as bulk per unit volume, **compactness**. Its unit in this text follows the SI, the kilogram. Because Density (originally defined by Euler as mass per unit volume) is derived from a distinctly different conceptual mode, CB avoids the term and uses in its stead the word **compactness** in terms of convergence power effect per unit volume. Density is defined as a composition of the dimensions $[M][L^{-3}]$, where M is mass and L is length, hence mass per unit volume. Compactness is defined as a composition of the dimensions $[B][L^{-3}]$, where

B is bulk and **L** is length, hence bulk per unit volume. This compactness sustains comparability limitations that correspond to bulk comparability. *Bulk only measures aspects of neighborhood CB interactions.*

Example: From the old definition, water at 4° (more accurately 3.98° at standard pressure) Centigrade (Celsius) is 1,000,000 grams per cubic meter or 1,000 kilograms per cubic meter. To reduce this ratio directly to one gram per cubic centimeter by arithmetical division assumes that measured weight is Eudoxian, an assumption not sustained in the Convergence Buoyancy reference-frame (shown next).

Accretion Weight. In CB, *accretion weight* is a special case of a measured weight in that it excludes any constraint impetus effects. The measured weight for debris resting on the surface of a planet is an effect of accretion weight and is composed of the debris *ex parte* weight and the host *impetus* weight (without constraint impetus). On Earth my accretion weight is 90 kg, which is my measured weight without constraint impetus. Similarly, my accretion weight on the Moon is about 15 kg. The next Chapter discusses the more general case of impetus weight that includes constraint impetus.

When debris sink toward the surface of the Earth this action reaches its isodynamic state only when the debris accrete to the Earth surface. Prior to accretion, the debris are always in a motion toward an isodynamic state. If the debris are not returning to an original state, then they are in a *qualitative* regression that is driven toward a new isodynamic state.

Impedance as Weight. In CB, impedance to acceleration is expressible in weight units such as kilograms because, in fact, the measure for weight is always a measure of impedance to acceleration. In the case of weights on Earth CB uses the notion of accretion weight to designate a weight based on restraints but free of constraints. To say I weigh 90 kilograms is to say that the impedance to my acceleration toward the center of the Earth is 90 kilograms.

Momentum has no dereference without a neighborhood because its units have no dereference without a neighborhood. It is a term invented by Newton and he himself never found anything useful when it is not converted to a force by a change in velocity. In current physics instruction it is used for demonstrating the validity of Newton's Third Law of Motion even though no one really makes any sense of the kilogram meters per second as a unit, a verbalization of **mv**. Impedance is measured in weight units. Accretion weight is a form of impedance. but impedance is not necessarily an accretion weight. Impedance has a larger scope of applications of which accretion weight is a subset.

To say that we would weigh in at different weights on the Moon and on Mars is to say that there is a variation in the magnitude of impedance.

If we fell off a building to a sidewalk below, the required impedance to halt the fall on the sidewalk would reveal the impact weight pressed against us when we land. The units for impedance are only weight units. Knowing the momentum at any given moment in the fall has no use as we can directly calculate the impact weight relative to any building height. If we are struck by a bullet and it goes through the body, the body offers some resistance to its flight and this resistance is an impedance to its flight. If at the same distance the bullet were fired to a wall that it could not pierce, this impedance to its flight is expressible in weight units and we would know the impact weight that the bullet applied to your body. Again, momentum has no use here.

Momentum Misconceptions. Momentum is often written as **p=mv**. Newton, who invented the concept, never found much use for it. A moment of acceleration is a velocity, and this is why he named **mv** momentum. Catechism physicists have been trained to believe that momentum is a force. For Newtonians, force in an inertial reference frame is written as **F=ma**. This implies that **p** is not a force until its velocity changes. A change in velocity is an acceleration (plus or minus). Many physics teachers think when they use collision experiments that they are demonstrating momentum. Collisions apply abrupt changes to velocity which convert momentum into a force. The potential force of momentum can only be revealed through a change in velocity in an inertial reference-frame. It is easy to show that

$$m\Delta v = m(v/s) = ma = F.$$

where Δv is a change in velocity. By using collisions they decelerate the vehicles, etc. In other words they are converting momentum into a force. Thus all they show is how certain instances of Newton's Second Law of Motion can exemplify his Third Law of Motion.

The term 'impedance' is a more general term that expresses in only weight units the measure of resistance to a motion relative to a given CB reference-frame, whether that relative motion is accelerating (plus or minus) or at a fixed velocity. This also implies that the weight units themselves always express an impedance (more in next Chapter).

WEIGHT DECREMENT FUNCTION. It is supposed that the *ex parte* weight quantification (bulk) of an object remains fixed for every neighborhood that the object participates in if its physical state, including volume, is conserved. Since the bulk quantity itself is Eudoxian (though not in the terms of Newtonian mass, since bulk is calculated from convergence buoyancy relations rather than through some absolute notion of quantity of matter) it seems feasible that a formula could be found that would render this graphically. To discuss how this works we can use an apodeictic model that displays the main features required for such a formula in a CB reference-frame.

Let **b** be a bulk, a Eudoxian variable. **W** is the accretion weight, a non-Eudoxian variable, With **(W≥ 0) & (b ≥ 0)**.
 a is a constant of proportionality.
 k is the bulk float-sink boundary, an aorist constant.

> **Note:** An *aorist constant* is like a snapshot of an event in a moment of time, barring changes in convergence power, this makes **k** a function constant that only varies if the pressure in its ambient neighborhood varies. The current use of specific heat is also an aorist constant as we know that specific heat differs according to the temperature we start at for liquids and solids. However, these differences in gases have very low empirical visibility. Velocity as a moment of an acceleration is an aorist constant.

Then with $W = a(kb - b^2)$ we map an apodeictic model, Exhibit 2.4, designed to display the main characteristics of accretion weight in CB arguments.

This is a **Weight Decrement Function** for accretion weights.

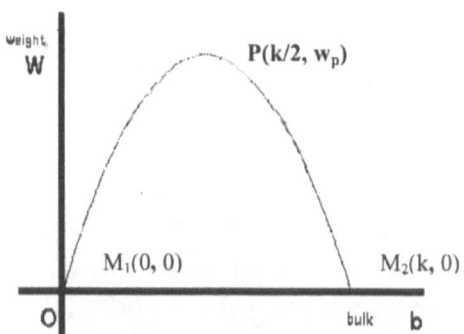

EXHIBIT 2.4: Weight Decrement Function $W = a(kb - b^2)$; where accretion weight is expressed as a function of bulk under a squelching convergence, with the peak weight, w_P, at **P**, and zero weights at M_1 and M_2. M_2 is at the boundary bulk, **k**. (Except for **k/2**, each weight has two distinct instances of bulk symmetric about **P**).

The graph in this Exhibit shows accretion weight as a function of *ex parte* weight, bulk. From the graph we see that accretion weight, **W**, is not Eudoxian and the decrease in material quantity maintains an increasing rate of change in magnitude (slowing down) that produces a reversal pattern as shown. More investigation is required to determine the actual curve. It will likely take the form of an ellipse segment that is catenaried upward, because for the most part it is a resultant of the repelling relations between debris and a host determined by their material distribution adjusted for distance. It is certain that a specific accretion weight, w_p, depends on the convergence power of the reference-frame. Thus it may vary from reference-frame to reference-frame.

Weight Decrement. For squelchable objects (debris), the apodeictic graph for the Weight Decrement Function when we add two equal and similar in kind bulks, shows that their combined accretion weight is smaller than two times the original bulk measure (this is a sufficient condition for showing it is not Eudoxian). More generally, if we combine any two debris bulks together to form one bulk, the resulting accretion weight is less than the sum of the individual accretion weights. This is why the result of adding or multiplying these weights is referred to as having a **weight decrement**. We can define the reverse process, subtracting or dividing, as **weight increment** for the accretion weight of the parts. Dividing an accretion weight into two parts, the measures of the parts sum to more than the whole.

CB provides a paradigm for describing in physical terms why atoms and molecules that combine into larger compounds weigh less than the sum of their parts. The smallness of scale renders a proportional increase in the comparable differences. In the Newtonian tradition this disparity in weights is thought of in terms of Newtonian masses and called the **mass defect**. In that same tradition this set of phenomena remains a curiosity with no generally accepted explanation. The CB model also provides a paradigm for understanding the differences among the electromagnetic 'force' and other 'forces' on the micro-particle scale. If it involves opposed polarities, then it is CA, else it is CB.

> **Note:** It is said that the strong force, electromagnetic force, and weak force govern the structures of atoms. In 1969 Steven Weinberg, Sheldon Glashow and Abdus Salam are credited with unifying the weak force with electromagnetism, which is CA. CB provides a different mode of analysis.

CB provides a basis for putting into question the interpretations in the use of electric polarity in regard to analysis of microparticle phenomena. It suggests that the microparticle phenomena observed in electromagnetic fields are sufficiently distorted by these fields to class the interpretations as possible misrepresentations. That anything effective might come of this sort of research likely stems from a parametric effectiveness (see Apology). That an electron is 'negative' and a proton is 'positive' is based on the simplistic acceptance that the powerful fields used in these studies do not change the apparent nature of the particles observed. Maxwell teaches us that a body charged positively will move toward a fixed position lesser positive charge *or* negative charge. Thus we may not know if the positively charged body is moving toward a lesser positive charge or a negative charge. This can lead to a misrepresentation in the interpretation.

Debris and Host Objects. In CB theory, {host, debris or their synonyms; large object, small object} are categories that are useful for distinguishing types of like-pole convergence interactions in a given scale of physics. The graph in Exhibit 2.4 helps to clarify the demarcation between 'large' and 'small' objects of like kind.

DEFINITION: The term **debris** (small object) denotes an object below boundary bulk (< k); the term **host** (large object) denotes an object above boundary bulk (> k). The term **boundary object** denotes an object that equals the boundary bulk (= k).

As noted before, the dereference for 'boundary bulk' varies according to the physics that dominates the scale of observation. (There are other details of the boundary bulk concerning positioning that are not considered here.) Therefore 'small' and 'large' also dereference according to scale {gravitation, surface tension, atomic}, they refer to quite different objects on each of the scales. In micro-convergence neighborhoods surface tension convergence overpowers gravitational convergence, etc.

Though the graph for the Weight Decrement Function illustrates the principle used for the demarcation between the small and large objects, it remains apodeictic. We have yet to find a decidable range of magnitudes for them.

Thought Experiment. As you add bulk to a debris object with surface convergence **e**, you are also increasing **e**. The major portion of the accretion weight is generated by the host convergence, **E**, and is expressed as an *impetus weight* (the residuum). As **e** increases, the debris object bulk and accretion weight augment at a slower and slower rate (the squelching power on **e** diminishes as the convergence of **e** augments). Finally, we can add a last bulk that stops all possible additional increase in the weight measure in its relationship with its host. This is at the **turnabout weight** point, **P**, the *peak weight* for debris convergence. All further weights added to this bulk generate an **e** that repels sufficiently to subtract from the total weight, beyond this point the weight diminishes toward zero as its bulk (in terms of the number bulk units) augments. This is what is denoted by 'turnabout weight'.

Note: Though the number of bulk units augment in a Eudoxian manner, the graph shows that the Weight change is non-Eudoxian.

When augmenting the bulk from the origin until accretion weight reaches zero again it is at the **e** boundary convergence level, M_2, the boundary between accretion and buoyance. At this point, M_2, adding more weight makes the object appear to have negative weight in a convergence neighborhood with higher pressure. The **e** is here large enough to generate a repelling to buoyance effect; in effect **e** goes to **E**. This repelling maintains a level with the host convergence that is isodynamic with it. When **e** goes to **E** the framework required for a direct access to accretion weight is lost. Float buoyancy is like a negative weight that buoys up to an isodynamic pressure where buoyancy and convergence are in equilibrium. Between the host and former accretion system there is now a reversal zone, a **Convergence Reversal Zone** (CRZ).

Note: This equilibrium between buoyancy and convergence is not equivalent to the boundary buoyancy. Boundary buoyancy is strictly defined to accord with the point M_2 on the Weight Decrement Function.

Convergence Reversal Node (CRN). There is an isodynamic Convergence Reversal Node between the Earth and the Moon. The difference between CRN and CRZ is that the CRN maps the point where this reversal occurs and the CRZ maps the locus of possible occurrence. The isodynamic CRN lies within a discus zone of repelling placed between the Earth and the Moon (CRZ). Today the isodynamic CRN in line with the opposed convergence lines corresponds to what we know as one of the Lagrange points, **L1**. Lagrange calculated in Newtonian terms that this is an equilibrium point and the net force is zero (note that empirical observation shows that it is not exactly at the Lagrange predicted place). The mean distance of this

Earth/Moon reversal occurs near 338,648.6 km from the Earth center, about 37,638.3 km from the Moon center. Using values from the Newtonian model, at this reversal node the mean acceleration is calculated at close to 0.0035 ms^{-2} for each of the two opposed directions. For the (Newtonian) Lagrange point it is a place of tension, for the CB Lagrange point it is a place of compression. This is a fundamental duality between Newton and CB. Note also that this reversal node does not represent a reversal of actual convergence in the manner of the CIS. The isodynamic CRN is a demarcation for the boundary between two external convergences (within the neighborhood they participate in). Finding the compression at a CB Lagrange point requires more considerations than the simplistic calculations required for a Newtonian Lagrange point (see below).

The isodynamic state generating the CRN allows us to analyze the CRN as between *convergence powers* and their maximum *generated repelling* (relative to the neighborhood pressure they participate in). The interacting objects would continually conform to and influence neighborhood restraints. Note that at an isodynamic CRN, the linear resultant magnitude seems too small, but the magnitude of isodynamic repelling centered at this node is never zero and spreads over the entire area formed by the repelling convergence lines (Exhibit 1.2A). This puts the CB Lagrange point closer to the observed location.

Reference-Frame use of Limits. The Earth and the Moon form an approximate ideal system of two host objects of different convergence fields that come into equilibrium without obtaining surface contact. It is approximate because there are other objects influencing this two-body system. In this regard all convergent systems we consider are condemned to approximation since we have no way to isolate any system that is less than everything. However, it is at their limits that convergence systems show their *ex parte* measure of participation. On the other hand, it is at their limits that the inertial systems fail. To have a purely inertial system we would have to turn off gravity as well as other influences different in kind that distort inertial effects. In CB for example, using limits we can artificially isolate the convergence powers of neighborhoods and convergence objects. These limits can be used for accurate estimation of *ex parte* participations. However, the galactic scale may introduce convergence influences based on matter and antimatter polarities, which are not a form of CB, etc.

The spin of an isolated Earth (at a hypothetical limit of isolation) would not exhibit as a spin because spin has no significance in a convergence-free neighborhood. In CB, kinematics and dynamics are functionally independent of each other, though a bound equivalence between a kinematics case and its dynamics may be possible with a constant or function of proportionality to correspond the two different modes of analysis with the exactitude required for the problem at hand or for scalability.

CALCULATING CB WEIGHT. The previous arguments set up a demarcation problem that requires operational definitions. If my accretion weight is the result of an interaction, how can we calculate how much of the weight is from me and how much is from the Earth? How can we reason to the *ex parte* magnitudes that determine the contribution of each participant. The usual numbers we associate with the Newton mass and the ways we talk about it suggest that it is me rather than the Earth that is largely responsible for the weight that I register. In fact, my contribution to the 90 kilograms I weigh is at least 22 orders of magnitude less than 90. In order to clarify the dereference for the *ex parte* CB weights, I shall first argue with Newtonian formulas in the manner used in Chapter 1.

Using Newton Formulas. Writing big letters for the big object (host) and small letters for the small object (debris) we have two Newtonian equations for expressing the weight relation with gravitational acceleration:

$$w = mG$$
$$W = Mg$$

My weight on the Earth is '**mG**' and the Earth weight on me is '**Mg**'. This weight is expressed as newtons. As vectors, **w** and **W** are opposed, but as we are only interested in comparing their scalar values we can write that $|W| = |w|$ and

$$m|G| = M|g|.$$

From this we have:

$$|G/g| = M/m.$$

Comment: This proportionality shows that two masses in a weight relation have the same ratio as the scalar values for their surface convergence accelerations. I argue that what Newton called mass is always determined by ratios and that this renders it as nothing more than a constant of proportionality that only seems unchanged from reference-frame to reference-frame. Expressed as ratios of same units they are independent of dimensions. These conclusions show one source of Ernst Mach's complaint. The proportion also shows why Newtonian mass may be used parametrically even though this use of the magnitude can mislead our physical interpretations.

From results in Chapter 1:
Using the derived proportions for weights on Earth
$9.81/g = 5.975 \times 10^{24}/90$, which yields
1.477×10^{-22} ms^{-2} for my **g**.

This is my *ex parte* gravitational acceleration due to me. The Newtonian weight is given by **W = mg**. If the mass is 90 kg and **g** acceleration is 9.81 ms^{-2} then the weight is 882.9 newtons. I use the ratio of **m/M** $= 90/5.975 \times 10^{24} = 1.506 \times 10^{-23}$, which is one part to 1.506×10^{23} parts. Multiplying this scalar result by 90 kg and we can derive two quantities that match this ratio: 1.3556×10^{-21} kg and $(90 - 1.3556 \times 10^{-21})$ kg.

My *ex parte* mass, 1.355×10^{-21} kg, is some 22 orders of magnitude below the usual Newtonian mass numbers for 90 kg. Note that this result is only about three orders above current values for the masses assigned to atomic particles. Note also that if we calculate the measure of these particles in this new way that they also become some 22 orders smaller in their *ex parte* calculation results. This also renders **E=mc**^{2} about 22 orders smaller. This renders the equation into a pseudo-equation insofar as the magnitude of its **E** (energy) depends on the **m** (mass) units, which in a Newtonian framework cannot be determined in any absolute way. If the new units derived from the proportion are chosen for **m** in this equation this makes the fission bomb outdone many times over by a firecracker. This is another way of saying that the **E=mc**^{2} equation (Olinto De Pretto's formula, 1903) using the current value for **c**, is submicroscopic in its predicted effect and therefore either **c** is wrong or the formula has nothing to do with the fission bomb. Furthermore for CB, energy is not an alternative form of substance, it remains a capacity to do work, a function of physical structures and arrangements.

Applying the same mode of calculation to physics on a different scale can be misleading if we do not account for the shift in what dominates the actions on this different scale.

Using CB Formulas. Except for the peak weight at point **P**, as shown in Exhibit 2.4, each weight has two distinct instances of bulk symmetric about **P**. The first problem in the general solution for *ex parte* weights is to derive the CB weight so that we have a Eudoxian representation of weights as bulk. With the CB weight there is no duplication, no pairings of bulk as found in accretion weight. This argument affirms that Exhibit 2.4 displays two modes of expressing weight; accretion weight vs bulk (*ex parte* weight).

Rather than taking my 90 kg as mass I shall now regard it as my accretion weight on Earth. I then reason from the fact that whatever weight is measured is the result of the deceleration from the Earth's surface, a form of impedance. If my accretion weight is 90 kg on Earth, the CB interpretation suggests a difference in the mixture problem. This time we conceive of two partitions such that the accretion weight has an impetus due to the Earth with a quantity removed from it. The removed quantity is due to the squelched repelling relation between the debris and host. If your accretion weight is the result of squelched repelling then it is reasonable to assume that your 'actual' weight is more than your accretion weight. We have left to determine a plausible quantitative effect.

In a relationship with accreted debris material, the effective repelling power that is exercised by the host can only be as great as the debris, **e**. The rest of the host convergence power sustains the debris to the host surface. Since my accretion weight on the Earth surface is 90 kg, this implies that there is an equal and opposite 90 kg repelling against me from the Earth. This repelling is squelched by the

Earth's gravitational convergence. As in the Newtonian argument, I could say that if I weigh 90 kg on the Earth, the Earth also weighs 90 kg on me. However, that weight on me from the Earth is determined in a manner different in kind from my accretion weight.

It follows that if the debris surface convergence power is **e**, then the maximum surface repelling power that can be expressed is also **e**. This is only part of the interpretation. The maximum repelling power to get expressed between two converging objects is the lesser of the two. At the surface of the Earth, the Earth **E** is greater than my **e**; therefore at the Earth surface my **e** expresses a repelling power equal to **e** and for Earth the maximum surface repelling power that can be expressed is also **e**. This implies that the measure of the sink acceleration power is cut in half by the repelling. This implies that **e/2 = 1.477x10^{-22} ms^{-2}** or **e = 2.954x10^{-22} ms^{-2}**. This implies a corresponding effect on the measure found in accretion weight. Therefore an accretion weight of 90 kg when adjusted for the repelling would be 180 kg. Since the doubling is by symmetry we need not yet consider the limitations of Eudoxian magnitude calculations. Here we note that it is the composite measured acceleration power that has been halved by the repelling and this is what cuts the accretion weight to half the CB weight. The **e** value for this example remains too small (invisible) for our measuring instruments.

We emphasize here that for CB, the measuring of the velocities of 'falling' bodies to determine the acceleration due to the Earth **E** yields a false measure. Their accelerations are due to squelching, which entails over-powering a repelling from the 'falling' bodies. And as **E** is determined through the use of debris objects, all of these determinations have to be half the actual value. Therefore, accounting for tidal effects, locale variations, and squelching, it follows that the operative convergence power between a squelched debris object and its host is accounted for with **(E − e)/2**. When these powers are adjusted for distance the composite extrinsic measure is half the derivable intrinsic value,

(E − e).

Note: The implication here is that Earth E = 2x9.78 m/s^2 at the Equator, which has a weight doubling .

Subject to the limitations of Lilliputian empirical measures we can adjust for the local convergence power, which includes rendering the measure independent of the circadian variations due to tidal effects. An example, Moon tidal effects on measured **E$_g$** can be calculated from empirically obtained gravitonic measures on Earth taken throughout the circadian Moon cycle, etc. The adjustment for the Weight Decrement is much more difficult. For exactness we need to find a workable Weight Decrement Function. This requires further investigation.

The prepartitioning adjustments (includes the Weight Decrement Function), transform accretion weight into CB weight. CB weight is Eudoxian and we can partition it into *ex parte* weight and *impetus* weight. If we know the debris CB bulk (here the debris *ex parte* weight), then we can divide the host impetus effect by this debris bulk and we obtain the corresponding host *ex parte* factor that we can use to obtain the host bulk magnitude. We multiply the CB weight by the host *ex parte* factor and this yields the host *ex parte* weight, which expresses the host bulk magnitude in like terms to the debris bulk magnitude. Note that with the conversion to CB weights that the symmetry about the peak accretion weight is no longer a consideration because the CB adjusted weights are Eudoxian and display a linear monotonic increase (conventionally shown left to right).

The debris *ex parte* weight is functionally independent of the CB reference-frame it is measured in. It is not a contrived choice (as if based only on an Earth standard). In this regard it is an improvement over Newtonian mass that ignores any *ex parte* partitioning.

Invisible Differences. Certain composite differences important to conceptualization and generalization make no difference on our measuring instruments. For a CB weight of 180 kg on Earth we can calculate that **e = 2.954x10^{-22} ms^{-2}**. This shows that two Lilliputian objects could vary in **e** by a factor of many hundreds of millions and though we could calculate the difference we could not measure this acceleration difference. We could not even measure the gravitational convergence difference between the Empire State Building and a common sewing needle though the *impetus weight* of these accretion weights on Earth differ by many orders. If we could drop these two objects onto the Moon from the same altitude at the same time they would appear to us as reaching the Moon surface at the same time.

This is why Galileo could argue for equal rate 'free-fall' with an accuracy that could not be contradicted by the evidence within the scope of his measurements and why Boyle's Tube did not reveal the difference between the sink accelerations of a gold piece and a feather. This is why the Eötvös, Dicke, Branovski/Panov experiments fail to show the difference required by CB (the last attempt to 10^{-13} cgs). This is also why the NASA STEP program under its current specifications will fail in its intent, unless its only intent is to increase the accuracy by a factor of 1,000,000.

Comment: Newton's 'mass' likely has no valid usage distinct from its possible function as a constant of proportionality, a parametric variable that is usually not recognized as one. Furthermore, any use distinct from a constant of proportionality is a misrepresentation and becomes a source of misconception. For this reason Newton mass is here relegated to an apodeictic fiction. The appropriateness of this conclusion is suggested in neutral buoyancy where, in the shuttle craft bay, using only one finger, a newly arrived astronaut can move a satellite that on Earth weighs several metric tons ('newly arrived' because in neutralized convergence the astronauts weaken with time). On Earth, one finger could not even make it wobble. If the satellite really contributed tons to its own mass, the definition of Newton inertia forbids this ease of movement in *any* reference-frame. Its CB *ex parte* weight shows that we can expect this when the impetus weight reaches micro-convergence levels.

We have no direct access to quantifying matter in material objects. We can only measure the composition of their accelerations and decelerations and attempt to decompose these compositions into their constituent parts. Using the CB calculation modes it is expected that accretion weight (on a Lilliputian scale) is in round numbers only half the true measure. *This eliminates the need to hypothesize a compensatory heavy core for the Earth.*

Impetus Weight Factor and *Ex Parte* Weights. For host objects, such as the Earth or the Moon, the calculated CB weight cannot be used directly to determine the *ex parte* weight of the host. In a squelching relation, the accretion weight (due to acceleration from restraint impetus but devoid of constraint impetus) is an effect of the host on the squelched object. This is another way of affirming that the debris *ex parte* weight (due to **e**) and the *impetus* weight (due to the acceleration invested in the debris by the host) are magnitudes that differ in kind in regard to how they are produced. With respect to a given host, the restraint impetus weight effect varies according to the squelched object bulk. The impetus weight effect generated by the host is expressible as a multiple of the debris bulk magnitude. For a given host **E**, this multiple (a scalar quantity), here named the *restraint impetus weight factor*, is invariant. If we can obtain the CB weight adjusted for the Weight Decrement, then we can obtain **e**. This adjusted weight is functionally independent of the reference-frame (e.g., any particular host).

Furthermore, we can include consideration of the dual argument for the weight of the host on the debris as shown in the following argument:

Let **(E-e)/2** be the effective composite convergence at accretion, which includes the repelling effect as already argued.

Here **e** is the *ex parte* debris convergence power and **E** is the *ex parte* host convergence power.

Let **K** be the scalar factor of the host object, i.e., the multiple of **e** that gives the CB weight, W_e, on the host in terms of **e**., or

$$W_e = Ke.$$

Next we let **k** be the scalar factor of the debris object, i.e., the multiple of **E** that gives the CB weight, W_E, on the debris in terms of **E**., or

$$W_E = kE.$$

Whatever the units for the magnitudes, for these weights in opposition we have $|W_E| = |W_e|$, and therefore:

$$K|e| = k|E|$$

or

$$|E/e| = K/k.$$

As ratios this result is homomorphic with what was obtained from the Newton equations though with some important differences. **K** and **k** are scalar factors that are not identified with bulk. An impetus weight factor is identified with the particular object that generates it. We have no direct way to determine the impetus weight factor. It must be derived from a systemic based algorithm, the subject of the next Chapter.

We now consider how to estimate the relative magnitudes of these variables in units familiar to us.

Estimating the Host Bulks. I could obtain a rough estimate for the bulk of any planet I can stand on by dividing the double of my accretion weight there by my bulk and multiplying the double of my accretion weight by this factor. The percent error remains small. Because we are relatively Lilliputian we can determine with impressive accuracy the bulk of any planet we can stand on.

At first it seems that the calculation becomes functionally independent of the actual shape or volume of the convergence object. This is not exactly true because cases larger than Lilliputian can introduce significant error. We have to remain on guard for these considerations in a manner similar to Newton's requirement of adequate distance when using center of mass calculations.

Note that my *ex parte* weight is the same everywhere on Earth. However, host latitudes can have differences in convergence power and this yields differences in the accretion weight. For CB, this variation in convergence power is likely due to the material distribution within the Earth. The debris bulk is functionally independent of these variations. However, our calculation method must take into account the local variations in accretion weight in order to obtain a more accurate overall *ex parte* weight for the Earth.

Fulcrum of Revolution. When two hosts move toward each other both objects have a repelling power effect that changes with distance. While this repelling power is below the neighborhood convergence pressure of the two objects the distance between them diminishes. In the two-body problem (artificial isolation) the two bodies generate the neighborhood. When the repelling power is equal to the neighborhood convergence pressure (which varies according to relative position) the two objects stabilize into an isodynamic state. There is always a CRN with distances to the two objects that are proportional to their respective surface **E** powers, and the two objects as they approach proximity begin to revolve around each other, though in a two body CB Universe this revolution remains invisible. This revolution conserves the convergence relationship for the given distance. The convergence relationship necessitates it.

This isodynamic revolution remains invisible to us if there is no neighborhood that is more comprehensive for relative judgment. The revolution of two bodies about each other requires a third body for its manifestation. The Earth and the Moon have obtained a mutually stable and continual convergence toward each other. Their fulcrum of revolution is not the CRN.

The CRN cannot be the fulcrum of revolution unless the two objects are of equal and similar bulk. The differences are compensated for by placing the fulcrum of rotation toward the dominating object in a manner that proportionally balances the power they each have over the revolution. In the two-body problem this is an inverse proportion with respect to their CRN distances to their centers.

This identifies their isodynamic fulcrum, but if there is no more comprehensive neighborhood then their revolutions have this invisibility and for observers this is equivalent to non-occurrence.

SUMMARY CONCLUSIONS. The CIS produces a tension driven agitation that has a heating effect with a radial symmetry distorted by tidal effects. The diachronic evolution of the CIS can also give rise to resonance effects (dampening or amplifying) that may account for cycles of heating and cooling in long term star evolution.

There are many modes of analysis for describing the building of stars. In terms of synchronic oppositions; systems and convergence are inextricably bound in mutual influence. In objects large enough for a well-defined CIS, it is proposed that the local mutual influence generates harmonics expressed as mechanical discontinuities within the spherical structure, which in turn affects the field structure. Material conflux is expressible in terms of sink or float buoyancy.

Barring resonance effects, the degree of heating is directly proportional to the bulk of the celestial sphere and relates to the proportions of the various elements that it contains. The proportions of the various elements are determinate in how compact the material system becomes for a given quantity. Theoretically, neutrons could cluster under CB relations to form the most compact material objects. Neutron stars, if they exist, could be the most compact of all material objects as they contain mostly neutrons, the 'heaviest' known material microparticle.

Star clusters are formed under the influence of gravitational convergence, but the formation of galaxies may have another convergence principle added to it. If galaxy centers are of antimatter, like-polarity relations no longer would dominate the convergence laws on the galactic scale. The much more powerful CA matter-antimatter actions would form the basis for galactic evolution. Galaxies may be formed from star clusters that have encountered spinning clusters of antimatter. This also suggests the possibility of anti-galaxies and a new set of laws for determining the possible forms of mutual influence for encounters {matter-matter, matter-antimatter, antimatter-antimatter}. It looks like we have some star clusters near our own galaxy that are about to join our galaxy. In this preliminary study, with its emphasis on Convergence Buoyancy, galactic dynamics is not discussed in any detail because its unlike-polarity relations do not generate CB. Even so, together the like-polarity and unlike-polarity relations form the foundation for a new comprehensive celestial mechanics.

Concerning certain intensional magnitudes, there is great difficulty in escaping the possible circular reasoning. To understand this problem let us consider an example.

To separate weight into its Newton components of mass and acceleration we need to know one or the other of these quantities. In order to find the mass of a weighed system we have to know the participant accelerations. In order to know the participant accelerations of a weighed object we have to know its participant masses. The way out of this circle attempted by Newton was to seek the acceleration through timed 'falling' experiments. His procedure ignores the composite nature of the empirical measurement of this acceleration. In these Lilliputian laboratory experiments the critical differences occur after the 22nd decimal place, a precision that is still (2004 AD) impossible to measure. The Eötvös, Dicke, Braginsky/Panov experiments are useless for this order of magnitude. None of them get us out of the circle and neither will the NASA STEP program under its current specifications.

The convergence acceleration at the surface of a host is our most common access to the measure of the material quantity, and therefore the **'g'** (itself the resultant of all 'attractions' interacting with the 'falling' body force of attraction) in Newton's equation, **W=mg**, is more related to the physical measure of a material quantity than '**m**' for a given object. The '**m**' for a body was intended as a constant, quantifying the unchanging mass of the body relative to different convergence reference-frames. I argue against this interpretation as **g** must be a composite value that varies according to the g-value of the falling body in a given reference-frame. This variance of **g** in a given reference-frame disallows scalability for **g** (see also Chapter 1).

The graph of the apodeictic Weight Decrement Function displays the main characteristics within the squelched convergence range of accretion weight. Accretion weight is here expressed as a function of bulk. Bulk is the calculated *ex parte* weight and is a function of **e**, its gravitational convergence power. The graph exhibits the **peak weight**, w_P, at point **P** (also called **turnabout weight**), and the **boundary bulk**, **k**, etc. It is at **k** that we note a demarcation for a difference in kind. This demarcation is the natural demarcation for the boundary buoyancy. Beyond (greater than) this demarcation the host and squelched objects become related as two host objects in float buoyancy. Thus **k** is the demarcation between the CB terms 'host' and 'debris' (debris < **k**, boundary = **k**, host > **k**). The CIS amplifies opacity and this reduction in transparency to gravitational convergence may contribute to a more forceful convergence and repelling relation. This in turn could narrow the scope of variations in magnitude for boundary bulks found in different hosts.

It is the weight decrement function that predicts the curious result that any accretion weights of material like-in-kind correspond to two possible distinct bulks except for the peak weight at **k/2**, which corresponds to one weight only. The two positions as accretion weight are equal in accretion weight, but as bulks they are not equal (different *ex parte* weights). The first problem in the general solution for *ex parte* weights is to derive the CB weight so that we have a

Eudoxian magnitude. With the CB weight there is no duplication, no pairings of bulk as found in accretion weight.

According to the CB model it is correct to think of the *ex parte* participation of the related objects, host and squelched object (debris), as artificially separable magnitudes. However, we are left with how these artificially separated magnitudes are to be interpreted. It seems reasonable to presume that if my accretion weight on the Moon is 15 kg, my CB weight will be close to 30 kg (since the magnitude is Lilliputian and left of the peak weight) and my *ex parte* weight will be the same as it was on the Earth. That my *ex parte* weight remains constant follows if my volume remains constant, etc. The only argument left to resolve is whether the suggested mode of calculation actually accesses the wanted measure. Until we have a precise Weight Decrement Function we have no general algorithm for composing or decomposing accretion weights.

Bulk is not characterized as an extensional magnitude in CB and we have yet to develop a method for determining a magnitude for the quantification of matter. Bulk is a magnitude associated with matter but our only access to its measure is through convergence interactions. The Weight Decrement function helps identify what is needed to develop a Eudoxian CB weight measure from accretion weight using accretion weights. Even so this does not achieve or validate the assumption that equal CB weights of different materials have the same quantity of matter. This Newtonian assumption for mass is left undecidable for CB.

The previous arguments for bulk roughly relate to Newtonian arguments concerning gravitational mass. This mode of argument suggests a further question. Can we determine the CB weight using some other procedure? The answer is yes and the procedure resembles the Newtonian method for determining inertial mass. In this artificial analysis, CB corresponds this to the general concept of impetus weight, the subject of the next Chapter.

END Chapter 2

Chapter 3: IMPETUS WEIGHT and the ILLUSION OF INERTIA

INTRODUCTION. *An artificial separation permits us to speak of the 'interaction' between an object under scrutiny and its neighborhood (restraints). An object that has its own power to change direction could only have effects of a magnitude proportional to the convergence relationship it has with its neighborhood. This proportional relationship varies according to the effective power of the neighborhood restraints on the object. All restraints on the constraint impetus vanish if the neighborhood vanishes. In this extreme case there would be no resistance to any self-imposed change in motion, a resistance that Newton's mode of reasoning would still require. It is this CB variability that denies Newtonian inertia. Even so, this extreme case loses the definatory reference-frame for dereferencing 'change of direction' that is so often considered for alleged inertial effects. Motion has no significance without a neighborhood. In CB terms, it is from the observed Lilliputian cases of impedance that help sustain the illusion of inertia, but it is the form of variability of this impedance that denies inertia.*

Since Newton assumes that inertia is a characteristic of material objects, it seems logical that the anti-Newtonian view would posture that it is rather a characteristic of the environment, i.e., it comes from without. This is as Ernst Mach expressed in what the Einsteins dubbed "Mach's Principle," which states that all inertial forces are due to the distribution of matter in the universe. Though this interpretation is a feasible reversibility of Newton's view, it sustains the concept of inertia. For Convergence Buoyancy this is not the case for it is inertia itself that is denied. CB asserts in its fullest generality that the universal condition for homeostasis is the regression of all objects toward an isodynamic state. The investment of a constraint impetus in an object invokes a potential or actual change from isodynamic movement. For our Lilliputian experiments on Earth, CB attributes the impedance for gaining or holding a constraint impetus as the squelching effects of restraints in a debris-host relation. For example, if an object in motion in a neighborhood is not isodynamic with that neighborhood there is a continual impedance to the constraint driven aspect of the motion.

Astronauts in neutral buoyancy trajectories experience the Earth neighborhood impedance to constraint impetus in a far smaller magnitude than that encountered on the Earth surface. In CB the Lilliputian invisibility of **e** suggests an alternative answer as to why in the shuttle bay the astronauts can move Earth-surface measured multi-ton objects with one finger, an observation that surprised catechism physicists (who like chameleons now assert there is no surprise even though they had not yet found a way of quantifying the surprise out of the first observation).

Types of Movement. Only three types of movement are characterized for Convergence Buoyancy, namely:

1. Revolution,
2. Rotation,
3. Translation.

It is with respect to neighborhoods that objects display revolutions, rotations, and translations. A continued convergence can be sustained in two objects by revolving around each other. However the rate of movement depends directly on the convergence power adjusted for distances and local pressure, dynamic considerations.

In a universe of only two objects, certain rotations can be visible, but a third object is required to render the possibility of visible revolutions (as Bishop Berkeley knew). When two objects sink to float buoyancy they conserve their sink energy toward each other by revolving around each other. Maelstrom induced revolutions can in turn induce rotations. However, in a two-object universe, if either object obtains a relative rotation, it may be undecidable whether a revolution generates the rotation or if it is generated by some action within the object itself.

Note that constraints may be applied to each of the three modes of movement (unnatural or foreign to the reference-frame, or interpreting Aristotle and Galileo, they call it, 'violent' movement). However, the translation mode through a neighborhood always requires some constraint influence.

CB reinstates circular movement as the foundational form of isodynamic movement. All celestial objects are restrained toward circular movement. This rehabilitates Aristotle's belief in the primacy of circular movement for the heavens, though with some important adjustments. Interactions of three or more objects distort these circular movements, etc. This necessary distortion from circular orbits is the CB approach to the 'three body problem', etc.

Angular Velocity. We start with an apodeictic fiction, a perfectly rigid object. This supposition allows us to create a model for the characteristics of the motion that are functionally independent of any deformation that occurs during the motion of that object. We neglect the deformation and further assume that the distances between particles of that object remain unchanged in their mutual distances. Let us take a rigid object of limited size, arbitrary in shape (but no holes), rotating about a fixed internal axis. Using infinitesimals expressed as differentials, if the object turns through an angle **dθ** in time **dt**, the angular velocity at any given moment is expressible as:

$$\omega = \frac{d\theta}{dt}$$

where ω is expressed in radians per second.

For uniform rotation, one revolution per second gives the angular velocity of **2π** radians per second. For uniform rotation this simplifies the formula to:

$$\omega = \frac{2\pi}{t}$$

a kinematics formula.

The velocities of the object-points are called linear velocities. These can be different, but the angular velocity, ω, is the same for each point. This is the first property of rigid object rotation.

In terms of infinitesimals expressed as differentials, if an object moves a distance **ds** in **dt** time we have:

$$v = \frac{ds}{dt}$$

In turning through an angle **dθ**, a point describes the arc **ds = r dθ**. If this is done in **dt** time we have:

$$\frac{ds}{dt} = r\frac{d\theta}{dt}$$

Since $ds/dt = v$ at a given moment and $d\theta/dt = \omega$ in that moment we find the simple relation between linear velocity and angular velocity:

$$v = \omega r.$$

which shows the kinematics generality of the uniform motion formula.

The Alleged Convergence Deficit. In Newtonian physics there are two forms of alleged convergence deficit interpreted as centrifugal force that we associate with the motion of objects. One corresponds to the rotation of objects; the other corresponds to the revolution of objects.

If you are standing at the equator of a rotating Earth, this implies you are continually sinking farther and therefore faster than if you were standing in Paris. Just from the geometry we calculate from the centripetal acceleration, v^2/r, an expected loss in **g**, i.e., 0.034 m/s^2, at the equator. It is a popular assumption that this alleged convergence deficit from rotation accounts in large part for the differences in **g** found at various latitudes. The Earth contributes about 9.832 m/s^2 to surface contact acceleration at the geographic poles, at the equator at least 0.034 m/s^2 of this is opposed to surface contact, which leaves 9.798 m/s^2, which is within two hundredths of the fraction in the measured 9.780 m/s^2 using Newton's mode of reasoning. But can these calculations yield anything useful for a non-inertial physics?

CB denies these Newtonian interpretations.

If we accept the current surface convergence power for Jupiter, $E_J \approx 2.528$ Earth **g**, then the *ex parte* convergence deficit at its equator is about 9 percent of this surface convergence power. If Newton's notion of inertia were correct and if atmosphere is held by the acceleration due to gravity then with this deficit of its equatorial gravity acceleration Jupiter should have more oblateness.

Jupiter's atmosphere is being held by something different from the Newtonian understanding of gravity and centrifugal force. This may also indicate a significant error in the mode of reasoning. We may infer from these results that we need to look for another way to account for these latitude differences in **g** that we observe on Earth, etc. A possible solution is suggested back in Chapter 2 in the discussion on a possible tidal redistribution of material at the Earth's core.

In CB, convergence reference-frames such as planets and stars can generate measurable tidal effects on one another when they are within sufficient proximity. It is the material distribution adjusted for distance and neighborhood pressure that determines the tidal effects that objects have on one another. The only role that movement has on tidal shaping is a change of position of the material distribution.

The alleged convergence deficit from the rotation itself is inconsistent with a non-inertial physics. If the Earth were isolated, how is the fact of its rotation made visible?

The only way rotation can be made visible is relative to a more comprehensive neighborhood (Bishop Berkeley). For an object in an environment independent of any neighborhood, v^2/r has no dereference for a gravitational convergence reference-frame. The formula is rendered useless as a determinant of inertial effects. Even if there were an *isolated* rotating large object capable of gravitational convergence, there would be no means to discover that it rotates. Its material convergence lines would have no outside influences to counter the rotation and its shape would not reveal any oblateness (tidal effects relative to a neighborhood).

These alleged convergence deficits intended to correspond to a kinematics quantity, a geometry in motion quantity, are calculated in disregard of the actual dynamics both in Newton and in CB. Newton tacitly presumes an equivalence between the kinematics and the dynamics. In CB this equivalence remains to be proven and cannot be accepted without a justification for each case in which it is invoked. In Chapter 7 some of these underpinning difficulties to Newtonian gravitation are discussed.

Drift Component. With visible planet rotation, an additional velocity component may be induced by the direction of rotation. Contingent on axis orientation the total velocity in the orbit could be higher on one side of the planet than on the other side. The planet convergence field may therefore generate a drift component in the direction of lower pressure (analogous to the Bernoulli principle). This complicates the analysis of celestial movement if it sustains validity.

IMPETUS WEIGHT.

For a given convergence power adjusted for distance we have two correlative reference modes convenient for the object motion, one normal (at a right angle) to the power, E_n, the other parallel to that power, and with these we can describe any of the possibilities in between. Normal to the power includes any motion parallel to the geoid (smooth virtual surface for mapping the surface variations in terms of variance from it). Parallel to the power includes any motion in line with a surface point to center of the host (normal to the geoid). Under constant acceleration in a pressure zone an object seems to gain the effects of an increased weight proportional to and in line with the motion and this *weight is sustained for the duration of the acceleration. It is as if for a fixed bulk that its weight changes directly with any change in acceleration. In CB it is not the bulk (ex parte weight) that can change but rather it is the weight effect that can change. In CB the changeable weight effect is termed* impetus weight.

Changeable Weight. For CB, weight is proportional to an impedance to an acceleration. If there is no constraint impetus we have left only the neighborhood restraint acceleration synergetic with the object impetus.

As various hosts may differ in the acceleration effect they generate on debris, the required impedance varies accordingly. In effect, for CB, the measure of impedance required for slowing or stopping a moving object relative to a gravitational convergence neighborhood is expressed in weight units.

For CB, on Earth when we swing a stone sling in circular motion it is the increased impetus weight due to our applied constraint that drives the stone sling outward (the outward constraint driven weight is exceeding the convergence driven weight). We see as the revolution increases in its angular velocity that the sling becomes more parallel to the Earth surface. The constraint impetus is proportional to the angular velocity. When the stone is released its constraint impetus immediately begins the regression toward accretion along the direction of its release. In contrast, an astronaut in microconvergence would find extreme difficulty in revolving a stone sling. The on-Earth possible exercise is synergetic with the Earth convergence power and pressure, etc. The weight of a body that is due to any acceleration (artificial or natural) corresponds to a measure of an applied impedance to that acceleration.

Accretion weight (no constraint impetus) is a special case of measured weight that is always specific to a given CB reference-frame (e.g., host). Any variation in impetus found in a measured weight is construed as a tidal effect or a constraint effect, as the case may be.

Impetus Weight and the Kilogram. For CB, the kilogram as determined on Earth can be regarded as another measure mode for a deceleration effect and therefore functions as an alternative expression for deceleration.

The measure of impetus weight is always relative to a deceleration or acceleration in a neighborhood. We can never determine impetus weight through direct measure that is independent of a deceleration or acceleration, which is to say that impetus weight is never absolute. On the other hand, as expressed earlier, **the *ex parte* acceleration, e, is functionally independent of any host used as a reference-frame.** This independence renders the *ex parte* weight invariant and allows for the following summary definition:

> **DEFINITION:** The CB *bulk* magnitude of a debris object is identified as its *ex parte* weight (its invariant weight).

Ex parte weights expressed in terms of their units of measure are always relative to a measurement reference-frame. This does not preclude that any Earth-bound standard unit of measure is arbitrary within the limits of usefulness and its use remains parametric.

All impetus weights measured in neighborhoods have restraint impetus. Impetus weight for debris has two partition modes; *accretion (unconstrained) weight* and *impact weight*. 'Accretion weight' is a locution for the composite of the *ex parte* debris weight and the partition of the weight due to the host restraints. Furthermore, accretion weight excludes any constraint impetus. On the other hand, 'Impact weight' is a locution for the composite of an accretion weight with an impelled weight that is generated by relative movement.

Impact Example: If you drop a weight onto a spring scale that has a live pointer and a dead pointer the dead pointer is pushed to the impact weight and the live pointer goes back down to the accretion weight. The difference between the impact weight and accretion weight is the *impelled weight*. This impelled weight roughly corresponds to Newtonian linear momentum at the velocity of impact (**p=mv**), with some important differences. As discussed in the previous Chapter, in Newtonian terms, momentum is converted to a force when its velocity is changed.

CB momentum can be expressed as **dp=bdv**, where **dv** represents a moment in its acceleration, a formula that has no dereference without an impetus relative to a CB reference-frame. In effect, momentum is only expressible as a relative moment of acceleration since in CB all objects are in relative states of motion changes. Maxwell knew this when he said that the momentum of the Moon relative to the Earth is not the same as its momentum relative to the Sun.

Note: Maxwell's book, *Matter and Motion*, contradicts Newton's concepts of absolute Space and absolute Time. In effect it establishes a theory of relativity for Classical Physics. He notes that because our understanding of motion is always based on relativity that there can be no absolute representation for a 'state of rest'.

When a meteor strikes the Earth, the impetus weight of this meteor strike is its deceleration expressed in kilograms at the moment of impact. Any impedance applied to this meteor in flight varies in its measure according to the moment and position in a neighborhood. The measure of this impedance is the most general expression for impetus weight. The measure of impetus weight for this meteor in flight is always relative to its relations with its milieu. We use the term 'weight' in all of these cases and their measure is expressed in units that are consistent with all of them, e.g., kilograms.

TERMINAL SINK ACCELERATION

Any object that sinks at an acceleration rate that is not isometric with the neighborhood restraints must regress toward that isometry unless its rate is otherwise sustained by constraints.

When the sink acceleration is isometric with the restraint driven acceleration this is the **terminal sink acceleration**. If we could run a pipe down past the CIS and drop a marble through the pipe, it must at first accelerate to reach its terminal sink acceleration and then its terminal sink acceleration should diminish to zero as it approaches the CIS. In the case where within certain limits the initial conditions have not regressed to zero at the CIS we expect an oscillation that regresses toward the isometric state. As the terminal sink acceleration itself is diminishing to zero as the marble approaches the CIS, the marble must finally achieve the terminal sink acceleration even if this occurs only at the CIS. Even so, the tension between the shell and the kernel volumes implies that an isometric state with the CIS state generates an oscillation even on the microparticle level. The marble will become hotter.

In so-called "free-fall" experiments we cannot immediately detect the resultant terminal sink acceleration with the neighborhood (at the position of measure). This terminal sink acceleration is the resultant of the object in its relation to its convergence reference-frame and differences in the objects bring differences to the resultants (though as yet Lilliputian differences remain invisible to direct measure). Note also that to be isometric with the terminal sink acceleration is not to say that the sinking debris is isodynamic with the host. Only when the debris has reached and settled on the host is it correct to say that it is isodynamic with the host. This suggests that regression may also apply to diminishing the initial conditions, which may either be artificial or natural. The natural can stem from large scale breaches of physical integrity limits such as result from a nova or collisions of matter and anti-matter.

We can reason that on the Moon the atmosphere is too thin to influence sinking objects (normal to the lunar geoid) measurably. I say that if we allow an object to sink to the Moon surface from a high altitude starting at a zero sink rate (a negative constraint) relative to the shortest path toward the Moon center, or starting at a greater sink rate (a positive constraint) than the ambient host-debris convergence rate on this same path; in either case the object sink rate regresses toward congruence to the ambient host-debris convergence rate. More generally, any accreting object that has an acceleration impetus with initial conditions that are not isometric with the ambient host-debris convergence rate regresses toward the ambient convergence rate. When this regression attains the ambient convergence rate (influence of non-isometric initial conditions nullified) we say that the sinking rate has reached the terminal sink acceleration.

Parallel to the Earth geoid we can reason that when a passenger plane accelerates for take-off, its power over the passenger **e** is not just a few orders. The passenger plane acceleration can exceed Earth **E**. It is because the passenger plane acceleration is above the lower perception threshold that there is a perceived feeling of increased weight (impetus weight augmented by a constraint impetus) in line with and opposed to the direction of acceleration. This shows an impedance to this change in motion. When the acceleration ceases, the weight in line with the plane acceleration seems to stop. The passenger then moves along at the new fixed velocity without perceiving any regression from restraints.

In a plane in flight at a constant speed, on a passenger the *regression* effect normal to **E** (adjusted for altitude) is proportional to the passenger **e**. Even though this regression rate offers impedance to the constant speed it is too small for our sense of motion to detect.

For small objects like people (debris) the restraint impetus effect would be invisible while in the Earth neighborhood even if the aircraft could exceed many times over the currently presumed speed of light *in vacuo*. We can calculate the impetus weight for any fixed speed. ***The impetus weight is a measure of the total impedance required to induce a specified speed relative to a given CB reference-frame.*** The impetus weight for a passenger in an aircraft at fixed speed normal to convergence (parallel to geoid) is invisible to the passenger, thus making possible the illusion that an aircraft in flight at a fixed speed fulfills the requirements of a spheroidal inertial reference-frame.

If the aircraft maintains a given angular velocity in a great circle above the Earth surface, for Newtonians this approximates an inertial reference-frame along the planar surface parallel to the geoid. This provides a Newtonian rationale for why the airplane passengers may move about ignoring the airplane motion. CB denies this interpretation.

> **Note:** At this point the reader inclined to mathematics should be able to calculate the fixed speed (velocity at a given moment) required in the Earth neighborhood to bring the impedance-to-movement to a perceptible level. Likewise, given a fixed speed, calculate the neighborhood restraint impetus required for bringing the impedance into the perceptible range.

INERTIA

For Newton the inertia of an object on Earth is the same as its inertia on the Moon. Insofar as it is the inertia from the same object, this conclusion is consistent with his system. CB denies this interpretation. An object cannot have an assigned direction or speed that is non-arbitrary if there is no neighborhood convergence. This is another way of saying that there is no inertia in the object. A space vehicle with a sufficient degree of isolation from neighborhood convergence could stop or turn suddenly with virtually no effect on its occupants. Accelerations have no effect on the occupants of a vehicle that is isolated from neighborhood convergence. Again, any movement change loses all significance without a neighborhood.

The Earth itself is a CB reference-frame. Its rotation could not be recognized if it were not in a more comprehensive neighborhood in which it is just a part. If there are rotational effects on it (such as pendulums exhibit), they may be classed with neighborhood tidal effects. Otherwise said, the only disturbance from rotation for the debris on the Earth surface comes from the neighborhood tidal effects and not the rotation. This argument alone, if it holds, accredits classing 'inertia' with 'phlogiston' as a contrived hypothesis, in which case it violates Newton's dictum "Hypothesis non fingo." With respect to system architechtonic it may be useful to regard it like Euclid's fifth postulate (parallel postulate), variations on which have produced other coherent geometries.

> **Note:** Today the movement by Earth rotation of an object at rest at the equator is over 1677 km/h. Ptolemy's argument against Earth rotation is related to this (see Ptolemy section in Addendum 6). It is current belief that inertia prevents objects from being swept away by this Earth rotation. That this inertia seems to work in a circular movement is why Galileo needed a kind of circular inertia. This is also why he is not credited with the 'modern' definition for inertia. In current physics the Earth surface is characterized with certain restrictions as an inertial reference-frame. In part this seemed required for the assumption that the Earth rotates.

For debris on the Earth surface (independent of air resistance, etc.) the regression rate is determined by the magnitude of the convergence relation between the object **e** and the neighborhood **E**. For a 90 kg human being walking or otherwise moving normal to the direction of Earth **E**, the regression driver magnitude is extremely small (on a Newtonian scale circa 1.477×10^{-22} ms^{-2}). This too-small-to-be-visible relationship between debris and neighborhood explains why the concept of 'inertia' has not been falsified by our Lilliputian experiments. The fuller qualitative analysis required for the CB quantification of debris regression is discussed back in Chapter 2.

Impetus Weight Guidance Systems. 'Inertial' guidance systems lose their effectiveness in deep space because deep space cannot support the illusion of inertia. For a constraint driven gyroscope, the pressure and power from the host drives the regression that allows the illusion of inertia, which is another way of saying that it makes it look as if we have an inertial guidance system. In deep space the lack of inertia becomes evident. It is more accurate to call these systems 'Impetus Weight guidance systems' because their function depends on Impetus Weight, which is to say that it requires a gravitational convergence reference-frame to assess it. Impetus Weight only exists relative to a neighborhood. *Impetus Weight guidance systems in deep space weaken or fail until brought within a sufficiently strong neighborhood*. A gyroscope invested with spin would spin longer on the Moon then on the Earth (in an equivalent vacuum chamber), because the Moon as the dominating neighborhood convergence for the gyroscope has less convergence power than the Earth. We can also reason that this spin has less effect on gyroscope stability on the Moon. Likewise for a given initial spin rate we can reason that the gyroscope would spin longer at the Earth equator then in other latitudes. Similarly, we can reason that for a given constraint impetus a billiard ball rolls further at the Earth equator than at an Earth geographic Pole. Note that gyroscopes are not the only form of 'inertial' guidance systems, all of which must weaken or fail in deep space (e.g., the Laithwaite system.).

There are places in deep space where a disturbed object could take eons to regress toward an isodynamic movement, if CB is the underlying dynamics, etc. For most laboratory cases the invisibility of **e** allows catechism physicists to believe that without friction that inertia impels a perpetual motion until otherwise interfered with. CB theory denies this.

This brings to the fore another contrasting correspondence between CB and Newton. It is easy to reason in Newton's system that perpetual motion is impossible, whereas in a CB universe with three or more objects, under definable conditions, a form of perpetual motion becomes a necessity. In CB, for a universe of three or more material host objects, the motion never stops. This derives from the interacting non-isotonic convergence fields.

Objects of the same type also have limits to size. Dust and gasses that accrete into stars must end in a nova, which limits the bulk that material objects may obtain to in a given pressure zone. Any two hosts that converge toward each other reach a proximity where repelling only allows the convergence to express as a revolution about each other. If the two object system is isolated then no revolution is visible.

CB Friction. On Earth, a gyroscope rotation is also a spinning convergence field that interacts with the Earth convergence. It is *not* in isodynamic convergence with the Earth. The spin is artificially induced, a constraint impetus. When the application of the constraint stops, the spin impetus regresses under the mutual influence of its field rotation and the neighborhood convergence. The restraint from Earth convergence resists the gyroscope spin and the spin impetus (a constraint impetus) regresses to zero from this impedance. In CB the neighborhood restraints are a form of friction to natural or artificial initial conditions that are not yet isodynamic. This is *CB friction*. This friction may also be characterized as the restraint impedance to non-isodynamic motion. In this view the friction drives the regression toward sustaining an isodynamic state.

Exit Velocity. In CB a material object is always in a convergence relation with other objects. Furthermore, all material systems have this synergetic mutuality in their convergence relations. There is also a convenience in regarding certain relations as dependencies. For a convergence center that dominates the relationship with the object under scrutiny it is convenient to regard the object as a dependent of that dominating convergence center. Thus we can say that the Moon is a dependent of the Earth, the Earth is a dependent of the Sun, etc. In ambiguous cases we can assert a mutual dependence, etc. In CB it is unlikely for an object to be independent of all neighborhoods. However, it is possible for an object to change its dependency from one convergence center to another convergence center.

The minimal velocity required for an object to leave its dependency on one convergence center to enter into a dependency on another convergence center (be taken over by it) is called *exit velocity*. For CB its exit velocity varies according to its synergetic relation with the convergence neighborhood. The path from point-of-origin to destination is non-isotonic, which may or may not have an adjacent proximity. It corresponds to Newton's escape velocity, but Newton predisposes us to consider a single escape velocity for the object, as if it were shot from a cannon. Lagrange refined this. For an object to exit the influence of the Earth's gravitation and be taken over by the Moon's gravitation in a direct way requires a velocity sufficient to pass the Lagrange point between the Earth and the Moon. This sense of 'escape' is only for objects that move from one dependency to enter another. If the Earth were alone in the Universe there is no escape velocity. Every unconstrained velocity comes to a halt and the object would return to the Earth.

CB LAWS OF MOTION (*ex parte*)

Law I: (Law of Regression). *An impetus is isodynamic or moves toward an isodynamic state unless otherwise constrained.* The constraint impetus portion regresses until the objects are in isodynamic convergence unless the constraint impetus is otherwise sustained.

Law II: (Law of Impetus). *In proportion to both magnitude and direction, impetus measures the total constraint impetus and restraint impetus invested in an object, i.e., impetus = (restraint impetus) + (constraint impetus), where '+' is the inclusive 'or'.* The constraint impetus is a measure of both the magnitude and the direction of the constraint effects on that object.

Law III: (Law of Isostasy). *Every investment of impetus generates an equal and opposed reaction impetus to both modes of investment (restraint, constraint).* The invested constraint impetus generates an equal and opposed reaction constraint impetus, etc.

With respect to constraints these are the CB Laws that are dual to Newton's Laws of Motion. When our concern is impetus and regression we limit the use to actions that involve constraints. That they are expressed *ex parte* implies important restrictions for their use in a systemic analysis. One has to remain vigilant against using *ex parte* concepts as if they dereference to something more than mere aspects of fact.

Misuse of *ex parte* concepts can generate artifacts that have no proper dereference in matters-of-fact (see Addendum 2).

1. CB Law I is the **Law of Regression.** It is equivalent to the statement: *Material objects regress toward a convergence that is isometric with the neighborhood pressure and power unless otherwise constrained.* Note that the sorting states of buoyance, accretion, and the boundary between them are driven by the effects of material distribution on convergence. They are manifestations that derive from this convergence. There is no stipulation here as to hardness (surface tension) or to compactness. If it is material, it converges to one of these three CB states. This Law of Regression is the anti-inertia law (for a non-inertial physics) that corresponds to Newton's Law I, the Law of Inertia (taken from Descartes), which states:

> "Every body continues in its state of rest, or of uniform motion in a right line, unless it is compelled to change that state by forces impressed upon it."

From this law of inertia we derive the concept of inertial reference-frames, an apodeictic convention that we associate with great advancements in Physics. Inertial reference-frames satisfy the following definition:

> **Inertial Reference-Frame:** A reference-frame with respect to which all systems which do not interact with other systems are at rest or move uniformly in a straight line.

A strict inertial reference-frame requires that we switch gravity off. Since all matter is presumed gravitationally attracted, an inertial reference-frame cannot exist for material objects. Therefore, in Newton's own terms the concept of an inertial reference-frame is self-contradictory and is found in nature only in various degrees of approximation. Catechism physicists readily admit that there is no such thing as a perfect inertial reference-frame. Every example is presented with its particular limitations for fitting the definition.

From the *ex parte* CB First Law of Motion we can derive the concept of **convergence reference-frames**. In such frameworks there is no such thing as an inertial reference-frame. There is only convergence. Objects are either isodynamic with a convergence reference-frame or there is interference to an isodynamic relation that derives from other sources of impetus investment (constraints). These investments may be construed in two forms, by field-to-field interactions or by surface-to-surface impact. When pushed to extreme this distinction fades because both forms can be treated as fields. Both forms may also be regarded from the point of view of *impetus*.

2. CB Law II is the **Law of Impetus**. For CB it corresponds to Newton's Law II concerning motive force, which states:

> "The change of motion is proportional to the motive force impressed; and is made in the direction of the right line in which that force is impressed."

Note that for CB the term 'force' is redundant, it adds nothing to intelligibility. In CB Laws, the roles of *constraint* and *impetus* replace this use of the term 'force' insofar as they are outside the domain of restraints. 'Constraint' is analyzed into interactions contrary to isodynamic movement and its one-sided dereference is 'constraint impetus'. Impetus is analyzed as an *ex parte* measure and as a view it artificially sustains one-sided dereferences.

A movement counter to isodynamic is against the restraint pressure. The restraint pressure interaction with the constraint can change the shape of the object. If the movement is regressed to isodynamic movement the object returns to its original shape insofar as the material memory allows in the isodynamic relationship, i.e., relative to the restraints. Otherwise it contributes to a new isodynamic state.

The quantification for *impetus weight* is not a fixed magnitude in the manner required for Newtonian 'inertia' (which is always a fixed magnitude). The magnitude of impetus weight is relative and proportional to the magnitude of the applied impetus effects (restraint, constraint). Any differences from the neighborhood restraint effects represent a measure of constraint effects.

3. CB Law III is the **Law of Isostasy**. For CB it corresponds to Newton's Law III, which is dual to itself. In Newtonian terms it concerns the opposition of action and reaction and states:

> "To every action there is always opposed an equal reaction: or, the mutual actions of two objects upon each other are always equal, and directed to contrary parts."

For surface to surface actions, this is the fundamental principle for billiard ball physics. For Newton, 'action' usually refers to direct surface contact. He admits that he could not account for the physics of 'action-at-a-distance' and this is why, at best, when he speaks of physics it is a *billiard ball physics*. This statement does not deny the effectiveness of his geometric analysis of gravitational phenomena.

Pseudoconstraints. Only CB reference-frames have physical validity for CB analysis. Any other kind of reference-frame can generate a pseudo-impetus, or pseudorestraint, or pseudoconstraint, just as any non-inertial reference-frame generates pseudoforces in Newtonian physics. CB reference-frames are not functionally independent of the physics that happens in them. All physical reference-frames here are non-inertial and participate in structuring the physics that occurs. This is *dual* to the Newtonian inertial reference-frame, which is said not to affect the physics that occurs in it.

HOMEOSTATIC CONVERGENCE SYSTEMS. *Suppose that our solar system were free of debris and that every object in it were in float, boundary, or sink buoyancy. Further suppose that every object subject to sink buoyancy is already accreted. If its neighbors were in the same state then this neighborhood system is in an equilibrium without growth. To this system add debris in the form of comets, asteroids, and interplanetary dust and gases. This debris feeds the system in that its various objects may grow in size, and this in turn may induce other modifications. With this introduction of the possibility of growth we introduce the concept of* **homeostasis**. *This is the mixture of steady-state with growth. A* **homeostatic convergence system** *is in a growth form of isodynamic convergence.*

In effect, homeostatic convergence systems better model the structures and functions of the celestial mechanics driven by CB in our locale. The isodynamic state is defined through limits. The lower limit of the homeostatic convergence system is the isodynamic convergence state without growth. At the limit, no growth is possible.

A neighborhood, by definition, is the convergence influence exterior to the object under scrutiny. Even so, the neighborhood itself may become the object under scrutiny. When we speak of a neighborhood as if it were an object, it is always from a convergence reference-frame. This way of talking is consistent with an *ex parte* analysis. With respect to the Moon's neighborhood, the Earth is the strongest contributor; the Sun is another strong contributor. Each celestial object influencing the movement of the Moon is a contributor to the Moon's neighborhood.

Mechanical constraints do not generate repelling. For example, the space-shuttle can circle the Earth at a rate that zeros out the sink acceleration due to Earth gravity at the given altitude. This has a superficial similarity with a convergence deficit (a mechanical concept corresponding to a Newtonian analysis), but it is not generated by centrifugal force (inertia). All that is required is a sinking at a rate that is equal to the Earth restraints at the given altitude while keeping a 'forward' movement that sustains the altitude. This can be synchronized so that the space-shuttle maintains a fixed altitude. This is a mechanically induced neutralization of the Earth restraint buoyancy effects for every position in the locus, but it does not neutralize the neighborhood directional polarity and it is not generated from centrifugal force (as the Newtonian system requires).

Objection: If we accept that artificial changes in the sink rate can zero out the sink acceleration due to Earth gravity at the given altitude, then it seems that the formula for the centripetal acceleration, v^2/r, could be used at the Earth surface where the altitude is zero.

Response: For CB, the formula, v^2/r, is functionally independent of host spin. Any apparent correspondence must be defended for each case. Differences are easy to find. Example, since pressure affects the geometric $1/r^2$ law, different pressures would modify this law according to these differences. The Earth E_n is not a constant for all surface points. This does not preclude its use to calculate the trajectory velocity required to match the centripetal descent acceleration for a given altitude from a given surface location. For CB the centripetal acceleration formula, v^2/r is equal to E_n. E_n is the shuttlecraft centripetal acceleration toward the Earth surface and is calculated using the shuttlecraft velocity, v, relative to the Earth surface. The CB argument starts with the supposition that if the Earth were isolated there would be no apparent Earth spin. No tidal distortions would occur on the Earth shape. The Earth could spin at any rate whatsoever and this cannot be known without a neighborhood. Therefore, we cannot use the **v** for the Earth rotation in the formula, $\mathbf{v^2/r}$. This sustains a consistency with the CB supposition that the disposition of any object in CB relationships (including on a host surface) is dependent on material distribution. Any effects that Earth rotation might beget can only be in terms of the tidal effects on material distribution.

Artificial Satellites. CB Law I implies that all *artificial* satellites in 'orbit' to date are in an accretion course with their respective hosts. These artificial satellites have insufficient bulk for sustaining the float buoyancy that CB orbits require. They follow trajectories that spiral toward the host surface. Their trajectories can never become orbits, and therefore it is incorrect to speak of these cases as 'decaying orbits'. They are ***accretion trajectories***.

One of the reasons it is easier to put an artificial satellite around Venus than around Mars is that for Venus, **E** is 0.91 of the Earth's and the Martian gravitational convergence is only 0.38 of the Earth's. Thus for Mars the repelling power is much weaker. This implies that it is much easier to overshoot the attempted trajectory around Mars, etc. It is also true that navigation devices based on 'inertia' are less effective in weaker pressure zones. So far, for Mars, failure is more common than success. In part, this is also why Moon artificial satellites are so short-lived.

The CB Lagrange points are compression rather than tension positions of gravitational convergence. Placing a convergence object at one of these points is a disturbance to the equilibrium of this compression. Debris located at CB Lagrange points disturb the equilibrium only slightly and therefore exhibit a location stability at these points that exceeds other locations.

SPIN OF PLASTIC SYSTEMS. The impedance to material deformation from neighborhood restraints corresponds to the degree of plasticity the material exhibits in its relation to the object field deformation. There are two general ways this deformation can be rendered, one is *not contained* and the other is *contained*.

1. *Not Contained*: The first case is the deformation of a plastic system that is not physically contained, i.e., it is suspended under conditions that render its shape independent of any physical container. In a shuttle-craft we could simulate this in large degree by setting a globe of water to float in air. Its spherical shape is a function of its own convergence, the surface tension of water and the neighborhood convergence (tidal effects). The shuttle-craft revolution around the Earth generates a spin in the globe. On plastic spheres with an isodynamic spin, the equatorial region has the highest angular velocity. This suggests that the spin of the Sun and the Jovian planets is from an external applied pressure that has its greatest effect at their equators (which spin faster than their polar axes spin).

2. *Contained*: **Newton's Spinning Vessel.** The second case is based on the plasticity of water in a spinning vessel (bucket) near the Earth's surface. A dual analysis of this Newton experiment inspired the origin of CB theory. Newton presumed that the flat water is the natural state and the spinning water going up the sides needed to be explained. An anti-Newtonian thought experiment is to assume that the water going up the sides is natural and the water staying flat needs to be explained. The spinning water surface is concave, and the lowest point is at the axis of rotation. The highest point of the water surface is toward the rim of the bucket and this highest point has a lower angular velocity than the axis. In the rough the spinning water follows the shape of a negative sphere and the drive of the water spin seems to originate from the axis of spin (the vessel spins from an unwinding rope). This is opposed in principle to the first case. Furthermore, as CB proposes that the water at its edge in the bucket is moving counter to an attraction to the Earth (the CB natural state) we are left to explain why the water is 'flat' when the bucket is not spinning.

> **Note:** Since the solar maelstrom drives the Earth rotation by pushing west to east, at the Panama isthmus the Pacific Ocean is on the high side and the Atlantic Ocean is on the low side. This difference is a permanent solar maelstrom tidal effect found on all longitudinal coastlines, etc. Earth tides driven by the Sun are thus of two types, solar maelstrom tides and solar isostatic gc tides.

The calculated measure of bulk (*ex parte* weight) is completely dependent on the convergence interaction of objects. *Movement in itself does not change the material bulk*; though it can change the impetus weight (see below Impetus Weight and the Equivalence Principle, also Addendum 5). However, induced movement in a neighborhood changes the shape of the object's field, which in time changes the shape of the object in accordance with its plasticity. When the induced movement is brought to a halt, the object might revert to its 'original' shape. For solids, this reversion is referenced as 'material memory' though this animistic term does not indicate any physical principle underpinning its occurrence. It may be more accurate to regard these effects in terms of molecular clustering. For fluids (liquids and gasses) that support more chaos the term 'material memory' fades in significance. However, we can account for their reaction states in terms of molecular clustering.

Examples: That the equatorial region angular velocity is augmented on the Jovian planets, etc., is observed. Their equators spin faster than their axes spin, as if driven by the Sun's maelstrom (Chapter 2 shows other examples). The Sun itself seems caught in the maelstrom of our galaxy and also spins faster at its equator. This suggests that the galactic maelstrom may drive the Sun spin. This galactic maelstrom, in turn, may be driven by matter/antimatter relations. That the driver of this maelstrom is not gravitational is suggested by the fact that the velocity for the galactic stars with respect to the galactic center seem close to equal regardless of distance from the center. This drives galaxies into a spiral form. It also violates Kepler's Third Law. On the Sun, the flattening of tidal effects with distance allows it to rotate in the galactic maelstrom. If the galactic maelstrom accounts for the Sun spin we can surmise the relative rotation direction of our galaxy from this spin.

There was a long controversy over whether the Sun is oblate or not. It is important to inertial Physics that its spin renders the Sun oblate. In CB there is no rationale for oblateness of the kind Newton requires. Celestial object oblateness in this non-inertial Physics is generated by tidal effects in its synergetic relation with a neighborhood convergence that is determined by material distribution adjusted for distance and not by a centrifugal force (inertia). On the Sun, the tidal effects from the planets and other stars do not render as easy-to-measure. Material objects that are large enough for significant influence on the Sun are too far away to convey this influence as a measurable tidal shaping effect. For gravitational convergence there is a calculable limit for measurable tidal shaping effects. There is a flattening of effect with distance as convergence lines approach parallel for subtended objects. This does not preclude rotation which itself is a tidal effect. The possible galactic matter-antimatter considerations are still too speculative for a detailed development.

The gaseous state attributed to the Sun encroaches upon a physics that is strange to Newton physics. Empirical evidence (Chapter 4, Russian Experiments) suggests that gases follow the radiation gradient. The enormous radiation gradient toward the Sun may compress gases into a metallic state at the Sun's surface. The granular surface pattern of the Sun is related to supercooling (more in Chapter 4).

CB INTELLIGIBILITY MODES. A convergence reference-frame can be any determinate convergence center (determinate for its dependents, the old reference-frame method choosing some zero point, here a convergence point), or everything is judged as relational convergences. The older mode is useful for establishing CB correspondences with Newton. A newer mode is modeled after object-oriented computer programming. Each convergence source is an object. The object is thought of as encapsulated with its methods and data. Its methods (convergence/divergence) are characterized as determining well-defined responses to other convergences (that result in sink, neutral, or float). Its data determine the parameters of its response. Thus, **E**, if correctly estimated at the surface of the Earth is a convergence parameter (a variable with a definable scope). Correct measurement is difficult because no physical *ex parte* measure for **E** can be direct. As we must choose an analytic system for solving this mixture problem we are exposed to the dangers of introducing systemic artifacts.

Regarding objects as monads is a defective characteristic of ex parte *thinking* (see Definition section in Addendum 1). It is also the underpinning of billiard ball physics. It is related to one of the extreme limitations often used in statistics, the concept of independent events. (Most science students only study these very limited special conceptual cases and only receive a token consideration of Bayes and other modes of analysis for non-independent itemization.)

In some respects this looks like we are back to billiard ball physics and *ex parte* analysis. If by distance we could dereference a monadic separation of objects, then we would have a billiard ball physics, but this is not the case. In a convergence Universe, nothing is entirely separate. Each object is within a convergence field and convergence fields 'interact' in accordance with CA or CB, as the case may be. Therefore, action-at-a-distance is not a problem because nothing has this monadic separation. There is no action-at-a-distance *per se* because there is no distance of this disconnected form. The fields are in contact and mingle. This theory rehabilitates some important aspects of the Cartesian vortex theory. Each object is the center of a convergence field and convergence fields interact. However, the term 'interact', like the term 'sunset', may predisposition us to think about the problem in a fixed way that could be defective.

If convergence is continuous (e.g., non-discrete waves) then all objects influence each other. If convergence is discrete (corpuscular, quantic) it may have a limited scope for its influence. Our notion of its physical presentation is highly influenced by our instruments of investigation. This includes the language it is expressed in.

Convergence Methods. Some convergence methods (response types) concern:

1. field shape,
2. movement,
3. material deformation.

In a neighborhood, there is no natural rest position for anything. Every neighborhood object moves unless constrained. This assertion is consistent with Galileo's belief in the primacy of motion.

To remain at rest, if a dereference for this can be found, requires constraints. In a previous thought experiment a celestial object was postured as isolated from neighborhoods. The conclusion was that movement no longer had a valid dereference. You might think that this implies we have a dereference for a celestial object at rest, but this is *non sequitur*. That 'motion' loses any possible dereference in matters-of-fact rather implies that 'rest' also loses any possible dereference. We have no power to determine whether the isolated object is moving or not. There is no available reference-frame for such a determination. This is another contradistinction to Newton's system.

This reference-frame requirement might be termed 'CB Relativity'. If we restrict the interpretations to kinematics then we have George Berkeley's relativity (*Motion*, 1721). Berkeley's relativity inspired Ernst Mach to develop this thesis into a relativity of inertial forces, a dynamics interpretation that is distinct from CB relativity. CB relativity develops a dynamics interpretation that is devoid of inertia.

impedance to a change in position *or* change in motion is the measure of the synergetic relation of converging objects. This reflects the mutual relations of their respective convergences. 'impedance' is a relative term. It is an interpretation relative to a reference-frame, relative to a context. For CB, without restraints or constraints there is no impetus to drive motion. This is why, for CB, inertia is an illusion.

The Earth rotation influenced Galileo's concept of inertia (also he never presented a term for the concept, i.e., he had no term for 'inertia'). Galileo argued for the Copernican heliocentric theory that requires that the Earth rotates. The problem he had to account for was the velocity of the Earth's rotation and why objects on its surface were not blown away by this rotational motion, etc. With host rotation, the closest thing to uniform motion of a point on the surface of celestial spheres is sustained in any surface circle normal to the geographic axis. Galileo's concept of inertia included this circular movement. This disqualified his concept from being the 'modern' one. Success in this argument would contradict Ptolemy's major objection to Earth rotation. The acceptance of CB principles invalidates Ptolemy's objections (see Addendum 6 for Ptolemy's objections).

FLIP-FLOP MECHANICS. Sink, boundary, and float are different modes of CB. Though CB is a physical principle, the ways for calculating it may be as kinematics or as dynamics. Under certain conditions CB can be modeled with inclusive flip-flop modes, like describing a glass of water as half-empty or half-full. They may each be used exclusively or both be used inclusively, without contradiction. However, this does not preclude that a constant or function of proportionality may be required for relating the two modes of analysis. The analysis of each dual mode takes, as basis, float or sink as the primary principle. With an appropriate constant or function of proportionality, the sink and float interpretations are as immanently reversible as the use of NAND and NOR. For example, the Earth and the Moon converge toward each other. In the sink mode, the sink buoyancy is determined from the rate of sinking toward a center.

Thus the Moon is sinking toward the Earth with an acceleration and the Earth is sinking toward the Moon with an acceleration. These accelerations can be calculated from the observed geometric relations and require no physical principles for this calculation. They are simply geometric relations. It is kinematics without established constants or functions of proportionality for relating it to a dynamics interpretation.

In the float mode, the Earth and Moon are regarded as floating around each other. The float buoyancy of each object is determined through calculations dependent on the physical state of buoyancy between objects. The analysis of this floating is based on convergence powers and pressures, the physics principles rather than only a geometric description. In effect, it is dynamics; though it also requires geometric considerations to determine the float buoyancy. In fact, it is the geometric considerations that open ways to applying the influences of dynamics. This does not preclude that these modes of calculation may not sustain a physical equivalence. Geometric quantities do not have the mixture difficulties found in physical quantities. In other words, geometric reasoning readily models *ex parte* analysis. Synthetic geometry was even founded on *ex parte* analysis. In physics an *ex parte* analysis requires suppositions based on a system of reasoning. The Newton and the CB systems of analysis are such systems of reasoning and herein are shown as incompatible.

All CB reference-frames are dynamics reference-frames. When we do kinematics calculations with respect to a CB reference-frame there is always an implied dynamics. Since spin has no dynamic significance outside a neighborhood, a geometric calculation (such as $\mathbf{v}^2/\mathbf{r}$) has no physical significance (dereference) outside a neighborhood. Recall the argument earlier in this Chapter (**Convergence Deficit**) about its use for the planet rotations. The intended Newtonian correspondence to centrifugal force fails. This failure is consistent with its dependency on dynamics for its existence. A case where the formula has coincidental accuracy to be directly useful is the one concerning the space-shuttle and adjusting its angular velocity relative to the center of the Earth to correspond to a sink acceleration that counters the shuttle-Earth convergence rate.

SCALE OF OBSERVATION. So far, the defined scale transitions for CB are; gravitation to surface tension, surface tension to atomic physics. These transitions have yet to be well understood in their detail and the role of convergence in these transition zones is not yet well defined. From the gamut of scale-sectors used for analysis of physical phenomena, we can separate the scale-sectors according to what physical principles dominate these sectors. Furthermore, we can characterize the transition zones that separate a conjacent pair of scale-sectors. That is to say that the gamut scale-sectors that show significant shifts in what principles dominate, are separated by transition zones.

It is when we cross over the transitions that convergence effects shift in scale, a shift that starts another mode of physics to dominate the influences on convergence effects. I argue that the two mentioned transition modes sustain a consistency in regard to the fundamental principles of convergence. A third transition mode may be from atomic physics to plasma physics and for now it seems that it may reduce the role of CB rules because the ionic influence of charged particles allows for different convergence rules (CA) dominated by the positive and negative charges. A fourth transition may be from gravitation to macrocosmic (galactic) scale. This may be another transition that changes dominance from CB rules to CA rules because galaxies may be driven by matter-antimatter relations (in which case applying Kepler's Laws is an error and the dark matter required from their use is eliminated). It may be that the galaxies reinstate a primordial plasma by the mutual annihilation of matter and antimatter at the galactic nucleus.

In the possible Physics based on matter-antimatter relations, the implied opposition in polarity suggests one of the most powerful macroforms of convergence yet conceived, a convergence based on an attraction to annihilation. This form of convergence follows a different logic from the like poles of buoyancy convergence. I note again that a simplistic bifurcation into poles such as {+,-} and its neutral form {0} may be misleading. A strong positive charge relating to a weak positive charge can appear to be a relation between a positive and a negative charge, etc., as Maxwell pointed out.

The underpinning logic here leads to a basis for all convergence host-debris accretions.

The discovery and development of a Physics principle starts with a scale of observation that dominates the interpretation of events under scrutiny. When we extend the use of the principle beyond the boundaries in which it was formulated we are either assuming its scalability or testing its scalability.

IMPETUS WEIGHT AND THE EQUIVALENCE PRINCIPLE

The following argument corresponds to the Equivalence Principle as it applies to the Einstein elevator argument. We conceive three reference-frames, K_0, K_1, K_2. An onlooker at K_0 can see that K_1 and K_2 are going in opposite directions (Exhibit 3.1, the 'Up' and 'Down' terms are relative to the onlooker at K_0).

K_1 is accelerating at rate **g**.
K_2 is decelerating at rate **g**.

The onlooker asks the twin occupants, one in K_1 and one in K_2, to report how much they weigh, each replies 90 kg. As they pass the observation point, K_0, the one-way transparent view shows them to the observer, K_0, as standing on the same ends of the elevators. The occupants have no way of determining, at that moment, if their elevator is accelerating or decelerating. The problem is not decidable for them. They each weigh in at 90 kg.

In K_1 the deceleration against the elevator floor is the equal and opposite reaction to the acceleration of the elevator. In K_2 the passenger is made to decelerate by a deceleration of the initial velocity of the elevator.

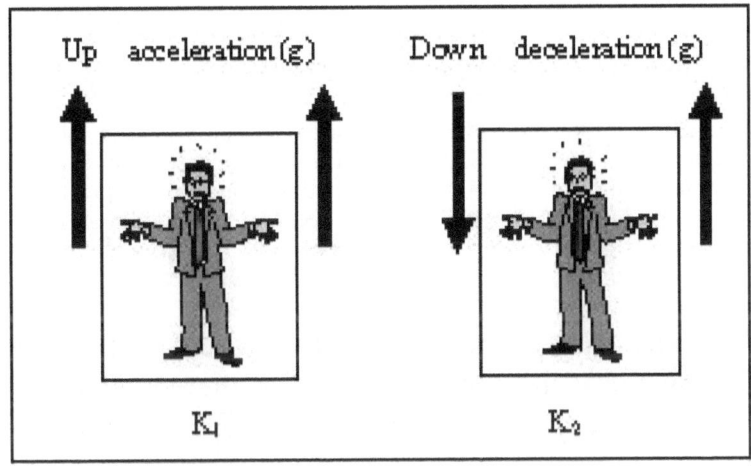

EXHIBIT 3.1: Elevators passing from opposite directions
(passenger image courtesy of Microsoft)

Comment: The Einstein elevator acceleration effect depends on inertia. From the CB perspective, there is no acceleration effect if there is no neighborhood reference-frame. Where motion has no intelligibility without a neighborhood, the acceleration also has no intelligibility without a neighborhood.

The Einstein elevator cannot duplicate the effects of host neighborhood pressures though in a sufficiently strong neighborhood their elevator can imitate a range of convergence powers. A more subtle point is that if *ex parte* weight derives from squelching then it has no possible expression where external gravity is eliminated from the elevator (as the Einstein thought experiment requires). The elevator cannot produce or imitate the required squelching effects for weight. This gives a fundamental reason for why the Equivalence Principle cannot be sustained in CB. Furthermore, the environment required for the Einstein elevator experiment denies Mach's Principle. This is because when gravity is turned off, Mach's source of inertia is turned off.

Next it is shown that the Equivalence Principle cannot be sustained within the requirements of either of the Einstein Relativity theories and in their own terms.

> **Note:** It can be argued that the occupant and elevator themselves make up a neighborhood. However, this neighborhood pressure is so small that the elevator would have to accelerate at rates many times the currently presumed speed of light (as shown in Chapter 1) for the occupant to weigh-in at 90 kg. Such speeds and accelerations are forbidden in Einstein Relativity and therefore the elevator can have no visible acceleration effect on the passenger.

A Time-Dilation Contradiction. Einstein Relativity carries the presupposition that weight produced by gravity is equivalent to weight produced by acceleration, and pointedly states:

> "The rates of the masses of two bodies is defined in mechanics in two ways which differ from each other fundamentally; in the first place, as the reciprocal ratio of the acceleration which the same motive force imparts to them (inert mass), and in the second place, as the ratio of the forces which act upon them in the same gravitational field (gravitational mass). The equality of these two masses, so differently defined, is a fact which is confirmed by experiments of very high accuracy (experiments of Eötvös), and classical mechanics offers no explanation for this equality." (*The Meaning of Relativity*, p56, Princeton University Press, 1956).

> **Note:** I have shown in Chapter 1 that classical mechanics does explain, namely the acceleration differences are too small for direct measurement.

It goes on to state how this relativity accounts for the equivalence.

Imagine being in an elevator in deep space and not accelerating. Under current modes of analysis, whether it has velocity or not is not determinable from within. For the argument we shall presume that in this elevator-frame the laws of physics are valid (recognizable). You judge that you are weightless. Next imagine that suddenly you move toward one end of the elevator with acceleration **g**. You are then asked to decide whether the elevator is in acceleration or in deceleration. It should be clear that we have no means for deciding from inside the elevator, using the Newton model or the Einstein model. At the moment that K_1 and K_2 pass K_0 we cannot even use the trick of dropping two balls and see if they approach each other (or not) since directional polarity at that moment is identical for both elevators, etc., (if convergence is present this is mandatory). Within reasonable limits you may determine **g** but you cannot decide what action generates it. These two cases are dual in their coherence and therefore one is not simpler than the other.

Now you are standing on the Earth with weight **W**. Is this due to an acceleration or to a deceleration, i.e., positive or negative acceleration (since **W** = **mg**, acceleration remains an intensional component of Newtonian weight)? With respect to the Newton system no debilitating consequence arises from the decision even if judged from another reference-frame.

With respect to Einstein Relativity there is a possible debilitating consequence that arises once the decision is made. This consequence arises from the necessary conclusion that is produced by how the Lorentz transformation formulas are used, and in particular, the reference-frame transformations that obtain the following sequence of reasoning.

One of the consequences of not recognizing the qualitative difference between acceleration and deceleration is the following:

Presuming acceleration (positive acceleration)

"The rate of a clock is accordingly slower the greater is the mass of the ponderable matter in its neighborhood. We therefore conclude that spectral lines which are produced on the Sun's surface will be displaced towards the red, compared to the corresponding lines produced on the Earth, by about 2^{-6} of their wave-lengths" (*op.cit.* p92).

The observation of this red shift has been claimed from two different sets of experiments. One series is the Pound-Rebka (10% within prediction, 1960); the other is the Pound-Snider (1% within prediction, 1965). These experiments used the Mossbauer Effect to detect the change in the frequency of gamma radiation in the Earth's 'gravitational' field. They showed that the rate is slightly slower in the Earth's 'gravitational' field (CB interprets this as a repelling effect). This redshift is not to be confused with the shift that astronomers use for estimating stellar distances or use for defending the hypothesis that the universe is expanding. This is a third use of redshift (for now we shall ignore how to decide which is which).

Assuming with the Einstein theories that gravity and acceleration have an equivalent effect on mass in terms of weight, we would say that in the case of occupants experiencing **g** in a decelerating elevator, the gravitational equivalence implies that our weight on the surface of the Earth is derived from a deceleration.

> **Note:** Some physicists like to say that an object resting on the surface of the Earth is not accelerating relative to the Earth surface. Of course in Newton terms this is a mistaken belief because it contradicts **W=mg**, etc.

When a force is applied to a moveable mass it will accelerate. When there is impedance to this acceleration, it is the measure of this impedance that allows us to quantify mass in Newton's system. If there is no impedance, there is no mass and measured weight equals zero. For accreted bodies, the Earth's surface resists the continued acceleration toward the Earth's center, thus the Earth's surface decelerates 'falling' bodies. (Close to the surface of the Earth, air increases this deceleration with resistance to movement from friction and from Archimedean buoyancy.) Weight can thus be defined in terms of deceleration.

Weight can also be defined in terms of acceleration. Newton defines weight in this way and the Einstein theories follow this. Newtonian physics leads to the equation: **W = mg**, where **g** is acceleration due to gravity and **m** is mass. The 'newton' units, **N**, are defined to eliminate the need of a unit constant (constant of proportionality). In this equation the deceleration is hidden in the concept of mass.

Note also for the argument that cyclic phenomena underpin the measure of time.

Question: If an acceleration slows down cyclic phenomena, does deceleration speed up cyclic phenomena?

Argument: Newton's system does not address itself to this problem, but Einstein theory holds Newton's system as a limiting case of Relativity. From Einstein theory we have definite arguments we can present.

Consider:

Case One, $v_2 > v_1$: The elevator accelerates and then stops accelerating. The elevator continues at the last velocity achieved. At this fixed velocity, time is dilated when compared to the elevator starting position.

An elevator is accelerating; it ceases to accelerate and is traveling at velocity v_1. It accelerates again in the same direction and again ceases at velocity v_2. We can say then, $v_2 > v_1$, and furthermore, according to the Einstein use of the Lorentz Transformation, time t_2, has dilated with respect to time t_1. This physical dilation of time t_2 implies that all cyclic phenomena slow down. This is what the Einsteins believed occurs in a manner directly proportional to the local **g**.

Case Two, $v_1 > v_2$: An elevator is decelerating and ceases to decelerate at velocity v_1, and then it decelerates again and ceases at v_2. We can then say that $v_1 > v_2$. According to the Einsteins, this implies that time t_2 contracted, the time at velocity v_1 is more dilated then at velocity v_2. This physical contraction of time t_2 implies that all cyclic phenomena speed up.

Comment: The physical phenomena are not independent of the Einstein reference-frames. From these two cases we see that they are opposite in principle and we infer that the choice is a critical one. The choice determines the nature of the predictions.

Question: Under acceleration, are cyclic phenomena slowing down or speeding up?

Response: The answer follows according to whether the acceleration is positive or negative. If acceleration is positive the cyclic phenomena slow down, if the acceleration is negative (deceleration) then the cyclic phenomena speed up.

Question: Can we decide which is the case for accreted objects (*e.g.*, at their host surfaces)?

Response: All gravitational bodies generate radiation (Chapter 2 and 4). The Earth gives off radiation on the low energy infrared side, invisible to the eye. If the vibrations generating this radiation were speeded up even more, the radiation could include visible light. The Sun, which we reason to have a much stronger gravitation, gives off visible light, but also includes very high energy light in the ultra-violet range, x-rays, and gamma rays (the x-rays may indicate a fissioning activity). This indicates that the vibrations that generate radiation on the Sun move on the average at a faster rate than on Earth. There is nothing subtle in this and the little red shift, that the Einsteins hoped to find, has no meaning in this context. We do not need this Einstein theory of relativity to tell us whether the cyclic phenomena on the

Sun are slower then cyclic phenomena on the Earth. The question has no subtlety, and the Einsteins are simply wrong. This also suggests that the Pound-Rebka, Pound Snider experiments need re-interpretation.

Comment: Catechism physicists may have a conceptual block to this conclusion because they do not consider the light from the Sun as convergence generated. We also observe that the metabolism of astronauts in microgravity does not speed up as the Einstein supposition would suggest. The metabolism slows down.

The Einsteins did not consider the deceleration case and therefore could not recognize the decision problem between the two cases required for completing their thought experiment. The speed limit ranges are the reverse of each other. The acceleration case must end before it reaches the speed of light and the deceleration case cannot begin at the speed of light, etc.

TWO PRINCIPLES OF CONVERGENCE RESSURE

Introduction: *Lucretius (c98—c55 BC) in his* De Rerum Natura *proposed that all sensed properties of visible bodies were determined by invisible 'seeds' (corpuscles). Sir Isaac Newton (1642—1727) proposed that heat consists in the internal motion of the body corpuscles. Ludwig Boltzmann (1844—1906) developed a quantitative model for this 'invisible' agitation in the form of 'corpuscular' vibrations that manifest as heat and his underpinning model strongly resembles a form of CB on the molecular level. In volume one of his* Lectures on Gas Theory *(1896) he states in the Introduction that for a molecule: "If it approaches a neighboring molecule it is repelled by it, but if it moves farther away there is an attraction." His description of this foundational principle of movement is defended on the basis of the* a posteriori *accountability that his derived theory provides for the phenomena under scrutiny.*

Corpuscular Vibrations. Convergence pressure affects corpuscular phenomena. There are two modes of corpuscular vibration and a common demarcation for them is that one mode is intracorpuscular (generates within the corpuscle) and the other mode is intercorpuscular (generates between the corpuscles). This is a theoretical supposition and other models may yet come into vogue. Cynthia Kulb Whitney (2004) has suggested that the periodicities we associate with intracorpuscular vibrations could be generated within corpuscle clusters, which would eliminate the need to complicate the inner atomic structure to account for periodicities (we know that the clustering modes of water molecules differentiate the physical states of water into solid, liquid, or gas, etc.). In either case, periodicity associates with a stricter cyclic vibration and connotes regularity.

> **Note:** These considerations suggest a study of the spectroscope to determine whether instruments can be designed to differentiate these varieties of intercorpuscular and intracorpuscular forms of radiation.

The definitions for vibration modes may extend to levels of higher complexity, e.g., certain biological processes may be regarded as complex periodicities, etc. These definitions may extend to larger scale astronomical observations as well.

First Principle: In convergence isopressure zones, corpuscular vibrations vary in their speed directly with the local convergence pressure.

Thus the greater the pressure the faster the vibration equilibrium rate. An increased vibration rate generates an opposed counter pressure. If the periodic rate did not increase under compression the material object likely offers less resistance to collapse (implosion). This principle functions only within a limited latitude of possible compressions. Sufficient pressure changes can change the structural characteristics of the material under the pressure change.

Second Principle: Material of sufficient bulk contracts along lines of convergence into a shape that allows the least impedance to restraints.

Systems exceeding a certain compactness and bulk must deform into spheres. An object of sufficient quantity and sufficient contraction becomes a fire-star and by shear compactness of convergence-pressure maintains the generation of light from heat. No artificial cycles are required to explain long term generation of light from heat and the sphere is the expected shape. The Carbon Cycle is an apodeictic fiction, etc.

SCHOLIUM FOR THE FIRST PRINCIPLE. There are two categories of convergence pressure opposed: compression and tension. In a material object, the highest tension is found in the interior at the CIS and its primary effect is an increase of the intercorpuscular vibration rates, an agitation augmentation of material corpuscles in that region. Intercorpuscular vibrations are the heat generators that Newton and Boltzmann talk about.

The highest convergence compression is found on the outer surface of gravitational convergence bodies. It has two primary effects. The first effect increases the intercorpuscular vibration rates that generate radiation from heat. The second effect increases the intracorpuscular vibration rates that generate radiation from internal periodicities. The effects this has on temperature have not been adequately studied. CB anticipates that a rise in temperature will also raise the intracorpuscular periodicity rate and the lowering of temperature will slow these periodicities.

Therefore, this First Principle contradicts the Einstein hypothesis that gravitational force slows corpuscular vibrations or periodicities (more in chapter 5).

Granularity. If we characterize gas, liquid, and solid in terms of material clustering we have a rough scale of increasing magnitude in the number of corpuscles in the clusters. These cluster sizes are herein referenced as the *granularity* of the material object. The intracorpuscular vibrations could be generated within the grains (mainstream says within the atoms); the intercorpuscular vibrations are generated exterior to the grains (or atoms). There is likely a correspondence between radiation frequency and the measures for this granularity. One complication in this analysis is the possibility of harmonics.

The stronger the host gravitational convergence effect, the more counter pressure is required for sustaining physical integrity for the zone or debris under scrutiny within or on the host. Faster periodicity rates for the corpuscles augment the counter pressure. In effect it provides a counter pressure that within certain limits reaches a genre of equilibrium with the gravitational convergence.

In a given convergence object such as the Earth the heat level associated with the 'size' of the Earth represents an effect from corpuscular vibrations that are proportional to the Earth's size. But these corpuscular vibrations and their relation to surface tension may produce other effects not considered here. The speeding up of corpuscular vibrations affects the radiation spectra. Thus in the Earth's interior, these vibrations can increase material pressure and this increased pressure increases the rates of periodicity. The Earth radiates. Likewise, all gravitational convergence objects radiate.

***Conjecture*:** When we attempt to measure the speed of light we are using devices that potentially mix modes. Corpuscular devices cannot be used to determine characteristics of wave phenomena, unless we want to 'know' the characteristics of wave effects on corpuscular devices. This is another way of saying that we cannot use corpuscular devices to determine if we are detecting waves or not. It would be like using a digital computer to determine if the data are analog or not. Furthermore, if we use material with periodicities slowing down, we may ask what effect this has on the presumed measure of the speed of light. We could put an atomic clock on the Moon and test its atomic periodicity relative to an Earth twin clock. Atomic clocks have been miniaturized (2004), which renders this experiment more feasible. The Moon could also be used for testing the long-term effects of temperature extremes on the clock periodicities.

If in presumed measures we found that the speed of light was constant *in vacuo* for every reference-frame, then with respect to light these reference-frames are at rest. This necessary conclusion is an absurdity that catechism physicists ignore. It may be that it is not the speed of light that is being measured. It is rather a mixture of the respond speed of the corpuscular device with what might be the speed of light that is being measured. It is assumed that light has speed and is Eudoxian, if light takes 4 minutes for distance, **d**, light takes 8 minutes for distance, **2d**. This is likely not found true in non-isotonic neighborhoods. This possible variable speed relates to the energy level of the polarization carrier and the time required for the photonic overlay phenomena carried on it. The corpuscular device used for the measuring is believed to be quantic in its energy level response. Thus a weaker polarization could take a longer time for the corpuscular quantum response. As the light beam may weaken with distance we could be using an illusory interpretation of light traveling. This puts the question of light traveling into a state of immanent reversibility that in one of its instances can deny that light travels. This could account for yet another interpretation for a redshift (more in Chapter 5).

It is doubtful that a laser beam 'reflected' off the Moon has the same energy level on its return as its origin energy level. The first problem is that there is no such thing as a perfect mirror. If on Earth we could filter the laser beam so that its energy level at point of measure has diminished from its energy level at point of origin equal to the Moon measure difference, would the measure of the speed of light be the same as if we were measuring with the Moon distance? If we cannot lower the energy level of light without 'slowing' it down we are left to wonder what we are actually measuring with our corpuscular measuring devices. Note that lowering the energy level of the origin is not a valid test for this question since it is the potential difference between the origin and the destination that must be sustained for the comparison. Note also that these considerations put the Einstein claimed analysis of the photoelectric effect in question. This does not preclude the fact that this Einstein analysis had a parametric plausibility. The fault addressed is the physical interpretation (more later).

SCHOLIUM FOR THE SECOND PRINCIPLE. The Sun is increasing its material quantity by increments. Over the eons this increase in bulk renders the CIS more active in the generation of heat. This generation of heat in the Sun may be of two kinds, the basal heating associated with the natural CIS tension, and the heating that generates from the breakdown of heavier elements under the CIS tension. We associate nuclear fission with the generation of x-rays. When the heavier element breakdown strongly affects the star, the star can swell for cooling and this cooling can render the star into a red or blue giant depending on quantity of heavier elements available. With a more limited supply of the heavier elements the star may become a red giant that can be subject to a sudden depletion of the heavier element fuel. The CIS can no longer support the star size. At this point, the red giant can implode, which generates a nova. In the case of the blue giants, it is possible that the accretion itself has created sufficient tension to approach a critical bulk, but it is also likely that the heavier elements are more abundant and fuel an amplification of the heat. This is why CB considerations predict that many blue giant stars are candidates for explosion and red giant stars are candidates for collapse.

If stars are hollow, the poles collapse toward the plane of the equator, following the tidal distribution of material in spinning hollow objects (thickest at the poles and thinnest at the equator), which would yield a torus explosion pattern about its equator with rings of material about the axis rather than a spherical explosion. When material stars are like magnetic bottles that leak at their poles we might expect secondary ejecta to form symmetric with the poles as a line of material along the axis.

> **Note:** The Sun's compactness is thought to be low because it is always calculated as if the Sun were not hollow.

There are other scenarios but not pursued here.

Note: If the center of our galaxy is antimatter then the attraction of giant blue stars toward this center should be greater than the attraction of giant red stars toward the center. The implication is that we should find more blue stars toward the center of galaxies. This does not preclude other possible reasons, such as availability of debris and how long has it been converging.

The planets are also accreting debris. This accretion must finally have evident effects on orbits in terms of speed and AUs. All celestial subgalactic material objects may be regarded as various stages of star evolution, all of them radiate. We are fortunate that the effects of this growth through accretion are relatively slow in our region. It has contributed to the time required to evolve life as we know it.

This evolution sequence for stars suggests in older terms a principle of anti-entropy. The universe is not "running down."

In convergence fields, heat increases with the strength of convergence pressure. This implies that all planets are pristine fire-stars. Heat and convergence increase correlatively. Higher pressure induces higher corpuscular vibration rates and 'heat' is another name for these vibrations. Convergence compaction also augments the rates of these material vibrations. These correlative vibrations function as a counter force to contraction. Theoretically, there are determinable limits of quantity of matter and contraction possibility in a given pressure zone. A limited side effect of the convergence pressure increase is an increase in the intracorpuscular periodicity rates.

This principle suggests that under certain conditions an artificial inducement of corpuscular vibrations in material objects may amplify their convergence power.

Thought Experiment. Conceive of what we shall name a 'critical quantity' of material in the form of homogeneous particles in space.

Next imagine all the particles accelerating along convergence lines to form a more concentrated system object that continues to contract all the while forming a system surface (tending toward a sphere) with an ever increasing surface convergence pressure.

Surface granulation develops for more efficient cooling. Growing toward a 'critical quantity' the critical pressure is reached where the rates of periodic phenomena exceed the possible limit for the material object to sustain itself in a stable form. Here the object begins to pulse to augment the cooling rate.

Finally it loses containment and implodes (like red giants, etc.) or explodes (like blue giants, etc.), a nova.

Our Sun 'attracts' debris that is rapidly accreted into its bulk. Over billions of years this increase in material quantity increases local pressure for further compaction of the Sun toward critical pressure at the same time the material quantity augments toward the nova-critical level. Aging is redefined in terms of change in state through accretion relative to a neighborhood pressure.

SUMMARY CONCLUSIONS. Note that the CB Definitions and Laws could be reduced in number in the manner of Ampère and Mach, but I see no advantage for this in a first presentation. Furthermore, the anti-Newtonian aspect would be less visible.

Any object that sinks at an acceleration rate that is not isometric with the neighborhood restraints must regress toward that isometry unless its rate is otherwise sustained by constraints.

The buoyancy convergence principles suggest a coherence for at least the three known physical scales for material convergence: gravitational, surface tension, or atomic. Some applications to atomic physics are discussed in later chapters. Special studies need to be done for the transition states. These principles may apply to matter-matter relations and even speculatively to antimatter-antimatter relations. Different principles are required for matter-antimatter relations and for plasma physics or any physics where opposed polarities dominate the action.

The Newton orbits are extremely fragile. The Convergence Buoyancy orbits are self-correcting and robust. With buoyancy convergence, the orbiting objects can bob and restabilize when disturbed. In *ex parte* terms, their isodynamic positions are specified to them by the synergetic convergence relations with the non-isotonic neighborhood.

The Newtonian celestial mechanics is self-destructive.
The CB celestial mechanics is self-constructive.

The question that is never directly addressed in the Principia is, "*How is change in velocity physically specified to the object that is changing velocity?*" For Newton we could say that it is inertia that specifies to an object that it is undergoing a change in velocity. However, as Newton is first concerned with mathematical principles, we are left with inertia as an axiom and no physics dereference. That he has named a resistance to change in motion as 'inertia' adds no comprehension to its physical nature. In other words, 'inertia' functions as an apodeictic fiction for Newton's physics. For CB, 'inertia' is filed in the same box as 'phlogiston' and violates Newton's rule: "Hypothesis non fingo."

Berkeley's relativity inspired Ernst Mach to develop Berkeley's thesis into a relativity of inertial forces, a dynamics interpretation that is distinct from CB relativity. In CB analysis we start with George Berkeley's kinematics relativity and develop CB relativity into a dynamics interpretation that is devoid of inertia. From this we derive that any change in velocity is physically specified to the object that is changing velocity through synergetic convergence fields. In certain respects this rehabilitates the Cartesian vortices. When the concept of synergetic convergence fields is understood, the answer to this question is understood. The thesis of CB, expressed in *ex parte* terms, is that neighborhood convergence 'communicates' with the affected object and this relation specifies to the object its relative status. The only way movement can be communicated to a 'moving' object is from a neighborhood. Expressing this in a language independent of *ex parte* concepts would require a major overhaul of our language and our attitude in its usage.

The sink acceleration of a material object that zeros out the Earth **g** (adjusted for altitude) is a form of neutralized accretion, thus it is negative. As argued in Chapter 1, this sink acceleration varies according to the *ex parte* e-power of the sinking object. These variations deny the traditional simplistic notion of 'free-fall'.

Material convergence generates revolution, rotation, repelling, and, furthermore, these generated movements correlate to the neighborhood restraints. The Sun maelstrom correlates to the Earth rotation, and the greatest speed at the Earth's surface is clearly found at the greatest distance normal to the spin axis. Movements not yet isodynamic with the neighborhood are under constraint or they are regressing to an isodynamic state. Movement is never independent of a neighborhood since in this mode of analysis we use the relation between the neighborhood and the object to determine the motion (Convergence Relativity). Translation movement is a constraint induced movement.

END Chapter 3

Chapter 4: RADIATION and HEAT

INTRODUCTION. *The first thermometers used material expansion properties that could be associated with an environment that influences this expansion. The measure of this expansion was usually in terms of a scaled subdivision between the freezing point and boiling point of water at a standard atmospheric pressure. Extensions were scaled to the practical limits of usage for the expandable thermometric material. With the theory of the corpuscularity of material the temperature came to be associated with corpuscle vibrations (Newton, Boltzmann). Because radiance can influence these vibrations, catechism physicists regard the radiance itself as another form of heat. The following arguments lead to another interpretation.*

TEMPERATURE. Temperature is not necessarily a measure of heat content. Also, as we go toward the extreme possibilities of temperature we find that our ordinary temperature measuring instruments based on antique technology (expansion properties) do not scale well to the current needs in science.

An International Standards (SI) Error. Fahrenheit was the first to establish a standard degree measure for temperature (1714). Celsius (1744) used a centigrade scale but used **100** for the freezing of water and **0** for the boiling of water. Independent of each other, Linnaeus (Sweden) and Cristin

(France) within a year (1745) set **0** as the freezing point of water and **100** as the boiling point. The International Bureau of Weights and Measures members voted to name the centigrade scale 'Celsius'. The voters were likely ignorant of the upside down scale used by Celsius. For these scales it was assumed not only that the thermometer materials expand linearly but that temperature itself is Eudoxian. We now know that this is not true. Progress in temperature measurement has been slow and though newer methods are available, antique methods are still in wide-spread use.

Basal Steady-State. Under standard atmospheric pressure, boiling water is a steady-state process. It can be regarded as a heating process or a cooling process. From the Newtonian perspective boiling is a cooling process in that even with the constant application of heat the boiling water temperature does not exceed a fixed amount. At a given atmospheric pressure, etc., the temperature of boiling water is fixed, because the water above the boiling point escapes as steam. For CB, all material objects sustain a basal heat level (also called its *ex parte* steady-state heat level) that can be influenced by its milieu and furthermore, on a host, this influence is an amplification that radiates in a signature pattern that can be used for identifying the material content (more in Chapter 5). The temperature for a debris object accreted to the Earth surface varies in a manner that follows the first principle of convergence pressure: *in convergence isopressure zones, corpuscular vibrations vary in their speed directly with the local convergence pressure.* This implies that for accreted debris the steady-state level that we measure has been raised to a composite level that corresponds its basal level amplified by the host surface milieu. When we measure a temperature it is always a composite of the temperature generated by the basal heat and the effect that the milieu has on this basal heat. We are left to determine the basal heat through systemic reasoning because it is invisible to direct observation even though it has determinate effects. To find the debris basal heat and the temperature that it generates requires finding the limiting value that its steady-state heat could take in isolation. If we cannot find a way to produce an isolation milieu our only recourse is by artifice, which is to say that it would require a systemic derivation to find the limit values. In theory, the Franklin Boundary State or Neutralization suggests modes of isolation that might one day lead to an empirical determination.

Our usual empirical temperature measurements consider temperatures in non-isolation and this includes non-steady-state as well as steady-state temperatures. Our usual empirical measurements deny the direct reading of *ex parte* values for temperatures and heat. An example using only hosts we can say that the Sun influences the Earth's steady-state heat. One of these influences is the permaheat found in tropical zones, which is a function of the CIS, Sun radiation, and tidal effects (among which; Earth rotation). For hosts the composite steady-state measurements are influenced by both the CIS and the neighborhood. All material objects have a basal *ex parte* steady-state. At present we have no well defined algorithms for finding these values.

That objects can have different temperatures in a given milieu can be described in a primitive qualitative manner in terms of the sensation of warmth or coldness felt from proximate or direct contact with them. This further implies that everything we learned about heat started with thermal changes that we noticed on thermal contact with them.

Parametric Heat Measure. Under a given convergence power, the quantity of heat is expressible as a function of at least three parametric measures; parametric because they are measures of effects and not a direct measure of the corpuscular vibrations themselves. Heat has a temperature, T_h, a volume, V_h, and a compactness, C_h. The geometry of a system includes volume and shape, but, except for extreme cases, thermodynamic properties are largely independent of shape. Therefore, most magnitudes of heat, H, can be expressed as an ordered triple of parametric measures: $H = (T_h, V_h, C_h)$, where H can be regarded as a point with coordinates or a vector with components. The order is strict by convention only so that we have the convenience of positional notation.

Thus temperature measurement alone does not indicate the magnitude (volume and/or compactness) of the heat and this quantity of heat does not indicate a fixed temperature, etc. It represents heat as a *vector attribute* and the calculation follows rules derived from linear algebra.

Note: On computer screens, color is a 3-dimensional vector attribute. The vector component variables are red, blue, green. Assigning values we obtain a unique color that cannot be gotten from any other combination. The resulting color is the vector attribute. Thus from a visible color on the computer we can deconvolute the vector attribute to its precise components, etc. The deconvolution of H is even easier.

Toward a Scalable Heat Formula (Heat measure using three parameters; a temperature, T_h, a volume, V_h, and a compactness, C_h.). In practice, the suggested linearity of these parameters is only useful on Lilliputian scales (narrow scope).

    1) Heat capacity can vary directly with compactness. When there are equal temperatures, same volume, homogenous material, then the heat capacity in a given milieu varies directly with compactness.

    Example: The material compaction level of the atmosphere in a sauna room, fixed at a certain altitude (atmospheric pressure), varies according to the quantity of water vapor it holds. If you remove the water vapor you are lowering the material compaction level, which lowers the heat capacity of the sauna atmosphere. In order to maintain the same temperature the sauna must pump more heat into the room. On the other hand, you can add water vapor to a sauna room in such a way that the compactness of the air with vapor is raised, the temperature goes up. The sauna must pump less heat or remove heat to maintain the same temperature as before.

2) The heat capacity can vary directly with volume. When there are homogenous materials, equal temperatures, equal compactness, different volumes, then the heat capacity varies directly with volume.

Comment: If we could ignore the physics of these volumes of material, CIS, resonance, etc., we would still have to contend with the volume and surface ratio. We could keep this ratio the same between two different volumes if we made the larger volume hollow or granulated its surface or some combination of the two, etc.

3) The heat capacity can vary directly with temperature. Starting with same homogenous materials, equal volumes, equal compactness, then the heat capacity varies directly with temperature.

Comment: To date we have yet to find a Eudoxian representation for temperature. For material indicators, it is not Eudoxian. Even so, physicists apply Eudoxian operators to kelvins that close on non-negative temperatures.

For scalability the quantity of heat expressed as an ordered triple is of the general form

$$H = (h_1(T), h_2(V), h_3(C)),$$

where the **h** notation indicates scales of adjustment for the non-Eudoxian aspects of the parametric magnitudes and the order is by convention for the convenience of positional notation. **H** as vector attribute can be used for representing the heat capacity of a given object for a given milieu or it can be used for the heat level of a given object. The former indicates the heat limit for that material object in its milieu and the latter measures the obtained heat level.

The role of chemical and physical composition in heat signatures is discussed in the next Chapter.

The quantity of heat exchange is analogous to amperage. There is no heat amperage unless there is a heat exchange. Temperature differences are analogous to voltages. Though stated as an analogy of thermodynamics from electricity, historically the electric quantities were modeled by analogy from thermodynamics (cf Lord Kelvin, etc.).

Thermodynamics. In early studies of heat the main concern was heat transfer. One of the proposed principles that came out of this work was the conservation of heat, and it is true that if heat transfer is the only activity, heat is conserved. However, the principle of conservation of heat became an obstacle to a fuller generalization of heat dynamics. It is Clausius who shifted the emphasis to the transformations that concern heat energy and work, where heat was regarded as a form of energy that could transform into work and work energy can transform into heat. This changed the emphasis to the principle of the Conservation of Energy in transformations.

> **CB Definition of Thermal Equilibrium:** Two systems in thermal contact are said to be in *thermal equilibrium* when the thermal balance shows no measurable difference between the corresponding heat vector components.

Using the definition of thermal equilibrium we formulate the following two Common Notions for the Thermodynamics of debris:

First Common Notion: *Two systems in thermal equilibrium have congruent heat vectors.*

Second Common Notion: *Two systems in thermal equilibrium with a third would be in thermal equilibrium with each other.*

The main use for these Debris Common Notions is the formal denial that a debris heat event is in thermal equilibrium. To guarantee that these Common Notions hold for debris in fullest generality, the objects in thermal contact must be of the same material and not destroyed by that contact, etc.

Heat Radiation. Radiation from corpuscular vibrations has a spectrum that has here been characterized as a combination from two modes for these vibrations; one either within the atom or generated by a clustering in groups of closer proximity, and the other either intercorpuscular or from the higher granularity of clusters interacting. These generators of spectra and their possible harmonics are associated with the granularity of their material drivers. Until this feature of granularity is better investigated we will not have a precise determination of the demarcation of these spectra. Whatever the case, it is useful to consider that the thermometer measure is based on a change in the corpuscular vibrations of the thermometric material. Thus it is more accurate to say that temperature measures some combination of the two modes of corpuscular vibration effects on the thermometric material. This is quite different from saying that it measures the temperature of another object or, more fictitiously, the 'temperature' of radiation.

Radiation is not heat, nor is it here conceived as having a temperature, however, it is here assumed that it has an energy level (a capacity to do work) that corresponds to the agitation levels in a material object and therefore may have a correspondence with temperature in a given material. Thus one approach to temperature measurement is to focus on radiation as the carrier of temperature data. Carrier radiation from sources other than the Sun can mix with the solar radiation to produce misleading results, as is common in upper atmospheric temperature 'measures' (more in Aeronomic section below).

At a distance a thermometric measure may convert the radiation spectrum from an object into a temperature reading from that object. Thus radiation from the corpuscular vibrations from an object may be the only data that is required for the temperature determination of the object. With appropriate calibrations we can use this radiation conversion to measure the temperature of radiation sources at great distances. Valid usage of radiation for temperature measure implies that we have found the "appropriate calibrations" that render the radiation spectrum into an adequate indicator of temperature.

Definition of Systemic Thermal Balance: Two systems are said to be in *thermal balance* when their thermal contact is in a steady-state.

The implication here is that the two systems remain intact, i.e., neither system is destroyed by the thermal contact.

There are many cases where systemic thermal balance cannot be achieved. For example, if a letter size paper is put directly in an oven hotter than 233° C, it cannot sustain the temperature of the steady-state heating and bursts into flame. It is easy to find other case types where the steady-state generator is destroyed or two steady-state systems destroy each other.

The following two case types exemplify some of the scope considerations for systemic thermal balance:

Case One: Systemic thermal balance does not imply equal

temperatures, though in many Lilliputian experiments the heat exchange drives to a state deemed as no significant difference.

Using contact measuring instruments to determine the temperature of a system, the effect of the measuring instrument on the system under scrutiny becomes an obstacle to exact temperature measurement. When two systems are put in thermal contact and neither one is in a *steady-state temperature* with the neighborhood and they differ in impetus heat, the warmer system becomes cooler while the cooler system becomes warmer until they reach a thermal balance. If one of the systems is a thermometer where contact has changed the temperature of the other system then we have no measure of the original temperature of the other system.

If the basal *ex parte* heats of two accreted objects (debris) differ, then the percentage difference in heat is reduced in a direct proportion with the host influence. It is also possible for the difference to be too small in magnitude for a direct observation.

> **Note:** CB analysis often requires considering a variation in magnitude that takes us outside the usual scope of direct observation. This brings to the fore the habitual objections related to testability and its use in scientific inquiries. For CB it is possible that these invisible magnitudes are revealed indirectly in a manner that could meet a testability criterion (Chapter 1).

Case Two: If a clinical thermometer is put in thermal contact with a patient and reaches the level of thermal balance, we can then read out the patient's temperature. This is significantly different from case one because the patient can be regarded as a steady-state heat generator and therefore when the thermometer is put in thermal contact any heat taken from the patient is replaced by the patient's metabolic heat and the final equilibrium brings the thermometer reading to the patient's temperature level (i.e., no measurable difference in temperatures).

Basal Levels of Divergence and Convergence (DC). All material bodies have their *ex parte* radiation divergence and gravitational convergence. As found in accreted material debris these effects are amplified by the host convergence power, etc. For matter, this accredits speaking of a basal level of divergence (radiation) and convergence (gravitation). For our purposes the basal level of convergence for debris is the debris *ex parte* **e**-power. We have yet to determine what this can be in terms of radiation from heat. Thus when we measure a temperature we do not know how much of it is basal temperature and how much represents amplification (First Principle of Convergence Pressure). If temperature measure were Eudoxian we could set out ratios in proportions as we did for gravitational convergence. For debris we can expect that the basal temperatures range down close to the kelvin **0**. Even if our technology one day is capable of measuring these small quantities associated with Lilliputian scaled debris, material neighborhood amplification on accreted debris prevents any direct measure. Chapter 5 suggests an empirical approach to this aspect of the problem.

Macrocosm Thermodynamics. Though heat balance can be defined between celestial objects, this balance does not have the homogenized notion of heat that we associate with the Lilliputian scale systems. The Earth and the Sun are two systems in a homeostatic stability and any notion of heat balance between these two objects clearly cannot follow the simplistic debris Common Notions. This implies important differences in the considerations required for defining thermal balance on the macrocosmic level. Thermal contact is redefined. This contact is a function of neighborhood pressure and radiation, both of which affect the corpuscular vibrations that manifest heat. For macrocosmic balance, these forms of contact are adjusted for the distances established in isodynamic relations. Therefore, in host-host synergy relations we have to make adjustments for 'contact' at a distance. For two hosts in isodynamic revolution about each other, their *ex parte* surface temperatures that derive from their own material, including the CIS as a steady-state heat source, are rarely equal (one to the other), e.g., {(Earth, Sun), (Earth, Moon), ...}, though we may speak of a thermodynamic balance sustained within a definable scope of variability.

Corpuscular Vibrations. Measures of corpuscular vibrations are relative to a reference system, i.e., aside from foundation assumptions, it requires translation equations for non-isometric zones and transformation equations for between measurement reference systems ('measurement reference systems' denotes various standards for quantification, e.g., Centigrade (Celsius), Fahrenheit, etc.). Corpuscular vibrations vary directly with convergence pressure (as argued in Chapter 3).

Radiation can amplify corpuscular vibrations in a thermometer. In other words, under certain ambient conditions, radiation induces heat in objects exposed to it. This does not imply that radiation has a temperature, which implies there is a quantity associative to 'radiant heat'. *Radiation can induce heat and heat can generate radiation, but radiation is not heat any more than magnetism is electricity.* In this context, 'radiant heat' (as self-reference) is a fiction that derives from a misunderstanding.

This does not preclude the fact that radiation can induce burning in objects exposed to it. Anyone with a convex lens and sunlight can raise the temperature of small material objects exposed to the focused light (the lens can even be made of ice). For this to work we have to avoid the possibility of a heat sink situation where an object is large enough or conductive enough to dampen (regress) heat at least as fast as it is generated. As related here, the only consistent dereference for 'radiant heat' is the corpuscular vibrations it induces that affects the objects exposed to it. Temperature is regarded now as corresponding to the average velocities of corpuscle vibration in a given object. It is a measure that is independent of volume. Heat is temperature relative to a measure of volume of a fixed compaction, or a variable compaction at fixed volume, or a mixture of these. Thus in laboratory experiments, a volume of twice the compaction and twice the volume of an object at a fixed temperature has about four times the heat capacity. Ludwig Boltzmann (1844—1906) quantitatively reduced the concept of heat to the 'invisible' motions of atomic matter. However, he did not divide these motions into the two modes mentioned above; one either within the atom (intracorpuscular) or generated by a cluster of corpuscles in closer proximity, and the other either intercorpuscular or from the higher granularity of clusters interacting.

Temperature is a parametric measure of corpuscular vibration. I say parametric because at present we do not ordinarily read temperatures as average velocities of corpuscle vibration. Narrow range temperatures near a standard pressure are still usually measured by association with expansion qualities of various materials, such as solid metals, mercury, alcohol, gases, etc.

The more scalable use of radiation emission/absorption (EA) theory is also parametric. EA theory associates temperatures with the electromagnetic spectra. In practice, this is used for estimating the temperatures of stars. This does not contradict the statement that radiance is not heat. Like graded litmus paper, it is an indicator and not the fact indicated. In recent times lisors (combined radar and laser technology) have been used for estimating temperatures in the upper atmosphere. However, I do not yet regard the basis for judging the data as well worked out (see below).

If temperatures record as different for each object exposed to a fixed level of radiance, this indicates that we have no way to dereference a constant 'radiant heat' as a well-defined fixed temperature. That radiance may amplify corpuscular vibration is consistent in theory with the effects observed. That it may amplify more vibrations in one material than in another (materials with different specific heats) is presumed through observed effects. However: **Heat amplification by radiation is not proof that radiation is heat.** Friction may amplify heat, but we do not say that friction is heat.

Interpretation: We already know that on Earth different materials have different specific heats. We also know that *a given material has different specific heats at different temperatures*. Specific heat for a material object is not functionally independent of its temperature.

What is claimed for the experiment here is that the variety of materials sustains different temperatures in equal volumes (and same shape) for the same level of radiance, etc. We know that a fixed form of radiance aimed at different materials can affect their corpuscular vibrations differently. This underpins how we class certain materials as insulation and others as conductors.

Furthermore, in neutral buoyancy environments the specific heats would have a measure different from those on the Earth surface (Chapter 5 considers this occurrence in detail). The host power of convergence directly affects the measures of temperature, the stronger the host convergence power the higher the temperature of the accreted debris, a form of heat amplification (Chapter 3, First Principle of Convergence Pressure). Many cases for these measures are within the limits of our current technology.

The thermometer may be used as an indicator of radiation levels but not all forms of radiation are measurable by a given thermometer. For the Boltzmann theory, the ideal thermometer would be sensitive to the radiation from corpuscular vibrations. In effect, all current thermometers are modes of radiometry insofar as radiation is interpretable as the carrier of temperature data.

Heat Impetus and Regression. For an object, the term 'heat regression' indicates a change in corpuscular vibrations toward an isodynamic state with its neighborhood. 'Heat impetus' that derives from a constraint indicates a change away from an isodynamic state. These constraint driven zone changes from an isodynamic state are found in three modes:

1. *Cooling*; the impetus slows the object corpuscular vibrations below the isodynamic state.
2. *Warming*; the impetus speeds the object corpuscular vibrations above the isodynamic state.
3. *Steady-state*; the object corpuscular vibration rates are such that constraint impetus = regression.

'Zone' may refer either to a part of the object or to the whole object, etc. Steady-state can be induced artificially or naturally. Natural steady-state is the isodynamic state (no constraint impetus).

CIS Heat. There are three major aspects to the generation of heat at the CIS.

- First, the *tension* relation between the kernel and the shell guarantees that there is an agitation in regard to material distribution between the two.
- Second, the *composition, quantity, and compactness of enclosing material* determine the opposed tension.

- Third, the magnitudes of the two modes of corpuscular vibration are directly proportional to the *opposed tension* level.

We associate corpuscular vibrations with photonic effects and gravitonic effects. Compare the Earth and the Sun. Infrared emission from the Earth is measurable. Clearly the Sun has faster corpuscular vibrations, else why does it produce such a quantity of visible light, ultraviolet, x-rays, etc., and the Earth does not. This assertion contradicts the Einstein belief that stronger gravitational convergence slows these vibrations. This calls for a re-evaluation of the Pound-Rebke, Pound-Snider experiments.

It is evident from the observation of effects that if we attribute vibrations to the corpuscles that there is a direct and proportional variation in corpuscular vibrations with the pressure of convergence. The greater the gravitational convergence pressure the faster the vibrations, constrained only by the limits of material integrity.

AERONOMY OF TEMPERATURE FOR EARTH

Cloud formations in the Earth atmosphere help demonstrate the transition between gravitational convergence and surface tension convergence. The water vapor, like a gas, is such that each droplet is small enough that it is subject only to microconvergence. That is, the angle of convergence is submicroscopic, small enough that gravitational convergence has a weak hold and surface tension dominates the cloud agglutination. The electrical properties also play a role in the surface tension. Strong electrical charges can augment impedance to water particle condensation. Clouds of water particles subject only to microconvergence are easily subject to billowing and blowing about in the wind. For a given atmospheric region and temperature, more heat capacitance is within clouds per unit volume then in the proximate atmosphere zones with less humidity. Therefore, if the heat level were everywhere the same in an isosphere concentric to the Earth center, this would imply that clouds have a lower temperature than the atmosphere they float in. If clouds are not isothermal with the atmosphere they float in, then there is a heat differential and the clouds will adjust to an isothermal position just as atmosphere molecules do (see Boltzmann, *Lectures on Gas Theory*).

The atmosphere is densest in the troposphere and this seems to be the ceiling for the water vapor that condenses on dust particles. In effect, cloud formations require dust for their formation and the troposphere altitude seems to be limited by the buoyancy level of fine dust. When an object is so small that the Earth convergence lines affecting it are nearly parallel, etc., the interference with squelching is easier. In this case the atmosphere agitation combined with Archimedean buoyancy delays or slows the squelching of dust particles. Water vapor condensation on these dust particles contributes to squelching them by inducing precipitation.

Going up from the Earth surface the air pressure falls. This lowers the atmospheric heat capacity. To maintain equal temperatures with the lowering of air pressure requires an input of more heat. The Earth gravitational convergence power diminishes with distance from the Earth surface and this also diminishes the amplification of temperature that derives from gravitational convergence. To conserve the heat level as altitude increases would require an increased corpuscle agitation, which, if this occurs, would finally augment corpuscle agitation to escape velocities. In other words, to sustain equal heats (isothermal) implies that temperatures augment as the altitude increases. This is about what occurs in the polar-regions.

Atmospheric Radiation. The atmosphere that blankets the Earth reaches over 560 kilometers from the Earth surface. For radiation passing through it, it functions as a filter and a resistance to its direct effects. Effectively, it is a form of shielding that modifies the state of the radiation that is passing through in two directions opposed. Other than the general cosmic radiation, there are two main sources for the radiation entering the atmosphere, one from the Earth surface (bottom up) and one from the Sun (top down). The Moon also radiates and it reflects solar radiation. The radiation from the Sun that reaches the Earth surface agitates the surface material, generating heat that augments the infrared radiation from the ground source. Thus the radiation from the Earth surface is composed from two main sources. One is generated by the CIS; the other is from the secondary heating of the Earth crust material from external radiation. Together they mostly emit infrared that is sent from the ground up. Thus radiation passes through the atmosphere from two opposed directions.

There is a mixture problem in determining the atmospheric temperature/heat pattern. On the thermometer readings, there is a mixture of source radiation that comes from the atmospheric corpuscular vibrations (which we wish to measure) and the other is generated by radiation that originates from the Earth and from the Sun, etc. It is difficult to differentiate the atmospheric source (corpuscular vibrations) from the other sources. On the Earth surface a crude effort to remove external radiation from this measure is attempted by using thermometers shaded in ventilated insulated boxes, etc. Most charts of atmospheric temperature variations with altitude do not bother to indicate any separation. The problem is further complicated by the specific heats of the measuring devices and the influence of varying gravitational convergence pressure on their corpuscular vibrations.

Ozone Hole. The stratospheric ozone layer above the geographic South Pole (which is the Earth's magnetic North Pole) is partially destroyed and then in some degree rebuilt annually. It reaches its greatest dilation in the southern hemisphere Spring (August and September). It is like an iris that alternately dilates and contracts annually. America is about 25 million square kilometers. On 12 September 2008, the Antarctic Ozone Hole reached 27 million square kilometers, the maximum size for that year. This was larger than in 2007, but smaller than the record size in 2006.

The quantification of ozone in the stratosphere is expressed in Dobson Units (DU), named after Gordon Dobson of Oxford University. In the 1920s he built the first instrument to measure this ozone from the ground.

> **Note:** One DU is 2.69×10^{16} ozone molecules per square centimeter, or 2.69×10^{20} per square meter. This is 0.447 millimoles of ozone per square meter.

The total ozone quantifications less than 220 DUs were not found in the historic observations over Antarctica prior to 1979. The average measure of ozone in the atmosphere is about 300 DUs. This would be about 3 millimeters thick if compressed into a single layer. The ozone hole average measure is about 100 DUs. This thinner presence of ozone passes an increase intensity of UV light that is harmful to surface life on Earth.

> **Note:** This higher reduction of ozone in the southern hemisphere associates with the increase of melanoma cases. Australia has one of the highest per capita occurrences of melanoma.

It is conjectured that excessive CO_2 in the atmosphere may produce a greenhouse effect, an impedance to heat dampening. The positioning of the micro-percentage of CO_2 in the Earth atmosphere does not resemble a greenhouse structure. However, the position of the Earth Ozone layer is well-placed for significant reduction in harmful effects from direct Sun radiation. Industries that create chlorofluorocarbons and other catalysts that destroy the Ozone layer are contributing to the destruction of an important part of our main shield from harmful effects of Sun radiation.

UV scattering through closed car windows pass onto the interior car furnishings, which in turn generate long-wave radiance (infrared), and by convection this heats the interior of the car to a temperature that is higher than the exterior temperature. It is the infrared that we feel as heat.

In terms of effects on global warming, the short-wave (UV) radiance heats up any Earth surface exposed to it, especially the vast deserts around the equatorial belts and the ocean waters. It also includes the polar ice deserts when they are in their Sun exposure months.

It is likely that a Carbon Dioxide greenhouse effect has very little to do with the surface temperature of Venus. It has been estimated that Venus had a surface meltdown about 500 million years ago and with its location nearer the Sun and its weak insulation (no Ozone layer, etc.) from direct Sun radiation resists heat dampening and this creates a strong radiation gradient toward the Venus surface. The atmospheric pressure is likely generated by this radiation gradient toward the surface (more below).

Permaheat. Radiation that reaches the Earth surface induces an augmentation of corpuscular vibrations at that surface. During daylight hours much of the visibility attributed to light effects on material objects is more accurately attributable to an effect known as fluorescence. 'Phosphorescence' is usually only applied to special cases of materials that after exposure to light in the human range glow in the dark within the human range of light sensing.

> **Note:** All material objects have a positive kelvin temperature and this implies that they radiate. In this sense we can assert that all material objects 'glow in

the dark', even though much of this glowing is beyond the range of human sight.

Impetus fluorescence lingers in all material objects exposed to radiance, but it commonly has a short half-life for light within the human sight range and time scales. The current term for the visibility of 'lighted' objects is still 'reflection' (based on an older way of thinking about it).

The Earth rotation is instrumental in distributing the received surface heating that generates from exterior radiation. With day warming and night cooling there is a variation in effects from the equator to the geographic poles. In temperate and tropic regions, as the Earth rotates, this heat never dampens completely. Like the permafrost that never melts in near-polar regions found in Alaska, Northern Canada, locations in Norway, Sweden, Finland, Russia, etc., the Earth crust in the temperate and tropic zones stores an undampened heat level that is greater than that communicated from the CIS alone. I term the surface composite heat level, ***permaheat***. More investigation is required to obtain the details and extent of this permaheat phenomenon. It would be interesting to determine whether there is any influence on **E** that corresponds to the variation of permaheat to permafrost (the next Chapter suggests a major reason why this could be interesting).

Without the current permaheat effects the Earth would be in a permanent Ice Age as we find in the polar regions. Even so, permaheat levels have seasonal variance. The further we get from the tropical zones, moving toward the geographic poles, the larger the percent of variance of the permaheat for the given latitude. In latitudes that have snowy winters and warm summers, the snowy winters are like mini-Ice Ages. These snowy seasons exhibit the low level season of permaheat. At the geographic poles, permaheat sustains its lowest level and the CIS has an enlarged percentage of contribution to its level with a variance that inversely follows the variable solar (and other exterior) contributions.

Consistent with these conclusions, if dust or ash blocked solar radiation from the Earth's crust, the permaheat would reduce sufficiently for the Earth temperate zones to enter an Ice Age. Furthermore, the demarcation for the Troposphere would be less clear, as it is in the polar regions. This implies further weather changes disruptive to life on Earth. Thus rain, which clears the atmosphere of dust, is a helper in maintaining large temperate zones and the tropics both by cloud formations (which require dust for condensation) and dust clearance through precipitation.

Corpuscular vibrations can be amplified or dampened. The Earth surface at the poles has the least amplification of surface heat. What is denoted is that corpuscular vibrations have very little amplification from sources other than the Earth itself. There the solar radiation effects are dampened by the angle of incidence and the presence of permafrost, deep ice, and snow increases the dampening effects. The dampening of corpuscular vibrations is usually incorrectly referenced as 'heat dissipation', which is more of a metaphor than a description.

This corpuscular vibration dampening helps keep a diminished surface temperature that maintains the polar regions in a perpetual Ice Age. On the portion of the Earth that is heated more directly by solar radiation we note aeronomic temperature inversions that are not nearly so marked in the polar regions. These inversions divide the atmosphere into chemical (ozone, etc.) and mechanical layers.

Since vertically the atmosphere can never be isothermal, it is permanently in movement, a state of perpetual adjustment.

Atmosphere Layers. Atmosphere layers are not so distinct in the polar regions. With the use of special instruments in space we have obtained a more detailed model for the atmosphere structure. Using recorded temperature changes, density, chemical composition, and movement, we identify four distinct layers in the temperate and tropical regions {Troposphere, Stratosphere, Mesosphere, Thermosphere}. The altitude for the outer limits of each layer varies with the seasons. The maximum daylight altitudes are reached in the summer months of the given hemisphere. These four layers are separated by pauses in temperature change {Tropopause, Stratopause, Mesopause}. The Exosphere is a fifth and not yet well-defined layer to designate the transition from the upper atmosphere to interplanetary space.

Under present modes of aeronomic temperature measure, there is no proper separation between the effects of radiance and the effects from atmospheric agitation. This precludes any absolute conclusions based on these layers.

1. **Troposphere.** Weather as we know it on Earth occurs in the troposphere. There are no clouds above the troposphere. The troposphere starts at the Earth surface and extends 8 to 14.5 kilometers high. In temperate zones, the temperature decreases with increasing altitude. Most long distance airline flights are in the upper side of the Troposphere, above the cloud ceiling. In temperate zones this air can be lower than $-52°$ C. This implies that the lowering of temperature in thermometer readings is indicating effects of the atmosphere on the measurement. Here we may surmise that there is a shielding. The atmosphere shields the radiation power for agitating the thermometric material. There is also the diurnal blocking of solar radiation with Earth rotation. The Troposphere inversion in temperate and tropic zones in part generates from the effect of *permaheat*. The Tropopause separates the Troposphere from the Stratosphere. Together the

Tropopause and the Troposphere make up the *lower atmosphere*.

2. **Stratosphere.** Starting just above the Tropopause, the Stratosphere extends to 50 kilometers up from the Earth surface. The ozone layer is within this layer. The Stratosphere has a reversal in temperature changes. The temperature readings gradually increase to $-3°$ C. 99% of the atmosphere is found in the Troposphere, Tropopause, and Stratosphere. The Stratopause separates the Stratosphere from the Mesosphere.

3. **Mesosphere.** Starting just above the Stratopause, the Mesosphere extends to 85 kilometers from the Earth surface. In this region the temperatures fall as altitude increases and can fall as low as $-93°$ C. The Mesopause separates the Mesosphere from the Thermosphere.

The *middle atmosphere* consists of the Stratosphere, Stratopause, Mesosphere, and Mesopause.

4. **Thermosphere.** The Thermosphere extends almost to 600 kilometers above the Earth surface. Also known as the *upper atmosphere*, it starts just above the Mesopause. The air in the Thermosphere is so thin that solar radiation obtains hardly any atmospheric resistance. In this region the temperature readings rise as altitude increases. Hardly a hundred kilometers from the Earth surface and the solar radiation can agitate Lead (Pb) into a liquid. Temperatures as high as $1,727°$ C have been claimed. However, this upper atmosphere is a near vacuum and it cannot hold a high temperature in terms of corpuscular vibration without the corpuscles reaching *exit velocity*, where 'exit velocity' references the velocity required for leaving the immediate neighborhood to enter (be taken over by)

an adjacent or a more comprehensive neighborhood. Thus the high temperatures recorded in these upper regions are for the most part the corpuscular vibration level of the thermometers amplified by ambient radiation. The ambient material vibrations have a much smaller part in these thermometric readings and it is not easy to obtain their *ex parte* temperatures.

The Thermosphere is the region of flight for the US Space-shuttles. The shuttle missions vary in altitude from 300 kilometers to a record altitude of about 600 kilometers. This record altitude for the US space-shuttle is the outer region of the Thermosphere where it was used for placing the Hubble telescope (mission STS-31R, 24 April 1990).

5. **Exosphere.** The Exosphere is a transitional zone from the upper atmosphere to interplanetary space. Its upper boundary is not well-defined but it is supposed that its upper boundary does not go beyond 1,000 kilometers.

Solar radiation coming down through the atmospheric layers meets with more and more scattering, which on the average leads to a lowering of the solar radiation power. When the solar radiation reaches the Earth surface, it amplifies the surface material corpuscular vibration levels. This amplifies the heat of the Earth surface and this surface heat gives off radiation that is sent back through the atmosphere. This radiation from the Earth surface is a composite of solar generated surface heat and the heat generated from gravitational convergence, which includes the CIS. This surface radiation also is subjected to scattering but here in reverse for intensity. It is subject to maximum scattering as it goes through the densest part of the atmosphere first. As over two-thirds (71%) of the Earth is covered with oceans with a deep blue tinge, this blue is subject to maximum scattering near the source, this renders the day sky blue over most of the Earth. Green skies have been reported near broad forest areas (Mars with its reddish surface has a pink sky, etc.). In the temperate and equatorial zones, the overall resultant radiation scattering in the troposphere between the solar radiation and the surface radiation is a lowering of temperature as the altitude increases.

The temperature inversions are a heat transfer phenomenon that is under an active investigation. The answer is likely to be found in the effects of the overlapping bi-directional temperature gradations in a possible harmonic distribution. And these variations vary according to latitude in a manner that seems consistent with the temperature inversions.

RUSSIAN HEAT TRANSFER EXPERIMENTS: Both Russians and Americans experienced surprises in regard to their expectations for experimental results in mechanically *neutralized buoyancy* environments.

Because current theories are not adequate to many of the space-lab observations, current experiment designs are often quasi-independent of any known theories (intelligibility reference-frames). In this respect the experimenters are put in the position of Francis Bacon when he performed his experiments without any clear theory in order to find foundation principles for a theory. This is the primordial *modus operandi* that Bacon accredits for scientific induction. The challenge is to find principles for a reference-frame that are parametrically consistent with the variations that are observable (see Addendum 1, Introduction, for more on Sir Francis Bacon).

The Convergence Buoyancy model generates a number of expectations that seem to match the new observations.

In particular, the Russians have performed experiments in neutral buoyancy environments that show unexpected results concerning heat transfer. Their experiments relate to the manufacture of materials on board spacecraft in mechanically neutralized buoyancy. These spacecraft provide specific conditions advantageous to the production of certain materials. The Russians state:

"The most important of these conditions is dynamic weightlessness which virtually eliminates buoyancy and gravity convection, but increases the significance of surface tension effects..." p.7 (from *Manufacturing in Space: Processing Problems and Advances*, Edited by V.S. Avduyevsky, MIR Publishers, Moscow, 1985)

In the interest of preparation of materials in weightlessness, the Russians made some important investigations into the physics of space flight environments. Among them were experiments regarding the problems of heat convection. They devised some very clever experiments revealing the role of surface tension in convection currents, in particular, the dependence of surface tension on the temperature, etc.

Of the experiments performed, one of the most relevant to the present argument is the Dreif experiment, p186 (op. cit.):

First Stage: The initial 10 experiments were done to estimate the magnitude of small accelerations aboard the ship. An isothermal liquid with gas inclusions of 3 to 14 mm diameter was observed.

"The bubbles of smaller diameter than the cell thickness (8mm) were spherical shape while one large bubble was oblated." (op. cit.).

Comment: This investigation adds rigor to the other experiments. Some of the noise from g-jitters and small accelerations were accounted for. No details were given for the oblateness mentioned. The oblateness was normal to the gradient (not specified in text, but visible from illustration). For the intent of these experiments the oblateness seems to have a dismissed relevance.

Second Stage: These experiments required a temperature gradient for the heat convection studies. The flow patterns were observed by using light 'reflecting' tracers. The tracers used were fine aluminum particles. The large oblate bubble was reported as not moving from its position. The small bubbles migrated toward the highest temperature (temperature range 22°—52° C).

Comment: The radiation toward the source of its divergence is here termed the *radiation gradient*. A temperature gradient in an isothermal material (does not significantly vary in volume with temperature) indicates a radiation gradient.

HYPOTHESIS: Gas corpuscles follow the neighborhood radiation gradients (lower toward higher).

The observation of the small gas bubbles migrating toward the highest temperature is consistent with this hypothesis. Even the oblate bubble is consistent with this conclusion since the oblateness is normal to the radiation gradient as if pressured toward the heat source.

If the medium were a fluid that significantly varies in volume with temperature we might argue that the density gradient was the driver. This was not the case.

Conjecture: This hypothesis suggests how Venus with a higher surface temperature but less gravitational convergence then the Earth can have a higher atmospheric pressure than the Earth. The Earth near surface average temperature is c22° C. The Venus average is c480° C. With such a high temperature we are taught to expect the atmosphere gases to expand and lower the pressure. Instead, at this temperature Venus holds the atmosphere in a compressed state that cannot be accounted for with our ordinary understanding of heat and gravity.

Conjecture: The hypothesis suggests that all host objects that hold an atmosphere have a radiation gradient toward the CIS that is sufficient for pressuring gas to the host surface.

Conjecture: An atmosphere on any celestial object indicates that it has an active Convergence Inversion Sphere generating radiation.

Objection 1: Catechism physicists have asserted that gases are held to planets by gravity.

Response 1: The hypothesis suggests a different emphasis. First, gas corpuscles are so small that the gravitational convergence angles affecting them are nearly parallel. This implies that gas corpuscles are only exposed to microconvergence directional polarity. The hypothesis asserts that gases follow radiation gradients (a physics reason is suggested in Chapter 5) and the Russian observations are consistent with this hypothesis. The Russian experiments may be sufficient for accepting this hypothesis as empirical fact.

Objection 2: The Moon and Mercury have extremely thin atmospheres yet their spherical shape implies that both have a well-defined CIS.

Response 2: They likely also have in common that the CIS has a distortion elongation along the distance to each host, a strong relation with each host. The Moon relation to the Earth actually prevents the Moon from rotation relative to the Earth. This could facilitate leaching of the Moon atmosphere.

Mercury only rotates about three days faster than its year. Two explanations offered for Mercury's thin atmosphere are: one, the solar wind strips it away; two, the Sun leaches it away. The leaching is consistent with the hypothesis that gases follow the strongest radiation gradients.

With respect to gas corpuscles both the Sun convergence and the Earth convergence are in nearly parallel lines, due to the relatively small corpuscle size. This has an effect on the gas corpuscles that resembles microconvergence, the form defined as convergence in nearly parallel lines.

Objection 3: On Earth the air is more compact at night than in daylight. It is further assumed that it is the cooler temperature at night that renders air more compact. This seems to contradict the assertion of following the Earth radiation gradient.

Response 3: It is true that on Earth the air pressure in daylight is less than at night. In daylight the Earth radiation gradient is significantly weakened by the Sun radiation, etc., which in turn weakens the atmospheric pressure. The Earth crust also stores heat generated by the Sun radiation and this impetus heat should have a calculable average half-life for the various latitudes. In temperate and tropic zones, due to the rotation rate of the Earth, this impetus heat cannot dampen to a zero effect on the CIS radiation gradient. The crust agitation level always exceeds the level that is isodynamic with the heat generated by the CIS. In effect, this incomplete dampening sustains a kind of *permaheat* that is amplified by solar (and other exterior) radiation and distributed by Earth rotation. It is this permaheat effect in the temperate and tropic zones that keeps the Earth from a permanent Ice Age. This is also the effect that changes the radiation gradient that affects the atmosphere

Atmospheric heat contributes very little to the radiation gradient that affects it, as is expected from such thin gases. The raising of atmospheric pressure at night is not due to the lowering of the temperature of gas corpuscles as catechism physics teaches us. It is rather due to the gradient increase toward the crust. This effect is maximized in the polar regions during their respective dark periods.

When daylight augments the Earth surface heat, this lessens the potential difference between the exterior (solar, etc.) and terrestrial effects on the atmosphere. The resultant lowering of the radiation gradient toward the crust lessens the pressure of atmosphere gases and the air thins out. This thinning would occur independently of the air temperatures (temperature range where air constituent gases remain in their gas state). This indicates a different mechanism for atmospheric tides. It also suggests that the Moon would have very little effect on atmospheric tides other than a kind of interference when it comes between the Sun and the Earth.

Gases have a very high dampening rate for heat. This is especially true for gases without closed containers such as the atmosphere.

Experiment. In a vacuum on a host, put atmosphere at a high pressure in a closed container on a suspension balance. Heat that atmosphere. Barring other influences, since the radiation gradient (potential difference) between the atmosphere gases and the host has lowered, the balance will rise on the side of the gases (taking care that the gas occupies the same volume, or in some other way avoids Archimedean buoyancy distortions). The rising of atmospheric gases is independent of pressure. It is the radiation gradient that determines the atmospheric pressure. The atmosphere temperature is a secondary effect of the radiation gradient. The prediction is that the 'weight' of the gas will diminish as the potential of the radiation gradient toward the host diminishes. Furthermore, if it is heated sufficiently to reverse the direction of the potential difference it seems that its convergence power would adjust accordingly.

Comment: Gases in space would compact at a rate directly proportional to heat differences between isothermal surfaces concentric to the intersection of the differential lines of radiation gradients. The driver is the radiation gradient. As they cool, these clouds tend to spheres (bubbles) about these centers of intersection, these points of convergence, in order to allow maximum heat dampening. This does not preclude billowing during the weak stages of convergence.

We note on Earth that volcanic clouds, which have high heat, are more granulated in their shapes than the usual condensation clouds. This effect is produced by the extreme difference in heat content between the clouds and the ambient atmosphere they enter; the greater the difference, the greater the granulation. We associate this form of granulation with rapid cooling. The rapidity magnitude between any two points in heat transfer is proportional to the potential difference in heat between these two points along with the adaptation required by any special conditions for the transfer between them. Increased granulation presents a larger surface area for divergence of radiation and the convergence effects of surface tension or gravitation, etc.

HEAT EXCHANGE. The theoretical dereference for 'heat exchange' depends on how we model the concept. An object with heat diminishing at a faster rate than its maintenance can be thought of as losing heat or gaining coldness. The problem in prematurely deciding on a viewpoint for this is that it predetermines any *ex parte* analysis, a bias that might blind us to important considerations. This *ex parte* analysis is likely why the concept of heat is often spoken about as if heat were a thing. In this sense, phlogiston is still with us. A dual theory formulation could easily be based on 'cold'. This gives us another pair of flip-flop concepts for Physics. My preference is DC (Divergence, Convergence), which corresponds to AE (Absorption-Emission rates) though AE has misleading terms inherited from past misconceptions (details in Chapter 5).

Steady-State Heat Exchange. Corpuscular vibrations are increased when accelerating against isodynamic convergence. This mode for rise in temperature accounts for part of the heat generated by a friction that is functionally independent of the atmospheric friction in take-off or re-entry. Thus astronauts under take-off accelerations should experience a rise in their body temperatures. After obtaining a mechanically neutralized buoyancy trajectory, their body temperatures should soon after register lower than on Earth. This reasoning also suggests a change in temperature of their vehicle that corresponds to the pressure impedance from the neighborhood convergence.

Radiation, like heat, can sound like a thing. In Chapter 5, photon radiation is presented as a field polarity and, in general, fields only travel with field generators, the convergence generators. Contrary to what Newton and others teach us that light corpuscles travel, Huygens defined a corpuscular theory where the corpuscles function as a medium for waves. It is not yet sure whether the photon polarity is clearly distinct from phenomena designated as microparticle radiation (alpha, beta, etc.).

Radiation participates in the generation of the shape of the radiating object, and large objects with a well-formed CIS take the shape that allows maximum radiation for the volume and pressure. In general, this shape is toward a sphere, which under E-power may also develop surface convolutions. CB conditions guarantee that all material bodies radiate a spectrum of light.

Radiation polarizes from a proportionally diminishing surface (r^2) as the volume (r^3) augments. This geometric relation implies that to maintain constant radiation polarization in spheres of different sizes of similar material that larger spheres must be hollow or form surface granulations that increase the per square area of surface or have a mixture of the features.

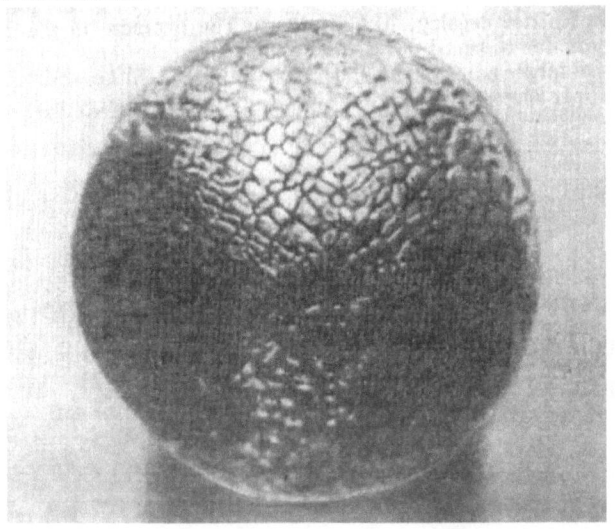

EXHIBIT 4.1: Convex grains on copper surface, from p124, *Manufacturing in Space*, editor Avduyevsky, 1985, courtesy of MIR Publishers.

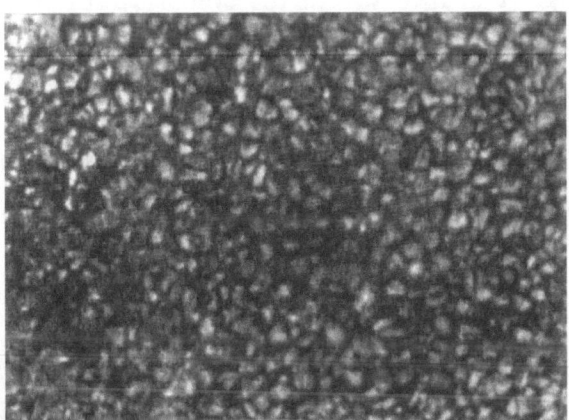

EXHIBIT 4.2: Convex grains on the Sun surface, from p280, *Astronomie Generale*, Bakouline, Kononovitch, Moroz, 1974, courtesy of MIR publishers.

Granulations are one of the effects of super cooling. It has also appeared in neutral buoyancy experiments done by the Russians and others. The granulated surface of the Sun increases the surface area for more efficient cooling. Hot volcanic clouds also show a marked degree of convolutions giving their surface a more granular look than ordinary clouds. In large part, the Sun looks like a steady-state version of a supercooled surface (compare Exhibit 4.1, 4.2). This suggests a plausible explanation for the Sun's granular surface.

Pulsing Sun. The radiation given off by the Sun is an effect of its 'steady-state' cooling system. The steady-state polarization required is a function of the bulk. When radiation falls below a steady-state for a given size, the star may contract to maintain the steady-state. If the contraction of a star is too violent the star may implode. On the other hand, the star may accumulate heat until it can no longer sustain containment of fixed size and it then swells. If this swelling is violent, the star bursts. Otherwise the star may spend a time pulsing. It is likely that all celestial spheres pulse and that this pulsing increases with accretion. The pulsing alternates with heating and cooling. Our Sun has a 22 year pulse cycle of accumulation and release.

Pristine Star Evolution. In the order of evolution from early to later we can surmise the order from younger to older as Mercury, Mars, Venus, Earth (where age is represented as a stage of development rather than as a fixed passage of time. Mercury is much like the Moon. One day, disregarding possible catastrophes, Mercury will accrete to the size that Mars is now. Mars has a solid ('frozen') surface. The surface shows signs of early volcanic activity, an indication of cooling. As accretion continues, the surface becomes hotter and hotter as the shell gains thickness and there is a concomitant impedance to heat exchange.

Venus is like an earlier Earth. When the plates form and the basins fill with oceans we reach the time that life molecules may form. The spreading and cooling can leave basins that become the ocean floors as Venus cools down and the rains come.

Earth swelled, the plates separated into continents and the continents spread with the swelling. The continents and their undersea plateaus may give a rough idea of the pre-tectonic surface area. This cracking and swelling may also indicate that the Earth is hollow. The swelling could thin the shell and quicken the cooling.

STEADY-STATE HEAT. A constraint driven steady-state system dampens at the same rate that it amplifies heat, and this maintains a fixed level of corpuscular vibrations. The diminishing of the vibrations matches the augmentation of the vibrations or vice versa, e.g., synchronic oppositions of impetus and regression in equilibrium

Lord Kelvin experimented with wires heated by direct current electricity and discovered that wires partly covered by insulation and partly uncovered became hottest where the insulation was removed. This result was not expected. If we assume that temperature is directly proportional to the rates of corpuscular vibration, we can infer from the empirical evidence that the greatest corpuscular vibrations are in the non-insulated portions of the wire. Insulated portions of the wire would have the lower magnitude of vibrations. Where there is no insulation there is least resistance to the influences of the ambient heat level. If the ambient heat is lower than the wire heat, then it seems logical that the wire would radiate more at the places of non-insulation because these are the locations of a higher difference in potential. There is a steady-state source for the corpuscular vibrations, the circuit current. The 'current' source remains constant for the circuit, though not necessarily the same constant throughout the wiring.

Note: CB characterizes 'current' as a circuit polarization. In a closed circuit nothing is flowing or traveling. The circuit is in a polarized state (more in Chapter 5).

The apparent greater vibrations in the non-insulated areas are sustained by the 'current'. *Here we may have temperature amplification through an augmented exchange rate.* When our senses reveal an object as hot we say that the object is 'losing' heat. Without the constant current the wire temperature would diminish. The non-insulated wire also radiates more than the insulated wire and this may confuse the description because radiation magnitude is dependent on potential difference between source and destination.

Under a given atmospheric pressure, when we heat water to a certain point it begins to boil. On the Earth surface this state of boiling is most often interpreted as an effective cooling process. It is the point at which the water does not get hotter. All parts that are hotter than the boiling temperature turn to steam. Convection driven by gravitational convergence conduces rapid heat distribution. This is a steady-state phenomenon (within definable physical limits, since the water is vaporizing). In steady-state systems, the faster the heat exchange the hotter the object appears to be. In dynamic weightlessness, the boiling bubbles coalesce and the water temperature likely registers somewhat higher due to the stronger influence of surface tension. Heat distribution is mainly by conduction. There is virtually no convection. This apparent homeostasis of boiling temperature suggests that by losing its surface tension at this temperature it turns to steam all parts that are hotter than the boiling temperature. In effect, boiling would again be a cooling process if the steam could be circulated out of the enclosed milieu.

Under a latitude of conditions playing on water surface tension, it is possible at a given atmospheric pressure to heat water to temperatures above its boiling point for that pressure. Disturbing the surface tension sets in the boiling and it cools down to a historically predictable temperature.

SUMMARY CONCLUSIONS. The components in the ordered triple for heat measure are expressed as functions so as to take into account their intrinsic non-Eudoxian magnitudes. What may be strange in their status is not this non-Eudoxian characteristic but rather the framework that is required to make the non-Eudoxian measures intelligible. For now it seems that this intelligibility mode can only be rendered by an apodeictic Eudoxian framework that remains a thought construct for reference. It is in this sense an apodeictic fiction without which the model that is regarded as closer to fact would have no dereference that we could understand.

During daylight hours much of the visibility attributed to light reflection on material objects for our visible range might better be attributable to corpuscular vibration effects that can be likened to a fluorescence effect. In general, within human visibility and time scales this fluorescence effect in material objects usually has a short but measurable half-life.

Solar induced crust agitation (corpuscular vibrations) has a half-life that for more than half of the Earth surface cannot completely dampen in one night. The lowest heat level reached in the rotation cycle dampening of this solar induced crust agitation is called the permaheat of that cycle. Furthermore, we can speak of seasonal permaheat, etc., just as we speak of permafrost in Northern regions.

Permaheat and permafrost are both fuzzy categories of heat levels, but a homonymic name would complicate conversation about them.

The Russian heat transfer experiments in neutral buoyancy environments show that gases follow the neighborhood radiation gradient from lower toward higher temperatures. This suggests a physical effect that accounts for certain atmospheric pressure variations found in the solar system.

The Earth atmosphere insulates us from the intense radiation from the Sun and that if this insulation deteriorates (such as ozone loss) we would be subject to global warming. The *ad hoc* hypothesis of the so-called Greenhouse Effect is ill-conceived and not required. It is a misdirection from the industries using fluorocarbons and other catalysts that are destructive to the ozone layer.

In the past 120 years there have been four scares with a promise of disaster; global cooling (1895), global warming (1920's), global cooling (1975), and now we are back to global warming again. However, the New York Times, 11 May 2007, reported that the Greenland glaciers are melting slower than geologists thought, implying a cooling down. The Serbian astrophysicist, Milutin Milankovitch, made a study of the relation between longterm climate change and Earth orbital variations. His calculations and predictions match very closely the known Ice Ages and agree with the geological evidence that the Earth is coming out of an Ice Age, which implies that despite the ups and downs the Earth is warming up until the next Ice Age descent begins.

END Chapter 4

Chapter 5: THE THIRD PERPENDICULAR

INTRODUCTION. *Michael Faraday (1791-1867) succeeded in showing a relation between magnetism and electricity, light and electricity. He attempted to find a relation between gravity and electricity. He kept an account of his experiments relating gravity with electricity in* Experimental Researches in Electricity.

In 1850 he wrote:
"2703. The long and constant persuasion that all the forces of nature are mutually dependent, having one common origin, or rather being different manifestations of one fundamental power (2146), has made me often think upon the possibility of establishing, by experiment, a connexion between gravity and electricity."

After many experiments he concluded:
"2717. Here end my trials for the present. The results are negative. They do not shake my strong feeling of the existence of a relation between gravity and electricity, though they give no proof that such a relation exists."
                                      Royal Institution, July 19, 1850.

There are two major decision problems that remain open in regard to the characterization of light. Both problems are left from classic physics. The first is whether light is discrete or continuous. The second is whether light has a medium or is its own medium. If there is a state of matter that actually loses its particulate nature we may have access to the determination of the first problem. Satyendranath Bose presented the theoretical foundations for such a state in his investigations and further development of Boltzmann's theory of gases. The second problem of whether there is a medium or not reduces to two general forms that may be referenced as the bootstrap and non-bootstrap models.

The bootstrap models are those that propose that light does not require a medium. For these models in a manner of speaking light provides its own medium. The non-bootstrap models require a medium. This decision is related to the experiments that attempted to determine the existence of aether, the proposed medium for light.

Determining the existence of aether became tantamount to a proper characterization of light. The Michelson experiments did not succeed in showing the existence of aether, but neither did they disprove the existence of aether, as Michelson himself reveals in his papers and this includes the work done with his student Morley relating to this subject. The most exacting experiments were done on a large scale by Dayton Miller (a former student of Morley) in 1922 through 1934. He claims to have detected the aether drift.

Einstein Particle of Light. In 1917, Albert Einstein (1879-1955) published the suggestion that quanta of radiation, introduced by Max Planck (1858-1947), should be considered as particles possessing a definite energy and a definite momentum:

$$E = h\nu, \qquad p = h\nu/c$$

The fact that he states they have a definite momentum shows he is not even up to Maxwell in the recognition of the role of relativity in representing momentum. We cannot even argue that the speed of light is functionally independent of any inertial reference-frame because the Einstein Theory of Relativity was not yet a mainstream catechism. In 1923 these particles became known as photons.

Note: The photon is a rehabilitation of Newton's corpuscles of light, but has some added detailed qualities with the feature of quantification. It also requires the re-introduction of Newton's concepts of absolute Space and absolute time applied to light. In effect, the absolute momentum for light implies that for light every inertial reference-frame is in a state of rest. This absurd conclusion derives directly from the stipulations of Einstein Relativity.

Dirac's Graviton. In 1959, Dirac (1902-1984) proposed the quantum of gravity, the graviton. He suggested that gravitons likely follow the same expression that gives the energy of photons. However, photons are regarded as carriers of the electromagnetic field, a vector field (which he prefers to call 'a vector space', following the mathematical bias that is functionally independent of the physics), and hence are regarded as vector corpuscles, i.e., spin = 1. Dirac, following Newtonian tradition, believed that gravity was fundamentally an attractive force, but he did not regard it as a vector (spin = 1). To regard the graviton as a vector leads to contradictions in accordance with the Newtonian model. He reasoned that the graviton is a tensor corpuscle (he did not use the word 'corpuscle') and assigned the graviton spin = 2. The energy of the gravitons is considered equal to Planck's constant, **h**, times the graviton frequency. This gravitational field equation provides a basis for a quantum interpretation for gravity.

Since the Dirac graviton spin is based on the Newtonian gravitational attraction apodeictic reference-frame, it has no direct basis in Convergence Buoyancy. The rationale for this is the main subject of this chapter.

Microparticle Spin. Spin is specified as a property of microparticles. It does not represent spin in any ordinary way. It represents an internal angular momentum. The projection of a microparticle in any direction assumes discrete values. In current theory this is interpreted as a manifestation of the specific angular momentum of the microparticle. The existence of spin in microparticles leads to the appearance of more internal degrees of freedom. In current theory, both photons and gravitons belong to the boson (named by Dirac after the Indian physicist, Bose) class of microparticles. In theory, both are stable particles.

A complete theory does not exist for quantum electrodynamics. There are fundamental contradictions between the corpuscular terms and the wave terms. With current theories, the interrelation between photons and electromagnetic field remains unclear. The interrelation between gravitons and electromagnetic field remains unclear, as well.

Light Impedance. In general, light phenomena riding on a carrier slows down in a proportion directly with the increase in media compaction (barring new alignments and resonance effects and within a limited scope of variability, and note also that a lowering of temperature may increase compaction). It speeds up at entering a less compact medium. This speeding up favors interpreting light as a polarization phenomenon riding on a carrier where the polarization effect travels, but the carrier medium may not (more below). Billiard ball physics with traveling photons handles this awkwardly with *ad hoc* compensatory assumptions not considered here.

In accelerations normal (perpendicular) to convergence fields (e.g., circling satellites) it is likely that the speed of carried light measured in the space vehicle will not vary for a fixed altitude. However, at various altitudes in the convergence fields (non-isometric zones), the speed will vary inversely with the field compaction. This implies that the greatest light impedance is found closest to the surface.

This suggests a possible solar optical effect on stars beyond the Sun. We have a star displacement set that was photographed during the solar eclipse of 1919 by Arthur Eddington (1882-1944). From his analysis he concluded that the Einstein General Relativity prediction was confirmed, which changed a ho-hum eclipse expedition into a widely publicized news item. The interpretation for this phenomenon by the Einsteins is that starlight beyond the Sun converges slightly toward the Sun when passing near its surface due to the Sun's gravitational influence. Looking toward the Sun this gives the effect that the star seems pushed out away from the Sun. Albert Einstein's first note to Eddington gave the above an erroneous halved result. He soon after posted a card doubling the value to the 'correct' result (in the nick of time before the eclipse event).

The CB interpretation of the observation is different. The Sun convergence is strongest at the surface of the Sun where we presume the Sun's atmosphere is most compact. This suggests that near the surface offers more impedance in terms of affecting the carrier direction and affecting the speed of carried light than at higher altitudes. This higher impedance is likely true even though arguments may continue on how much of it is from stronger convergence or from higher atmospheric compaction or some combination of the two. The impedance to polarization of the light grazing the Sun limb lowers with the distance from the Sun. Polarization follows the path of least impedance, this **principle of least impedance** *(least effort),* here proposed for physical optics, corresponds to the usage of Fermat's principle of least time in geometric optics. It seems logical that light passing near the Sun surface would follow the path of least impedance and thus move away from the Sun surface. Thus the images of the background stars are physically moved out from the Sun surface. Both explanations (physical, geometric) fit the observations, but their diagrams are different.

The Einstein model, though claimed as physically caused by gravitational influence, is consistent with *geometric optics*. This is why a number of physicists claim that it can also be regarded as an optical effect, and may not even be a gravitational effect. Since the light bends in, it appears that the stars move out. This proposes why on Earth the optical sunset takes longer than the physical sunset (an optical effect of the Earth atmosphere).

The CB model uses a form of *physical optics*. Using the polarization model, the polarization follows the path of least impedance, therefore the physical image moves out. Both of these optic modes of analysis also predict that a large gravitational body with an atmosphere could produce two separate images of a body behind it. The Einsteins postured this as a gravitational phenomenon.

> **Note:** In atmospheric optics the phenomenon of a double sunset mirage is well known and is regarded as an optical effect associated with temperature inversions. As the Sun's atmosphere also has temperature inversions we may expect to find corresponding double star mirages, etc.

For CB the non-isotonic variation in atmospheric compaction generates a corresponding and direct non-isotonic variation in impedance of light polarization. This non-isotonic variation in impedance likely holds the simpler underpinning if light has no substance. The following arguments lead to another interpretation.

PHOTON-GRAVITON CONJECTURE

We may say with respect to the Sun that its radiation is divergent and its gravitation is convergent. This opposition suggests a kind of polarity relationship between radiation and gravitation. Conjecturing on this opposition we may logically conceive of another type of system opposite in principle, a whitebody that is convergent for light and divergent for gravitation. For now this remains an apodeictic construct that has no agreed upon confirmatory observations.

As blackbodies are *material objects* that have gravitational convergence and radiation divergence, it is left to find a consistent underpinning physical matrix for whitebodies. If this matrix were material, then following Newtonian logic we might accredit the Pierre-Simon Laplace (1749—1827) conjecture that describes a gravitational body having such powerful convergence power that light cannot escape it and therefore the body would appear black. However, if we regard radiation as a polarization, then the more powerful gravitational convergence, rather than squelching the radiation, would intensify it. In effect, for CB, this conceptualization of the whitebody denies that it can be matter. Since the whitebodies are conceived as opposite in principle we may find a consistency in assuming that they also have an opposition in their foundational composition, which is to say that if they exist they are in some form of *antimaterial bodies.*

Looking into space from a photonic emitter such as the Earth, the whitebodies would be invisible or difficult to see. If we could view from a whitebody in the antimatter universe, the 'night' sky would be white and the antimatter stars would be sparkling black spots. From this point of view it is the blackbodies that become difficult to see.

Recent observations suggest the existence of many whitebodies, which are currently baptized 'black-holes'.

This discussion leads to the following conjecture as a thesis:

> **THESIS:** Photonic and gravitonic phenomena are polarities of the photon-graviton (**pg**) field orientation (also called the *Aetheric polarities*).

For quantum models this implies that in contexts where the spin factor actually sustains a valid dereference for photons and gravitons that they must have the same spin. In CB they can keep the photon spin value without contradiction.

Note that declared features of the photon-graviton, such as spin, or quantum, are apodeictic constructs at best. Though we may talk about a packet of the photon-graviton polarity and treat it as a corpuscle, this in no way implies that it actually exists as a packet. The packet phenomena of the photon-graviton polarity are usually related to interactions with material objects. This implies that we should expect decision dilemmas when we use a materially corpuscular measuring instrument to read both discrete magnitudes and continuous magnitudes. We have no critical test to determine whether the photon-graviton polarity is structured as continuous or discrete. This difficulty in the data interpretations is mainly due to our current inability to eliminate the discrete aspects of our measuring instruments on what might be continuous magnitudes that are being measured.

Prince Louis de Broglie (1892-1987) in 1923, while still a graduate student, proposed that 'microparticles' also display wave characteristics, a thesis that led to the development of the electron microscope.

The possible variations in how we model light contribute to complicating our considerations. We can allow our model to be corpuscles that travel, as Newton thought, or to conceive of it as a field of corpuscles that carries waves, as Huygens thought, or a field that can be polarized without committing to the corpuscle or some bootstrap form of the wave model, or etc. The bootstrap models require that light be its own carrier, which implies that light has substance. I find more explanation power in fields that can be polarized (a form of conduction) with the possibility of field movement, i.e., the field moves with the object that generates the field. Of course this stance re-opens the problems we associate with the carrier of light. Whether these fields generate convergence systems or convergence systems generate these fields is not a decision required for the present discussion. In other words we can leave it in a state of immanent reversibility. But if a system moves, its field moves with it, etc. This gives a rationale for the nil results in the Michelson-Morley experiments. *This consideration is expressed in Michelson's original conclusions.*

The **pg** (photon-graviton) field polarities suggest a model for explaining why all shining systems, that is to say blackbodies, have the possibility of being seen. At the same time it offers an explanation for why interstellar space appears black. Standing on Earth looking into a clear night sky we see black and when we look toward a shining object (photonic source) we see its light. This suggests an explanation for why space seems transparent. All black-bodies are visible to us if our instruments are sensitive enough to the photon-graviton field carrier of the polarization that is associated with the presence of blackbodies. In this argument, the speed of light is nothing more than the conduction speed of a light polarization riding on this field. This implies that the speed of light is not independent of the physical state of its carrier. (The bootstrap models require that light be its own carrier, i.e., light has substance.) Even so, we note the possibility of light participation in a kind of bulk convection insofar as the source of polarization (e.g., blackbody) is moving through space.

Another Redshift Interpretation. Polarization of this kind also suggests the mechanism that allows seeing objects beyond the Convergence Reversal Node (CRN). This CRN is a node in a compression zone for convergence and this compression spreads the convergence lines (see Exhibit 1.2A). All material objects in photon-graviton fields are associated with graviton convergence and this spreading of convergence lines at the CRN suggests the possibility of redshift amplification phenomena. Redshifts may be amplified by the repelling spread at the CRN. This amplification may increase directly with the repelling action taking place. This may lead to a method for measuring the gravitational pressure between two hosts.

Experiment: With respect to the CRN between a planet and the Sun, compare the Sun spectrum from the planet side, then from the Sun side of the CRN. The sensors in satellites going toward the Sun should pick up a redshift relative to Earth observations until the satellite sufficiently passes the Earth-Sun CRN, etc.

Flying Frogs, Andrej K. Geim, Michael Berry, etc. In the work of A. K. Geim and team at Nijmagen University, Netherlands; frogs, grasshoppers, strawberries, etc., were suspended in a field sector generated by a powerful magnet (a Bitter magnet 400,000 times stronger than the magnetic field of the Earth). Michael Berry, UK physicist, found similarities with the levitron toy and joined Geim to extend the current levitron explanation to include Geim's observations. They developed an electromagnetic model. However, the Third Perpendicular conjecture for a **pg** polarity suggests that we examine the sector of the field that levitates the objects for any evidence of a photon-graviton field that generates perpendicular to the electric and the magnetic fields. The cylinder end that levitates objects may be exhibiting a Franklin neutralization of the graviton polarity. The heat-generated radiation (photonic polarity) from the magnet appears to be perpendicular to the electric and magnetic fields.

This discussion also suggests that we could obtain a Franklin neutralization for heat-generated radiation. If the Third Perpendicular conjecture holds, then we should be able to obtain a neutralization of any charge (quantified polarity) associated with a photon-graviton field.

EMISSION AND ABSORPTION. From early interpretations of instrument readings it was assumed that an object emitting a certain wavelength at a given temperature would also absorb the same wavelength at the same temperature. This assumption may not be the case.

It is not a necessary conclusion that the color of an object is the complement of the color given off. Assuming these complementarities denies a participation of the object in the color given off in the following way. It may be that if a given object is exposed to the full spectrum that it emits what it cannot absorb. This renders the relation between absorption and emission as complementary. However, it is also possible that the light excites the material in a way that shifts the color emitted or absorbed.

This sort of 'reflection' is observable when we put ultraviolet light on fluorescent objects. Different objects can generate different colors *even from very pure ultraviolet*. If it were a question of reflecting or absorbing only ultraviolet, these other colors should not appear. This indicates that the object material participates in what color is radiated. Furthermore, this indicates that the term 'reflection' should have a restricted usage. Here 'emission' is more accurate. It also gives grounds for regarding all material objects as in some degree fluorescent.

We have left to consider how objects participate in the determination of the emissions they show. A first question is whether the object introduces a change to the polarization between the source (object) and destination (observation).

Let us say that perfect reflection is a perfect mirror and white shall denote that the reflection is scattered. Total reflection, be it scattered or not, is the state of perfect opacity. No claim is made for its existence in nature. For this argument it is an apodeictic construct. We do not know of a material that can' reflect' the entire spectrum.

Its opposite is perfect transparency from which it logically follows that black is scattered transparency. This represents a significant re-interpretation of black. When I look through a glass window, this transparency is a form of black. It is unscattered black (highly reduced scattering) for the spectrum that passes through the glass. Note that I am speaking of the window itself as unscattered black and not the spectrum that is passing through.

Ordinary glass scatters any ultraviolet that can pass through it. This is why ordinary glass appears as a scattered black on ultraviolet film. This film is a case where from indoors the window photographs as black and from the outdoors it photographs as white. Though the ultraviolet is scattered, it still can generate infrared from materials within a closed room or a closed car, etc.

Comment: As some UV is essential for healthy plant growth this fact draws attention to one of the many defects in the Biosphere 2 experiments (Arizona, USA). If they had used at least some panels of crystal glass that allow some unscattered passage of UV, this could have killed the surface aerobic bacteria that were robbing the system of Oxygen, etc. Plants grow long and thin in infrared; they grow fuller in ultraviolet, etc. An entire book could be written on the misconceptions of Biosphere 2. I informed the experimenters of these and other defects. A few months later they announced that after carefully researching the Oxygen problem, they discovered that it was aerobic bacteria that were using it up faster than it was made. My name was not mentioned in the announcement, etc.; *Bacteria steal oxygen from Biosphere 2*, Science News, Feb. 12, 1994, p. 110.

Blackbody Radiation. Max Planck (1858—1947) in 1900 suggested that light emission is a quantum phenomenon. He did not seem aware that it was his particulate measuring instruments that require a quantum approach. On his quantum assumption he was able to obtain a formula for the energy density of blackbody radiation more consistent with observation.

Ultraviolet Catastrophe. Paul Ehrenfest (1880-1933) applied this term to the failure of the Raleigh-Jeans Law. Without discussing details, this light-wave theory Law implies that for a blackbody both the power at a given frequency and the total radiated power is *unlimited* as higher and higher frequencies are considered. This implies that an ideal blackbody at thermal equilibrium will emit radiation with infinite power. The energy conservation law is violated. Lord Rayleigh (1842—1919, born John William Strut) and Sir James Jeans (1877—1946) found this discrepancy between observation and theory and it destroyed the generality of their own Law. The "ultraviolet catastrophe" was their catastrophe.

Basal Polarity. The photon-graviton model does not generate the problem in the manner conceived by Rayleigh and Jeans and suggests a different interpretation of Planck's solution. We have not yet determined if light is a wave or of a corpuscular structure. We do believe that the detection instruments for light phenomenon are corpuscular. Objects in a milieu entail an interaction between the basal state of the object and that milieu. The older terms, absorption and emission, can be misleading. This follows on the assumption that light is a polarization effect rather than a traveling corpuscle or wave.

> **DEFINITION**: The polarity *basal state* of an object references its polarity properties *ex parte* (functionally independent of its milieu).

For a material object, its basal polarization is such that the photonic polarity diverges from the object, while the gravitonic polarity converges to the object. In the older language one might say that in its natural state the photonic polarity is 'emitted' and the gravitonic polarity is 'absorbed'. If we restrict this to indicate *only* the directions of the polarities we can reduce the distortions in the semantics of polarizations. Using this restriction, this emission of photonic polarity can interfere with any milieu photonic polarization directed toward the material object, in effect, it generates a repelling effect that we associate with all like polarities.

Polarization Conductivity. The question arises whether under this repelling influence of the basal state any photonic polarity can actually penetrate the material object in opposition to that basal state. The response to this question requires consideration of the relative energy levels of the milieu polarizations and their interaction with that basal state polarity. Basal polarization frequencies that differ from the frequencies of the milieu polarizations can be more or can be less effective in the repelling. With respect to the milieu polarization, with respect to any levels that are not effectively repelled, we can speak of the polarization conductivity in the object.

Under perfect conductivity for a given polarization energy level, the object would function like a window that passes the polarization through with minimal distortion and scatter. Such an object is said to be transparent to this frequency of milieu polarization. The object may also scatter the polarization with various levels of completeness. This scattering represents stronger interactions. Such scattering can make the object appear at various intensities of black and white and produce interference modes that to our sight can produce the illusion of colors. These effects are still regarded as modes of transparency. Though these are the effects that CB corresponds to photonic 'absorption', in effect, *no absorption actually occurs*. The milieu polarization is rather thought to pass through the object in these varieties of transparency modified by interactions with the basal polarization of the material object. The only *convergence* polarity for a material object is the gravitonic polarity and this references only the basal gravitonic polarity of the material object.

Photonic permeation into material objects can raise the agitation level of these objects (periodicity rates of corpuscular vibrations). Microwave ovens are thought to heat food this way. The concomitant effect generates an increase in heat and temperature, which in turn affect transparency. This represents an invested impetus and can only be sustained in a milieu that supports the impetus. No photonic permeation can be sustained without a supporting milieu.

Bulkless Polarity. When referring to the polarities, to say that the photon and graviton (as polarities) have no bulk directly follows from assuming that these polarities are determined by the particular polarization of the photon-graviton field. That is to say that if the photon-graviton field is substance, it must have bulk (something the Newtonians would call 'mass'). But this does not imply that light or gravity have bulk since these are only the polarities of this field and not the field itself.

Radiation Sink. Interpreting from the older physics, the ideal radiation sink would absorb all and emit none of the radiation sinking on it (and without generating a temperature that would destroy it). Common materials that are proposed to approximate this abstraction in *STP* are things that appear flat black, such as soot, black suede, velvet, or flat black paper. In our Lilliputian laboratories we attempt to represent this abstraction with an isothermal enclosure. This can be a spheroidal enclosure that is blackened inside and is completely closed except for a narrow slit through which light or other radiation entering becomes trapped inside by multiple defective "reflections" from the walls. Even our best 'black' radiates some light but by containing the black walls, whatever is emitted gets diminished by multiple emissions (defective "reflections") from the black walls. These walls may be any material resistant to the temperature generated by the radiation (note the influence of the 'reflector'). This light or radiation sinkhole appears intensely black. Note also that a well-made radiation sink is as destructive to the human retina as staring at a bright light. This has yet to be adequately explained. The photon-graviton polarization suggests this darkness is the graviton polarity of the photon-graviton model. In effect, we can regard this as a form of darkness that is just as energetic as light on the retina since it is the other pole of light. This may also explain why in part creatures that inhabit dark caves lose their eyesight.

Historical Confusion. Here we have a confusion that generates from historical precedence. The old belief that an object's apparent color is the color emitted and the object's actual color is that which is absorbed led to calling the Sun a blackbody. It gave off all colors and therefore had no color, and white was regarded as all colors and black was regarded as no color. For CB the Sun has photonic divergence and gravitonic convergence, thus for short in CB the Sun is a *gc-body* (arbitrary emphasis on convergence).

Michell-Cavendish Experiment Revisited. CB analysis disqualifies the Michell-Cavendish experiment for applications to celestial objects. Earth-squelched objects have convergence fields highly deformed by the Earth neighborhood.

The Michell-Cavendish torsion balance (Exhibit 5.1) can only test a relation between two Earth-squelched objects. The spherical field shape is turned into a field shaped like a Lobachevski space. In this context the convergence field of the squelched object, **e**, is in a flattened convexity generated by the synergetic relation with the squelching Earth convergence, **E**. There is evidence against assuming that the so-called big-G constant is a magnitude that is functionally independent of the squelching field in which it is tested (Stacey et. al., experiments in Australia show evidence that suggest big-G is not a constant).

The squelched field shape of a debris object also has a very strong influence on how as a squelched object it responds to other sources of impetus (e.g., constraints) as the gyroscope readily shows. This puts the scalability of **G** into doubt when its applications are directly extended beyond the testing scale of squelched objects. Furthermore, if the host gravitonic convergence is passing perpendicular to and between two squelched objects (Exhibit 5.1b), it may in large part be more accurate to regard the big-G result as a measure of a photon-graviton Bernoulli effect.

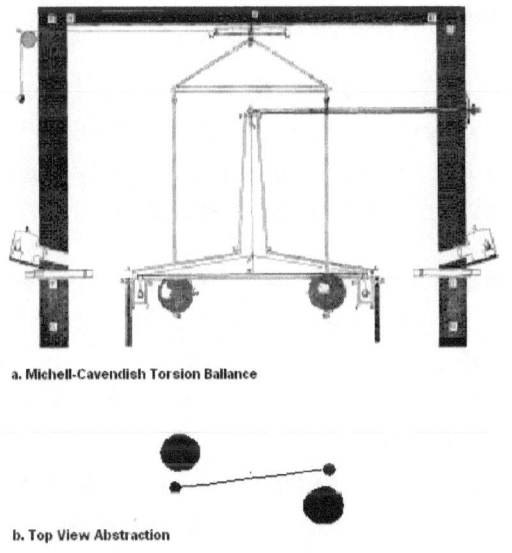

a. Michell-Cavendish Torsion Balance

b. Top View Abstraction

EXHIBIT 5.1: The Michell Torsion Balance Adapted by Cavendish

Note: Frank D. Stacey and G. J. Tuck found that measurements in deep mineshafts show big-G is not a constant (variations up to 0.7% higher than the SI value), and the deeper the location, the higher the value. See "Geophysical evidence for non-Newtonian gravity" (*Nature*, v. 292, 1981, pp. 230-232.)

EMA (Electric-Magnetic-Aetheric) Power Variation. There are three distinct modes for augmenting or diminishing the physical power of EMA polarities from a given source, one with variations in source, and two that have variations with distance.

1. **Variations in Source.** Changes may originate with variations in the source. Thus, as bodies accrete more material, the CIS becomes more well-defined and augments in its generation of heat and introduces the possibility of resonance effects. In general, increasing the quantity of material increases the power of EMA. Growth periods with resonance effects render the variations more complex. Compactness of the material adds another variable to consider.

2. **Variations with Distance.** One of the concerns in describing effects of EMA polarities is the possibility of augmenting or diminishing its physical power with distance. Variation with distance has two forms:

    a. **Kinematics:** Hooke's inverse square law follows from the geometric divergence from a center in a Euclidean field. Applying this Law to EMA propagation, the further you go from a central source, the wider the spread of the EMA power. The widening of this spread weakens the power per unit area. In a Euclidean field this diminishing of power is simply expressed as $1/r^2$, where r is the distance from a center to a surface point of an EMA. Depending on context, this Law may require an embedding adjustment for other spatial models and influences.

b. **Dynamics:** Physical fading is distinct from the power loss through divergence. Functionally independent of the kinematics considerations, it is possible for EMA to fade with distance from its source. This is a rationale for assigning variations in distance according to the degree of redshift. However, this does not account for other possible neighborhood effects on the carrier from source to receiver.

Redshifts could occur in a stable Universe that neither shrinks nor expands, as Halton C. Alp (b. 1927) and Sir Fred Hoyle (1915—2001) teach us. Edwin Hubble (1889—1953) associated the redshift with distance. Others convinced him that it indicates an expanding universe. A stable Universe contradicts what is now called the Hubble velocity-distance law of recession. In mainstream science the presumed expansion has become a catechism belief and in popular science this redshift is likened to the Doppler Effect. Nearby galaxies show a blueshift. This is taken as an indication that they are moving or accelerating toward our galaxy. These spectra shifts are also used for determining galaxy rotations, etc.

In combination, the two forms of fading with distance, kinematics and dynamics, suggest that embedding considerations are required for both geometric and physical conditions. These embeddings could actually augment physical power rather than diminish it, etc., according to the material disposition, spatial orientation, and possible resonance effects, etc., example: laser generators amplify the intensity of light. This can be done in a variety of embedded resonance configurations.

Toward a Universal Law of CB. First, we use reasoning in a CB simulation that corresponds to Newton's attempt to find a universal law of gravitation, which in terms of CB power, better corresponds to Newton's reasoning concerning force due to gravity. Convergence Power, P_c, in isodynamic relations can be determined in the following way. Following Hooke's suggested use of the inverse square law and using it *ex parte* for convergence objects we have:

$$P_c \sim 1/r^2$$

Furthermore, for the two objects under scrutiny, treat the convergence power in each of them as a charge, Q_1, Q_2. The possible polarity modes are; *electric* (negative-positive), *magnetic* (north-south), or *aetheric* (photon-graviton). The two quantities, Q_1, Q_2, must not only be expressed in the same mode for polarities but must also have the same polarity in that mode (for CB). Without knowing yet how to quantify Q_1, Q_2, if we divide the charge in half in one of the two quantities we expect that P_c will be assigned the value that is less than or equal to $P_c/2$. It is equal to $P_c/2$ if they are already isodynamic (sink or float buoyancy), not equal to $P_c/2$ if they are not (this follows the inverse square rule). Following a similarity to Newton's reasoning, this implies that for sink buoyancy:

$$P_c \sim Q_1 Q_2$$

However, we cannot then conclude that:

$$P_c \sim Q_1 Q_2/r^2$$

If there is no physical fading other than from geometric spread over distance and no adjustment is required for neighborhood pressure, the actual power of Q_1 on Q_2 reduces by the inverse square rule. Likewise, the power of Q_2

on Q_1 reduces with distance, following the inverse square rule. For spheres, holding the charge, to express this reduction independently of any measure system standard we can in each case use mean radius values as unity. Thus the radius of Q_1, as its own unit for radius, may be different from the radius of Q_2 as its own unit for radius. The *ex parte* convergence power for Q_1 on Q_2 is expressed as $P_1 \sim Q_1/r_1^2$ where r_1 is expressed as the distance from the surface of Q_1 to the center of Q_2, in terms of the Q_1 radius as unit. Similarly for the *ex parte* power for Q_2 on Q_1, it is expressed as $P_2 \sim Q_2/r_2^2$, where r_2 is in terms of the Q_2 radius as unit. Furthermore, for float buoyancy, we assume that the objects interacting are sufficiently removed from each other to be regarded as convergence points.

Independent of any system units for the dimensions of Q_1, Q_2, we can use a constant of proportionality, **k**, and express:

$$P_c = kQ_1Q_2/r_1^2r_2^2$$

Since we have not yet decided the CB units for Q_1, Q_2, we may choose them such that:

$$k = 1,$$

for each of the three cases (electric, magnetic, aetheric) and such that **1** is without dimensions. This gives us:

$$P_c = Q_1Q_2/r_1^2r_2^2$$

Notice that the terms Q_1 and Q_2 are regarded as charges and not as bulk or anything that corresponds to Newton's mass for any of the polarity modes. The concrete CB quantity-of-matter unit **[B]** is a unit of bulk (adjusted from accretion weight) that can introduce factors different in kind and different in effect, e.g., the CIS, and it is through the use of charges (polarity power) that the quantification of bulk can be identified.

It is better to avoid units that require constants of proportionality, as long as it is practicable. Without the constant of proportionality we have no tempting hanger for factored dimensions and this may help keep the dimensions assigned to their proper place.

From Coulomb's original law for electrostatic charge, his original unit is referred to as the *absolute electrostatic charge*. It is the *absolute electrostatic charge* that renders **k** = **1** in the electric polarity mode using Newton mass. Bulk may be more than 22 orders smaller than Newton mass units when we consider squelched bulks. However, Coulomb's use of a torsion balance on the Earth's surface requires that the initial conditions for the measures of electrical charges be shown as functionally independent of the other two EMA perpendiculars. Otherwise an effort needs to be made to establish compensatory adjustments. Studies in neutralized buoyancy could be indicative. In CB we can choose units for EMA such that for:

E: The *invariant* **electrostatic charge** renders $k_e = 1$.
M: The *invariant* **magnetostatic charge** renders $k_m = 1$.
A: The *invariant* **aetherostatic charge** renders $k_a = 1$.

For the photon-graviton polarity, by changing from the use of Newton's mass to a concept of an aetheric charge, (by design) the systemic constant known as big-G disappears. Thus any possibility of its apodeictic use vanishes. Once big-G is eliminated from the formula, the interpretation for the Michell-Cavendish experiment changes.

Calculating Q. This observation introduces a required complication into the measure of convergence interactions. In order to achieve full generality, a function must be found that accounts for the gamut of material bulks from small to large. If both objects are large and a CB relationship holds then the two objects converge to a proximity (not accretion) where convergence and repelling are in balance supporting a distance. If the second object is small then the two objects converge to accretion. If we paraphrased Newton's reasoning we would have the formula:

$$P_c = Q_1 Q_2 / r_1^2 r_2^2$$

The **Q** magnitudes required for this calculation are difficult to obtain. The measure for each **Q** cannot be direct. It can only be known through interacting convergences. Even if the appropriate data is obtained the formula works only when there is adequate distance for treating their respective centers as the sources for the **Q** magnitudes.

If **[Q]**, **[L]**, and **[P]** are the concrete units of EMA poles, length, and power, then following an argument structure adapted from Maxwell we have:

$$P_c[P] = (Q_1 Q_2 / r_1^2 r_2^2)\,[Q^2]$$

Note that length is subsumed in **[Q]**, which is evident from the denominator magnitudes, **r**, in the expression for P_c. From this we can derive a bulk interpretation where the brackets indicate the units of the magnitudes:

$$[Q^2] = [P] = [(BL/T^2)],$$

where **B** is the concrete unit of bulk (volume and compactness are only arbitrary within a latitude determined for practical experiments). This gives us:

$$[P] = [BLT^{-2}].$$

The dimensions of the unit-power of convergence are therefore, **1** for CB bulk, **1** for length, **2** for the reciprocal of time. This is the ***invariant* charge unit.** The scale difference found in the CB bulk unit was already noted as some 22 orders smaller than Newtonian mass magnitudes.

The squelched objects require a special function, **f(Q)**, which accounts for the effects on their field shape modified by the overpowering host.

Universal Sorting Law for CB. As remarked before, Newton's reasoning is not consistent with CA or CB phenomena. CB cannot properly represent the interactions of these charges in the form $Q_1 Q_2 / r_1^2 r_2^2$. For CB interpretations we must account for the sorting that occurs for the variations of *ex parte* $\mathbf{P}_1 = Q_1/r_1^2$ and $\mathbf{P}_2 = Q_2/r_2^2$ of the two objects. If $\mathbf{Q}_2$ is small and $\mathbf{Q}_1$ is large in their isodynamic position, then the power between them is expressed in terms of weight. If both are large or both are small the power between them is calculated for its expression at the CRN. In both cases of similar scale, the neighborhood pressure (independent of the objects under scrutiny) affects the isodynamic distance: the greater the pressure, the lesser the distance, etc.

Thus if $|\mathbf{P}_1 - \mathbf{P}_2| = 0$, then for $\mathbf{P}_1$ and $\mathbf{P}_2$ the sorting is complete as float or neutral buoyancy.

If $|\mathbf{P}_1 - \mathbf{P}_2| > 0$, then for $\mathbf{P}_1$ and $\mathbf{P}_2$, either they have not reached an isodynamic state or they have accreted.

From this we can show the more expressive form for CB Sorting (in a universe of two objects):

$$S_c = |Q_1/r_1^2 - Q_2/r_2^2|.$$

This formula applies to each and every polarity of EMA in a CB configuration. A more comprehensive formula can be developed from Boltzmann's work. Boltzmann underpins his theory of gasses with a model of molecular behavior consistent with CB. However, he does not offer any physical reasons for this phenomenon type. Even so, his mathematical model may be scaled to gravitational convergence and it can be adapted to include the major physical principles related in CB Chapter 1.

High precision thermal measurements remain among the most difficult to obtain. External radiation and host body influences may contaminate the temperature measures. Errors on the order of 5×10^{-3} % occur even in the most accurate measurements. We need a definition of temperature that does not invoke any of the species properties of particular substances.

GC-BODIES (blackbodies, gravitonic convergence), PC-BODIES (whitebodies, photonic convergence), and TEMPERATURE.

Introduction. *When Kirchhoff and Boltzmann were working on their emission-absorption formulas for radiation the only graph sector open to them was the positive one. Kirchhoff's Law only treats these magnitudes in the first quadrant using the Kelvin scale. The belief in the absolute zero limit denies the use of any negative sector.*

Kirchhoff Law of Radiation
The Kirchhoff law of radiation (1859) states:

> *The ratio between the emissive and absorptivity magnitudes is the same for each kind of ray for all objects at the same temperature and is equal to the emissive magnitude of a blackbody at that same temperature.*

The spectral absorptive magnitude is represented by A_{vT}, where A is absorption for a given frequency, v, at a given Kelvin temperature, T. The absorption is the ratio of radiation at frequency ν absorbed in the system per unit time per unit area to the radiation incident to it. By definition, a perfect blackbody has no absorptive power effect for radiation and therefore we can use its spectra to determine a measure for emissivity that is independent of temperature, as described next.

This interpretation of emissivity of material objects approximates the blackbody emissivity distribution set up by Planck. However, every distinct molecular structure has its own characteristic frequency of vibration. This implies that the molecules emit at energy levels that depart from the Planck distribution.

This distinct departure can be used as an identification signature. Furthermore, the radiance level of this signature varies directly with temperature. In order to compare materials independent of their respective temperatures the temperature effect must be removed. This is represented arithmetically as dividing the emissivity of each material body by that of the blackbody emissivity at the same temperature. Thus it may be expressed as $E_{\nu T}/\varepsilon_{\nu T} = A_{\nu T}$.

The spectral emissive magnitude is represented by $E_{\nu T}$, where ν is the frequency, and T is the temperature in kelvins, where $\varepsilon_{\nu T}$ is the emissive magnitude of a blackbody at the same temperature, and $A_{\nu T}$ is absorption, etc. This results in an emissivity spectrum without dimensions (since the 'division' of like dimensions, in a manner of speaking, reduces the dimensions to the division identity element, **1)**. These are the emissivity spectra that may be compared independently of temperature.

The wavelengths where an emissivity spectrum has a value less than one are in the regions of absorption. Since mixtures of materials linearly produce a convoluted spectrum we can use a catalog of known emissivity spectra to deconvolute the mixture spectrum, etc., a subtraction procedure through a kind of filtering.

Rearranging $E_{vT}/\varepsilon_{vT} = A_{vT}$ we get the original formula, $E_{vT}/A_{vT} = \varepsilon_{vT}$, formulated by Gustav Robert Kirchhoff (1824-1887) in 1859.

Negative Kelvins. Matter and antimatter are conjectured as dual to each other, and it seems therefore that their temperature modes would also be dual, one to the other. That is to say that in antimatter the photon-graviton polarity is the reverse of the photon-graviton polarity in matter. This reversal implies that for antimatter the photon polarity converges and the graviton polarity diverges. If we accept the possibility of a polarity reversal we are left to interpret how this affects the temperature measure. This duality implies that we reconsider the simplistic notion that agitation or motion only generates heat. In the material world this motion assumption seems to work as Newton and Boltzmann teach us. We do not understand how this works in an antimatter world. The suggestion here is that increasing antimatter periodicity rates for corpuscular vibrations increases gravitonic divergence, a kind of anti-radiance, and radiance is conceived here as a carrier associated with heat transfer. Therefore it seems logical that gravitonic divergence be conceived as a carrier associated with cold transfer. Antimatter is cold. This is consistent with a photonic convergence.

> **Note:** At this point note that photonic convergence does not imply photons are moving. Like a magnetic field, photon convergence only indicates the direction of the polarity and need not imply any convergence movement.

The antimatter object appears colder than absolute zero since not only is there no radiance coming from the antimatter object, all radiance in fields weaker than the passing antimatter photonic convergence will polarize to photonic convergence. This requires a redefinition of temperature to include the negative scale extension for kelvins.

Accepting negative photonic convergence (i.e., 'absorption' or photonic convergence) introduces the possibility of taking the resultant below what has been called 'absolute zero'. For CB this is the transition state referenced as 'modpoint zero' between a reversal of polarities. This passes the classic barrier of absolute zero to an extended range of kelvins. Absolute zero becomes the transition CB temperature, modpoint **0**. In the older language this is zero kelvins from which we expect no radiance (neither divergent nor convergent).

With this definition of temperature measure that relates temperature directly to a physical theory (emission & absorption redefined in terms of Divergence & Convergence) we can envisage that the gamut for temperature may hold a symmetry that extends as far in the negative as it does in the positive. If this temperature opposition exists, then the laws of thermodynamics in their large scale applications need to be revisited. This opposition redefines the role of entropy in the Universe.

'Emissivity' and 'Absorption'. If we rewrite Kirchhoff's Law for material objects (positive Kelvin temperatures) in strict terms of photonic emission and gravitonic absorption we could write:

$$\mathbf{E}_p/\mathbf{A}_g = \varepsilon$$

for a given photon-graviton frequency. For the basal level, this formula is trivial since the photonic and gravitonic levels must be equal for any material object in isolation from neighborhoods and therefore ε always equals **1**. This is significantly different from Kirchhoff's original formula (and therefore his conceptualization) as his formula was conceived for a positive light absorption, which **pg** as polarities denies for positive kelvins.

From the photon-graviton polarity we conclude that the basal level of divergence-convergence for any object isolated from a neighborhood is $|D/C| = |C/D| = 1$. This renders Kirchhoff's Law useless for the basal state.

We are left with determining the effects of the neighborhood on the debris in a manner that sustains the original usefulness of Kirchhoff's Law. For this, CB introduces the concept of *permeation* of the Basal State of debris.

CB Permeation. CB rejects the idea of light emission and absorption as used in nineteenth Century physics. CB associates *emission with divergence* and *absorption with convergence*. This changes the discussion from the usual bumper car corpuscle theories into discussing a theory of polarities functionally independent of a carrier. When accreted material debris are exposed to gravitational convergence of the host the debris has a certain level of transparency for gravitational convergence that varies according to the nature of the debris and its interaction with its host. It is expressible in terms of the degree of transparency to the host convergence or otherwise said the impedance to transparency, which we call *opacity*.

This debris opacity to host convergence depends on the debris material structure and could include a degree of scattering of the photon-graviton polarities. This degree of opacity combined with possible scattering allows the debris under steady-state conditions to hold a fixed quantity of the convergence polarity pressure of the host.

> **DEFINITION:** In CB the debris holding of a fixed quantity of the convergence polarity pressure of the host is called ***permeation***.

For material objects CB introduces the notion of gravitonic permeation and how it may be used for measuring ***gravi-***

tonic permeability in material objects (positive kelvin range). For material debris we need a representation of the gravitonic permeation that accreted debris are subject to on the surface of a host. Any staying power for this permeation in accreted debris represents a restraint impetus from its natural neighborhood (especially the host) and may include constraint impetus from other sources. A regression toward basal polarity begins as soon as the debris is distanced from the restraint and constraint influences from the neighborhood that sustains the permeation.

It is through selective gravitonic convergence due to the characteristics of the material debris that renders the color frequencies that are deemed divergent in the basal state.

> **Note:** From This conceptualization we surmise that CB denies that the visibility of objects can ever be based on reflection. It remains convenient to say a mirror reflects, but it is the selective permeation of gravitonic convergence that displays this so-called mirror image, *not* reflection, etc.

The measure of photonic divergence therefore determines the measure of gravitonic convergence (basal and neighborhood) of this state. Basal state gravitonic divergence does not occur in the positive kelvin range.

Basal Properties. It is feasible that the basal convergence ('absorption') properties are determinate as a limit problem, where basal measures are associated with the object *ex parte* (functionally independent of the neighborhood).

From the photon-graviton conjecture, *pc-body* (photonic convergence, white-body) implies gravitonic divergence in the basal state. This is one of the dual conditions for the extended form of Kirchhoff's radiation law.

With respect to divergent photonic polarity, every point on the extended Kelvin scale has a relative gravitonic convergence ('absorption') value, A_g, which is positive for matter and negative for antimatter.

The photonic divergence ('emission') value, E_p, is positive for matter and negative for antimatter. To redesign Kirchhoff's Law from this perspective we first keep the Lilliputian scale that we can work with in the laboratory. For material debris we need a representation of the gravitonic permeation that accreted debris are subject to on the surface of a host. This requires that we change A_g to represent a mixture of the basal gravitonic convergence, B_g, and the gravitonic permeation in the debris attributed to the host, P_g.

We can express this as $A_g = B_g$ & P_g, using the ampersand to indicate that this is not merely a numerical addition. It is a physical addition that requires we consider possible effects that these physical magnitudes have on each other (e.g., resonance, harmonics). Next we not only consider E_p positive for photonic divergence ('emission') in material objects, but a mixture of the basal divergence level and the photonic divergence level generated in accreted debris by the host. Again we must consider this a physical addition that requires we consider possible effects that these physical magnitudes can have on each other. Adapting Kirchhoff's Law to these terms we have an extrinsically homomorphic formula:

$$E_p/A_g = \varepsilon_p$$

for a given frequency at a positive Kelvin temperature. However, the variables do not dereference to Kirchhoff's original conceptualization

The divergent temperature range for ε_p varies from modpoint **0** through all the possible positive modpoint values for ε_p. These temperatures all depend on gravitonic convergence with its augmentation due to permeation. The removal of gravitonic permeation in this range implies that only the basal convergence remains.

With respect to this formula and its range, the modpoint **0** prevents photonic divergence ('emission'), basal or permeation, and no photonic divergence implies that we have a temperature that is at least down to modpoint **0**. Any material in this state regarded from the material side would appear as an intense black. This provides a means for determining how close we are to modpoint zero. As material beings, if we have an object pass into the negative kelvins we would not survive the observation without a bottling especially designed for the state.

To extend the generality of the formula to include negative kelvins, CB considers photonic convergence and a ***photonic permeability***. If we grant that we have photon-graviton polarities, then it is feasible to conceive of a reversal of polarities. CB renders photonic permeability as an impetus conjectured to occur in antimaterial objects though the feasibility of the polarity reversal does not depend on this supposition. With respect to photonic polarity, every point on the extended Kelvin scale has a relative photonic convergence ('absorption') value, $\mathbf{A_p}$, that is negative for matter and positive for antimatter. The gravitonic divergence ('emission') value, $\mathbf{E_g}$, is negative for matter and positive for antimatter. For antimaterial debris we need a representation of the photonic permeation that accreted debris are subject to on the surface of an antimaterial host. This requires that we change $\mathbf{A_p}$ to represent a mixture of the basal photonic convergence, $\mathbf{B_p}$, and the photonic permeation in the debris attributed to the host, $\mathbf{P_p}$. We can express this as $\mathbf{A_p} = \mathbf{B_p}$ & $\mathbf{P_p}$ under limitations corresponding to those required for gravitonic permeability.

Modifying Kirchhoff's Law for these terms we have:

$$\mathbf{E_g}/\mathbf{A_p} = \varepsilon_g$$

taking care to use the indicated domain restrictions for a given frequency at a given Kelvin temperature. We note that this gravitonic divergence ('emission') formula is dual to the photonic divergence formula. The two formulas are homomorphic and the generalized divergence formula that follows uses **D** for divergence, **C** for convergence, and keeps ε for the predefined domain restrictions. The domain restrictions are stated verbally as:

> *If ε is photonic divergence it is equal to the divergence magnitude of a blackbody at that same temperature; if ε is gravitonic divergence it is equal to the divergence magnitude of a whitebody at the same temperature; else ε is taken as in the transition state which has no divergence or convergence (modpoint zero).*

The generalized formula may be written as:

$$\mathbf{D_n}/\mathbf{C_m} = \varepsilon_n$$

where **n** is (**p** exor **g**) and **m** is from {**p**, **g**} that is not **n**, using the required domain restrictions for a given frequency at a given Kelvin temperature, etc. This generalized form modeled after Kirchhoff's Law includes the domain range {ε_g, **0**, ε_p} where as unsigned magnitudes $\varepsilon_g = \alpha_p$ and $\varepsilon_p = \alpha_g$, and this also indicates that the **modpoint temperature** is now in two parts with a transition state. For CB purposes this range is an absolute scale, which is physically independent of any chosen mode for scaling. For current convenience CB follows an extended Kelvin scale for T in the determination of ε_n.

The antimaterial case with photonic convergence is naturally below *modpoint zero*. If this is a natural state it represents the absolute range of cold. Antimaterial debris on an antimaterial host would have a steady-state photonic permeation that increases the average gravitonic divergence ('emission') level and this obtains an even colder temperature signature for the debris. If we could set up an array of lasers around the antimaterial object we do not increase the photonic convergence ('absorption') level but we could increase the steady-state photonic permeation level, which concords with the host photonic permeation under photonic pressure from all directions. This concordance in polarities could have a mutually enhancing effect, the balance of which would depend on the magnitudes that are interacting and their possible resonance effects (resonance can amplify or dampen). The expected overall effect is an increase in the average gravitonic divergence level. The antimaterial object obtains a signature that indicates a colder temperature than before the laser generated photonic steady-state permeation.

The blackbody case represents absolute warm on the surface of a material body that generates photonic divergence. Material debris on a material host would have a steady-state gravitonic permeation that increases the average photonic divergence level of accreted debris and this obtains an even hotter temperature signature for the debris.

If we use our current capability of an array of lasers around material debris we do not increase photonic convergence because photonic convergence cannot occur in material objects. However, we do force a highly volatile photonic insurgence that contradicts the material matrix and its host gravitonic permeation, though the photonic divergence level is not affected (unless the debris cannot sustain material integrity under the exposure). As long as the array of lasers holds a steady-state it allows a gravitonic divergence that mixes with the natural photonic divergence. This artificially introduced gravitonic divergence mixed with the natural photonic divergence (basal and permeation) weakens the effects of both polarities though their individual steady-state natures disallow any physical weakening of either member of the mixture as an individual participation. This mutual contradiction in polarities is destructive only to the effects of their individual powers, the balance of which depends on the magnitudes in conflict and their possible modes of interaction (synergetic response). The continual conflict is in a steady-state as long as material integrity is maintained. This provides a physical model for comprehending the destructive power of lasers. It is also useful for comprehending how solar radiation warms the Earth.

If we could use an array of anti-lasers (projecting gravitonic polarity) around the material object we do not increase the gravitonic polarity ('absorption') level but we could increase the steady-state gravitonic permeation level, which concords with the host gravitonic permeation under gravitonic pressure from all directions. This concordance in polarities could have a mutually enhancing effect, the balance of which would depend on the magnitudes that are interacting and their possible resonance effects. The expected overall effect is an increase in the average photonic divergence level. The material object obtains a signature that indicates a warmer temperature than before the anti-laser generated gravitonic steady-state permeation.

Note that in terms of the physical theory related to CD (convergence-divergence), which correspond to AE (absorption-emission) that divergence and permeation give a corresponding physical distinction between hotter and colder and this includes both absolute (i.e., gc is heat divergent, pc is cold divergent, Modpoint 0 is neither divergent nor convergent) and relative measures (e.g., any two distinct temperatures on the extended Kelvin scale). For accreted debris the basal energy levels are augmented by the host through permeation. This difference from the basal energy levels is a form of host impetus that becomes part of the debris energy level as measured on its host. This can create the illusion of 'absorption' when here it is regarded as a forced permeation from host restraints and possible constraints from other sources. As the host is regarded as part of the natural neighborhood of the accreted debris we can regard the host impetus present in the debris as generated by restraint from the host. Other neighborhood contributions may also be present.

Here we note that the capacitance for permeation varies in material objects in accordance with the basal CD level of the material permeated. All materials have their own vibration levels and frequencies relative to this capacitance. It is these differences under permeation that allow a continued use of Kirchhoff's Law modified for CD.

The Stefan-Boltzmann Law

Emissivity and Temperature. We have a proposed means for obtaining the total 'emissive' magnitude. In 1884 Boltzmann deduced theoretically (Stefan 1879 empirically) that the total emissive magnitude of, or the total radiation from, a blackbody (photonic divergence, gravitonic convergence) is proportional to the fourth power of the Kelvin scale temperature of the blackbody. Today we can derive this law from Planck's law of blackbody radiation. This is the Stefan-Boltzmann Law:

$$\varepsilon_T = \sigma T^4 \qquad \text{ergs/cm}^2 \text{ sec,}$$

Temperature, **T**, here is understood in a usual parametric mode on the Kelvin scale, where σ is the Stefan-Boltzmann constant,

$$\sigma = 5.672 \times 10^{-5} \text{ (cgs units)}.$$

σ functions as the constant of proportionality between the total energy radiated per unit surface area of a blackbody in unit time and the fourth power of the thermodynamic temperature. It is a derivable physical constant equal to R/N_A, where R is the molar gas constant 8.314570(70) JK^{-1} mol^{-1}, N_A is Avogadro's number.

In practice, the International Temperature Scale of 1990 (ITS-90) serves as basis for high-accuracy temperature measurements. CB interpretation uses ε as a modulus temperature. Furthermore, for the purposes defined herein, the given units could simply be read as "ε modulus," or "modpoint ε."

The formula, $\varepsilon_T = \sigma T^4$, represents a conversion relationship between the modpoint temperature and the conventional parametric Kelvin scale units. It also shows that the modpoint temperature in this parametric representation is not Eudoxian. In simple terms, if you double the temperature of a blackbody you multiply the radiant energy by 16, a non-linear relation, etc. It is interesting to note in this context that more blackbody emissive magnitude is produced from 100° C to 101 ° C then from 99 ° C to 100 ° C, showing that our ordinary treatment of temperature readings hides a non-Eudoxian scale for emission magnitudes.

In this perspective, the Stefan-Boltzmann Law qua law is a misnomer. It is not a law any more than

$$X \text{ inches} = 2.54(X) \text{ cm}.$$

It is rather a rule for converting or deriving the modpoint temperature units of a blackbody from the Kelvin unit indicator. If the Stefan-Boltzmann formula is accurate then we can say that the modpoint temperature of an AE system is quantifiable and completely determinable by its Kelvin scale measure.

Stefan-Boltzmann Law Extended. There are special difficulties in any attempt to extend this Law into the negative kelvins. We might use an apodeictic construct that for now is better named an apodeictic fiction and rewrite the Stefan-Boltzmann Law for photon absorption (i.e., photonic convergence).

From $\varepsilon = |\alpha|$ and $\varepsilon = \sigma T^4$, we correspond: $|\alpha| = \sigma T^4$

For this speculative argument we use a supposed symmetry, where σ symmetrically corresponds to the antimatter components. Here **T** uses the fourth root of $|\alpha|/\sigma$, and we have $\alpha = \varepsilon_{-T}$, and **T** is assigned a negative value for absorption. This allows that if we know **C/D**, then we can calculate the equivalent negative kelvins. Furthermore, σ remains as a constant of proportionality that is here used in an artificial reconstruction to correspond to antimatter. Whether or not there is such a correspondence or whether or not this correlates to an extended Stefan-Boltzmann Law does not vitiate the arguments that underpin the **C/D** and **D/C** reversal interpretations. Kirchhoff's Law has a more direct adjustment to the terms of Convergence-Divergence.

COOLING PROCESSES. The radiation given off material systems with little fluctuation suggests that like boiling water the material system reaches a certain temperature and maintains the minimum temperature allowed by the local conditions (constraints or restraints). It could be regarded as a steady-state heating process, but the current catechism prefers regarding this as a steady-state cooling process. The choice depends on what you want to emphasize.

'Radiation' is an unfortunate term; much like 'sunset' it can mislead the interpreter. Though radiation is not itself a loss of heat, it can be used as an indicator for the state of corpuscle agitation and periodicity rates. Separating these two sources of radiation {agitation, periodicity} is still difficult and the solution may be in terms of the material granularity associated with the assigned 'wavelengths'.

> **Note:** With CB constraints we can surmise that the connection between corpuscular vibrations and radiation is concomitant and therefore any impetus (restraints or constraints) invested in that radiation will also affect the corpuscular vibrations.

Earth-Bound Explosions. It is mistaken to think that because high temperatures are recorded on the periphery of an explosion that the interior temperature is proportionally higher at the same moment. Logically, at the moment just before the implosion (reaction that follows explosion) the center of the explosion should be at its coldest. This moment before implosion has the potential of having a much colder center than its pre-exploded temperature. Furthermore, the heat generated by the implosion is only a fraction of the explosion heat loss. This assumes an explosion that is not completely contained (Earth surface explosions represent partial containment).

In violent heat explosions it is expected that the center of the explosion cools far below pre-explosion temperature (CB Law III). Unlike steady-state cooling processes, a heat explosion has a reversal in its source. The explosion begins with a sudden expansion and as such represents an accelerated cooling process. Furthermore, the heat source for the explosion is usually not a steady-state source. This is why under certain controlled conditions an Earth-bound explosion can generate ice.

History has it that as early as May 1941, Tokutaro Hagiwara, in a lecture at Kyoto University (Japan) postulated that a thermonuclear reaction between Hydrogen atoms could be triggered by an explosive fission chain reaction in Uranium-235. Even so, it is likely that the American Hydrogen bomb was not intended to be a fusion bomb. Espionage reports found in Russia verify this. Materials acquired by the Soviet Union about the 'Super' bomb show that it was regarded as a boosted fission bomb rather than a fusion bomb.

The two main enhancements for fission explosions are neutron containment and the speed for achieving 'critical mass'. Improvements in these enhancements render the fission explosion more complete. An alternative design now known as the Teller-Ulam configuration created a double bomb (designed for Plutonium). The interior could be a cylinder or sphere stuffed with Lithium Deuteride (laced with Lithium-6 isotope, another neutron 'container') in concentric fissionable materials. The outside bomb ignites the inside bomb by making the interior fissionable material implode at a highly accelerated rate. This addressed the second problem, speeding up the formation of a 'critical mass'. The Hydrogen Bomb originally referred to the use of heavy Hydrogen as a neutron container, and this was not originally conceived as thermonuclear fuel for fusion as the aftermath publicity would have us believe.

It was observed that there was a significant increase of Helium in the aftermath of the first explosion. The quick conclusion (wrongly made by Teller) was that fusion had occurred. The public information is that the 'heavy' Hydrogen was fused into Helium by the heat of the fission bomb and that this was the designed intent. Theories were invented to rationalize that fusion of Hydrogen produces more energy than fission of Plutonium or U-235. Fusion of Hydrogen is supposed to be the manner in which our Sun generates heat (CB denies this).

More than one interpretation is possible here. Either the fission material broke up so completely that it produced Helium, or Hydrogen, in fact, is fused by the interior temperature of the explosion, or both events occurred that relate the two distinctly different processes. The question I open here is whether it was a cold temperature or a hot temperature. The Los Alamos officials announced to the public that fusion had occurred and that it was caused by exposure to high temperature and that this was intended.

This denial that Los Alamos scientists originally intended a fusion bomb is consistent with the original intent as expressed in the text materials acquired by the Soviet Union.

It also follows from these arguments that a nova produces the lower temperatures that are here hypothesized as required for fusing into elements heavier than Hydrogen or Helium. Note that 'thermo-nuclear' was introduced in place of H-bomb when non-Hydrogen isotopes were used for confining the neutrons. Furthermore, as mentioned before, these later bombs provided no direct source for Hydrogen to fuse. Any presence of Helium in the aftermath would support the belief that it was created from the breakdown of heavier elements or possibly from the fusion of Hydrogen that was first produced from a breakdown of heavier elements.

Note: Edward Teller (1908-2003) is mistakenly called the "father of the H-bomb." Stanislaus Ulam (1909-1984) was the director of the enhanced fission bomb project and Teller worked for Ulam. Even so, we may credit Teller for mistaking the enhanced fission bomb for a fusion bomb and thus he played a key role in the propaganda surrounding the tests for these bombs.

Speculations on Cold Fusion. It is suggested in this text that near modpoint zero from the material side may be conducive to fusion. Fusion in process (of material atoms) does not explode. It is more like a mode of condensation that occurs at near modpoint zero. This condensation may drive the fusion of matter into heavier elements. Modpoint zero is here conceived as a transition reversibility state. It is neither matter nor antimatter, but possibly it is a form of plasma. A central speculation here is whether this state temperature could be pushed down toward conducing the formation of antimatter. In theory this is possible and this may be how an exploding star of sufficient violence could produce antimatter. Again in theory, if a bomb could produce even a little antimatter this would source an increase in intensity of the explosion as matter and antimatter annihilate each other.

Theoretically, all elements heavier than Hydrogen are fissionable (though not in the local primitive easy way that 235-Uranium is, etc.). Uranium$_{235}$ has a material fragility that conduces chain reactions even using our currently primitive methods. The fusion process for matter is not a source of energy. In the material world, fusion is a mode of building potential energy. It requires energy of an unexpected kind to build the heavier elements.

The current First Law of Thermodynamics elicits the Principle of the Conservation of Energy, namely, the conversion of heat energy to work energy and the reverse in the positive Kelvin range. Scalability for this Principle would include the negative kelvin range when we consider what converges and what diverges. This compels a consideration of the concept of cold energy.

Using Light to Lower Temperatures. Bringing a material temperature close to modpoint zero (as a transition state) requires methods for reducing all photonic divergence short of switching the object to gravitonic divergence. CB denies the possibility of generating a gravitonic divergence in a material object. From the photon-graviton hypothesis we conclude that the opposite pole to a laser beam is the graviton polarity. In material objects on the modulus scale (absolute), if photon divergence communicates heat, then graviton polarity communicates cold (an equal and opposite reaction). Is it possible to lower temperatures using light?

If we surround a solid material object with strong counterpoised lasers we expect that the solid material will vaporize. However, if we could accurately estimate the energy level of the photon divergence it is feasible that counterpoised lasers with wide pencils could be adjusted to an energy level out of phase with the object under scrutiny. This would have an interference with the photon divergence. If we could coordinate a set of laser frequencies to equal but contrary amplitudes, then like sound waves they would cancel each other out. A total cancellation would achieve modpoint **0**. The nearer the object is to modpoint **0**, the blacker it will appear to be (from the material side).

> **Note:** The blackness associated with scattered transparency and the blackness associated with graviton polarity generate from different principles.

An Experiment. The tails of comets nearing our Sun always point away from the Sun and Johannes Kepler suggested as early as 1619 that this shows a possible pressure effect. A. Ashkin at Bell Laboratories, Holmdel, NJ, in the USA, suggested bending and focusing atomic beams and trapping atoms in focused laser beams. In 1984, S. Chu and co-workers (among them Ashkin and J.E. Bjorkholm) at Bell Laboratories produced an atomic beam of sodium atoms. This beam was fronted by a laser beam aimed in opposition. The effect was to slow the atoms down. This slower beam of atoms was then directed into an array of six pairwise orthogonal counter-propagating laser beams. Chu called this array "optical molasses" as its effect was to slow the atoms rendering them even slower (and by implication, cooler). The overall effect was that no matter which direction the atoms moved they were slowed down and pushed back into the array of six laser beams. These atoms were then trapped by shooting a powerful beam through the "optical molasses." In 1985, the group reported cooling a dilute vapor of about 10^5 neutral sodium atoms in a volume of 0.2 cm^3 to a temperature of about 0.2 mK. This work is described in the original publication (*Phys. Rev. Lett.* 55, 48 (1985)). Chu, Phillips, and Cohen-Tannoudji working independently in different laboratories received the 1997 Nobel Prize in Physics for work associated with this problem.

The temperatures were calculated under the assumption that the mean speed of the atoms was directly proportional to the temperatures. This assumption is consistent with associating temperature with the *intercorpuscular* agitation of atoms as Newton and Boltzmann taught us, but it ignores the temperature effects of periodicity within the atom (*intracorpuscular* agitation). In effect, it does not consider the variable effects of granularity. That an atom slows down from 4,000 km/hr to less than 25 cm/hr does not imply that the *intracorpuscular* activity has a corresponding slowing down (which we would expect from atoms near modpoint **0**).

> **Note:** The fact that the Chu team did not test for changes in periodicity vitiates the rigor of their claim.

Furthermore, though there is a claim that we shall be able to study these slow atoms, what we likely will learn about these atoms is the unnatural physical state of atoms subjected to light pressure from laser sources. The 'bottles' for these trapped atoms have physical effects on the captured atoms, etc.

Small Atomic Engines. If we could actually compress Plutonium, etc., it could be an alternative way of igniting a nuclear chain reaction using much less fissionable material and a less complicated bomb structure. A screw compressor of the type invented at the University of Chicago is compact and easy to construct, and the required quantity of primitively fissionable materials is sufficiently small in volume.

> **Note:** This screw compressor displays two main compression effects on corpuscular vibrations. The first is a rise in the heat capacity per unit volume. The second is that the material emits higher frequency colors that we assign to the intracorpuscular vibration rates. Materials that fission under self neutron bombardment have a third effect, because the new pressure facilitates this neutron bombardment and fission generates x-rays and heat.

Knowing the explosive aspects of this screw bomb defines the constraints required for using it as a heat source. It can generate much more heat with much less material, though we expect a reduction in half-life. It can also generate heat that passes through a shielding while holding the radioactivity in shielding. This renders small atomic engines feasible. These high-pressure screw compressors may also render certain materials fissionable that are currently not believed fissionable.

Large Atomic Engines (Star Heat Augmentation through Fission). That some combination of heat and pressure might facilitate fissioning elements that are heavier than Hydrogen suggests another way for stars beyond the first generation to augment the heat that is generated at the star's CIS. It may also suggest why lighter elements are so prominent on the surfaces of stars. These may be the 'ashes' of the heavier elements that are breaking down at the star's CIS region.

This implies that for stars that condense in regions with heavier elements that there is a store of fissionable fuel that can augment the heat generation in the star. For a condensation of sufficient size the star would have a CIS that is further fueled with heavier elements. This fissioning of heavier elements would render the star hotter than a steady-state CIS heat support. This condition has two general outcomes. The star either remains extant while its heavier elements fission down to Helium to Hydrogen and this progression leads to a Red Giant stage, or the heating that is supplementary to the CIS heating makes the star so hot that it swells (for steady-state cooling) into a Blue Giant stage. We also have to consider that Blue Giants may occur in locations with high accretion rates. Blue Giants can obtain to a heat level that cannot sustain their physical integrity. This may amplify pulsing. When this is no longer sufficient for cooling it goes to nova. Red Giants expand less energetically until the supplement heating derived from collapse starts. If the fission heating is brought to a sudden halt, the Red Giant implodes. The fission activity in any star is associated with the emission of x-rays. This could provide a rough measure for this activity.

> **Note:** After Robert Oppenheimer was forced to retire, Stanislaus Ulam directed atom bomb research at Los Alamos and remarked to the public that x-rays were the highest radiation form in quantity from fission bomb explosions.

That Plutonium is found in a natural state on the Moon surface and not on Earth suggests that higher convergence pressure is conducive to fission break down. This is consistent with the supposition that the stellar CIS is conducive to fission break down. With respect to testing the Moon's age, the lack of sufficient surface pressure for breaking down Plutonium suggests another cause that degrades our determination of the age for Moon rocks. This extended half-life would render Moon rocks to seem younger than they are, which implies they are older than now thought. They are already calculated older than Earth rocks and this discussion implies they are even older than calculated (using Earth standard half-life values).

Matter and Energy. Matter in motion displays energy. In 'Modern Physics' the distinction between energy and matter has become blurred. Today most physicists have been taught to believe the notion that energy has substance. The classic "capacity to do work" definition is corrupted in the formula, $E = mc^2$, proposed by Olinto de Pretto (1857-1921) and published twice (1903 & 1904) before Albert Einstein claimed it without reference (1905).

> **Note:** Albert Einstein finished his high school courses and bac examinations in Italy where he lived because his father ran a business there. On his examinations he obtained a grade of 5 in Italian and in German. His knowledge of two Swiss languages and a Swiss PhD in Physics qualified him to work for the Swiss patent office, where he was required to scan even the most obscure Italian engineering journals, including the journals where de Pretto's work is available.

This formula shows energy as a form of matter and matter as a form of energy, in effect, two forms of the same substance. The conservative view (which CB holds) is that energy is a capacity to do work and should not be confused with states of substance (the states named at the beginning of Chapter 3). Furthermore, due to misconceptions hidden in the Newtonian concept of mass (including conceptual repairs attempted by followers), it is ignored that mass is better represented by a number 22 orders smaller than the numbers in current use. This implies that De Pretto's **E** does not come out as big as the Einsteins thought it would be. In regard to the current magnitudes used for the variables in $E = mc^2$ the formula can be misleading.

Photon-Graviton Bulk. The converging photon-graviton field is a reference to the shape of the field and furthermore, this shape is one of the main condition variables for measured weight. To speak of the bulk of a photon or graviton has no immediate sense in this apodeictic reference-frame. The photon-graviton polarity form is the matrix condition for measuring weight. The measure of weight is one of the access modes we have to assign some measure to bulk, etc. It is deemed correct to assert that the photon polarity and graviton polarity are without bulk since they are aspects of field phenomena, polarizations. The wave-particle contradictions remain intact for as long as our experiment instruments have the granularity of corpuscular materials. For now a measure of the bulk assignable to the photon-graviton field is not determined.

Magnitude of Convergence: In terms of the photon-graviton conjecture it is still useful to speak of quantity of matter as follows. Convergence is locally defined according to polarity type and orientation. The measure of the magnitude of gravitational convergence is numerically identical to the measure of photon-graviton power adjusted for distance.

There are two components to this power. One is the compactness of the carrier field and the other is this field shape.

The magnitude of convergence power adjusted for distance is interlaced with the neighborhood pressure (that for calculation purposes can be regarded as functionally independent of the objects under scrutiny). Thus in a convergence isopressure zone the object (system) is compressed from all directions. From such a model we may imagine that objects placed in different convergence pressure zones will stabilize at different sizes and speeds of activity where speed of activity is judged at system boundary levels (intracorpuscular and intercorpuscular, where corpuscle = {atom, molecule, cluster}). This does not preclude the influence of the intruder on the pressure zone. As we near neutral buoyancy or microconvergence environments the importance of surface tension increases in its influences on shape and heat distribution. Note that this pressure model contradicts the Einstein equivalence principle for inertial and gravitational mass. Their thought experiment elevator only has *one direction* of pressure. Convergence in its gravitation mode has *omnidirectional* pressure (see Chapter 4 and the Einsteins in Addendum 5).

Speed of Light. We have been long-time habituated to thinking that light travels, though not knowing the nature of this traveling. Galileo made the first recorded attempt to measure the speed of light, but he failed. The first accepted argument was from the Danish astronomer Olaf Roemer (1644-1710) who in 1676 first estimated the 'speed' of light calculated from observing occultation of Jupiter's moons. Maxwell's electromagnetic theory shows the speed of light in a vacuum as a constant. The Einsteins for their theories assert as axiom that this speed of light is the same constant no matter what reference-frame is used. For CB the speed of light is dependent on a potential difference between the source and destination, as well as the physical state of the carrier that conducts the polarization.

Photon-Graviton Fields and Convergence Zones. Light as dynamic is already represented in classic physics and the Einstein work. There are still problems in the dynamic models (bootstrap models) and some of these problems are eliminated in a semi-static model, an elastic field model. When light slows down through a certain medium, it speeds up again when it goes back into the original medium. This can be expected for light conduction as a field polarization. It is not expected for traveling particles. *Michelson's original interpretation of the aether experiments was that while the Earth is moving the aether does not remain at rest.*

If photons are little objects that travel, then it follows that the emitting objects are losing material quantity. Catechism physicists 'fixed' this by saying that the photons have no mass. This also took care of another Einstein contradiction because for the Einsteins any mass that achieves the speed of light has infinite mass. All we have to do is believe that energy is a form of mass and that light is energy and that light has no mass. CB denies this mode of reasoning. Polarization does not entail material loss. If the carrier field moves with the reference-frame, then the experiment needs to be redesigned for convergence fields. Experiments done only on Earth may not be adequate for testing the 'existence' of aether.

> **Note:** Even so, this author believes that the Dayton Miller experiments need to be reconsidered. His experiments were subject to a criticism led by his former student R. S. Shankland, viz., R. S. Shankland, S.W. McCuskey, F.C. Leone and G. Kuerti, "New Analysis of the Interferometer Observations of Dayton C. Miller", *Reviews of Modern Physics*, 27(2):167-178, April. This criticism is the only noted contribution to physics that we have from this hostile student.

There are convergence fields, but the convergence itself is not a field. It is a field phenomenon (the shape of the field). Therefore, we can speak of **photon-graviton polarities** and

convergence zones in those fields. Though CB is consistent with treating convergence as a polarity orientation in a field, *the photon-graviton conjecture remains functionally independent of CB theory*, etc.

Nullification of Gravitonic Convergence. CalTech technologists, in the early 1950s, were the first to produce two sounds at a fixed frequency that were 180 degrees out of phase with each other. At the same amplitude they canceled each other out and no sound is heard. From this research much progress has been made to cancel work-station steady-state noise that is harmful to human hearing.

The light generated by the Earth (mostly infrared) can be regarded as steady-state. If we could find a way of generating a 180 degree out of phase light of the same amplitudes and frequencies this would nullify the gravitational convergence corresponding to the correlative photon-graviton polarities.

SUMMARY CONCLUSIONS. The Universe is locally cyclic. Each star has a life cycle, each galaxy has a life cycle and from our locale it seems these cycles are not in any discernible synchronization. Grouping these local cycles into larger more encompassing scales allows a form of statistical averaging that permits regarding these cycles as participants in a steady-state Universe. If this argument holds then the generality of entropy can be sized back down to where it came from, a Lilliputian laboratory phenomenon.

When Kirchhoff and Boltzmann were working on the emission-absorption formulas for radiation the only graph sector open to them was the positive one. The belief in absolute zero denies the use of a negative sector. The reversibility of photon-graviton polarity opens the possibility of taking the resultant below what has been called 'absolute zero'. For CB this passes the classic barrier of absolute zero to an extended range of kelvins. Absolute zero becomes the transition CB temperature, modpoint 0.

In electromagnetic field theory we have two polarities that are perpendicular to each other, the electric and the magnetic. Herein is proposed a third perpendicular, the photon-graviton polarity. There is already evidence that this photon-graviton polarity has a correspondence with electromagnetism. When a current passes through a Tungsten wire *in vacuo*, the wire glows. All wires warm up when we pass current through them, and this radiates. If the photon-graviton polarity is the third perpendicular then this radiation is evidence of the photonic pole of this third perpendicular. This may be more apparent when passing a spark between two electrodes in a vacuum. Rather than traveling electrons we could conceive that we have an amplified aether polarization. Maxwell agreed with Faraday that light was an electromagnetic phenomenon and we know that the speed of light *in vacuo* falls out from Maxwell's equations as a constant. This put him on the edge of recognizing a third perpendicular. Instead, he proposed that light is electromagnetic. But if we can generate photonic phenomena from manipulation of electromagnetic fields, the photon-graviton polarity suggests that gravitonic effects might also be looked for in these manipulations. It is the line of photon-graviton propagation that is the third perpendicular. The suspension discoveries and experiments of Andrej K. Geim and team at Nijmagen University, Netherlands may in fact be manifestations of the third perpendicular.

Further evidence of the possible interrelations for three perpendiculars is revealed in the polarities of the Earth. The Earth can be characterized as a photon-graviton bottle that leaks magnetism at the poles near its geographic axis. The surface of the Earth is electrically charged. The photon-graviton polarity is normal to the Earth's surface, the electric polarity covers the surface like a static charge, the magnetic polarities leak near the rotation axis points. As graviton polarity is convergent to the Earth, the Earth diverges (emits) light in the infrared range. The intensity of this radiation correlates to the intensity of the graviton polarity. The gravitonic spectrum corresponds to the photonic spectrum.

It is probable that all microparticles and material systems are structured variations on these fundamental polarities. That is why it is likely that all celestial objects will be found to have evidence of all three polarities. This also provides a systemic construction of the possible arrangements for these poles in objects. Theoretically, there are six configurations.

For accreted debris in terms of divergence, the total divergence from the permeation and the basal divergence together must be equal to the composite divergence. The absolute measure references an absolute scale such as the extended Kelvin scale and the relative measure is between any two objects of different temperatures.

If the arguments in this Chapter hold then they clarify why, after over fifty years of research, no one has been able to harness fusion as a source of energy. The basis for fusion has not been understood.

This discussion suggests that we take a more careful look at the artifacts of Newtonian based reasoning. For example, physicists have to say that the blackbody can 'absorb' any frequency that it can 'emit' (in the genre of Newtonian action-reaction). In CB this is not necessary because we are talking about polarizations and not little bumper car corpuscles (wavicles) that are moving in and out of a material object. It follows by necessity that since photonic and gravitonic are two poles along the same line of polarization that they likely have the same frequencies. From the symmetry about the transition state of the extended Kelvin scale we can anticipate the existence of equal frequencies with opposed polarities.

When we approach the transition state between blackbody (gc-body) and whitebody polarities (pc-body), matter may go into a kind of plasma state. Satyendranath Bose (1894-1974) suggested this possibility and it finally developed into the model known as the Bose-Einstein Condensate. Bose originally worked out his line of thought from the ideas of Boltzmann, and Einstein had discussions with Bose and helped him get published. The CB perspective suggests a different scenario. The transition state (modpoint **0**) of any material is highly volatile as this represents the reversal point of the photon-graviton polarity. Material objects cannot subsist 'below' the transition state and antimaterial objects cannot subsist 'above' the transition state. If it were possible for an object to cross the transition state from the material side, it would have to be transformed into antimatter to subsist there. Unshielded from the material environment this would generate an implosion that would react as an explosion greater than any known fission process as matter and antimatter neutralize each other. This may in part explain why when the Russians improved on the enhanced fission bomb that it was more than twice as powerful than predicted.

END Chapter 5

PART THREE
Science-Systems Implementation

Chapter 6: A DIFFERENT INTERFACE for our UNIVERSE

INTRODUCTION. *The foundational use for science-systems is to render our surroundings more intelligible to us through the use of conceptual framework interfaces (intelligibility frameworks). It is through our designs and systems of thought that an intelligibility mode is rendered to us. This "Different Interface for our Universe" is such a framework. Its design changes our mode of access to the way things are, the data-field. It does not change the data-field, but rather changes the way we regard it, the way we study it, the way we apply our science-system to it. This chapter encapsulates some of the new design and poses some of the new questions it generates.*

Mechanics is usually subdivided into three parts: statics, kinematics, and dynamics. *Statics* considers the laws of equilibrium of bodies. In *kinematics* the motion of the bodies is treated as functionally independent of the underlying physics that generates the motion. This frees the motion description from physical principles that could be faulty and for the development of new physical frameworks this freedom becomes a salient advantage in the investigation of the motion dynamics. *Dynamics* considers the laws of motion in terms of the underlying physics that generates the motion. In current physics it is assumed that there is a coherent correspondence between kinematics and dynamics, an assumption that is rarely faulted. CB introduces considerations in celestial mechanics that deny an exact correspondence and indicate the need for proportionality adjustments between the two modes of thought.

Most of the effort presented in this work focuses on qualitative analysis. The quantitative aspects depend on this qualitative framework developed for the intelligibility of Convergence Buoyancy (CB).

Scale of Observation. There are several physical scales of analysis that suggest the possibility of domains where a different dominating physics is required for understanding our observations. These are likely cases of *shifting* dominance rather than a separation of different forms of physics. For CB, the identified transitions are; gravitation to surface tension, surface tension to atomic physics. These transitions have yet to be well understood in their detail and the role of convergence in these transition zones is not yet well-defined.

Even though a physical principle might function throughout the entire spectrum of physical phenomena, the proportional magnitude of its influence on physical phenomena may vary according the scale-sector of observation. This suggests that at different points in the gamut of physical phenomena that there can be a shifting of dominance for the principles that influence these phenomena at various points. To say that a physical principle is scalable is to say that it holds validity when we change the scale of observation. When we extend the use of a principle beyond the boundaries in which it was discovered we are either assuming its scalability or testing its scalability. The boundaries of scalability for a physical principle measure the scope of that principle.

The recent (1997) 'discovery' that the center of our galaxy has antimatter suggests that galaxy dynamics is influenced in a manner distinctly different from matter-matter relations. This discovery implies that the convergence that dominates galactic spirals is a polarity opposition (CA), which is different in kind from matter-matter convergence that CB principles describe. The convergence action between matter and antimatter is much more powerful than the convergence of like polarities. This may explain why converging stars in galactic formations follow a modifying influence on their gravitational convergence. It seems that the stars in a given galaxy maintain a close to equal velocity independent of distances, a contradiction to Kepler's Laws. But antimatter may well have other scale influences not yet known. The subject of antimatter still remains high in speculation.

EMA Dipoles. The EMA field is conceived as holding three polarity modes perpendicular to each other.

An electric conduction through a wire helix generates a magnetic dipole field and this helix surface also emits a quantity of heat, a form of aetheric divergence. This dipole is commonly amplified by wrapping the helix around an iron rod or pipe.

This suggests the question: Can a helix of some kind generate an electric dipole or an aetheric dipole?

Optic fiber can be arranged to conduct light, but in CB this aetheric conduction implies the photon-graviton poles are both present. If we made an optic fiber helix conducting a laser beam, it is doubtful that it would be a magnetic dipole as this is already claimed by the electric helix. The Pigeon-Hole Principle suggests that mutual exclusions apply, else much confusion is introduced into the order of things. The only choice left is the electric field dipole. This dipole effect may be amplified by wrapping the optic fiber helix around a conductor such as a copper pipe or rod. This also implies that we have an electric field generator with no moving parts. The dipole field stands much like a battery. It will not deliver the electricity without first embedding it in a circuit that sustains this mode of polarization. We may also expect a measurable magnetic charge on the outer cylinder of the helix. To activate the helix requires that the optic fiber is itself part of a closed circuit of light 'conduction'. This conduction in CB is characterized as a polarization rather than a current. Refinements could render the helix active under any light source, including the Earth infrared divergence.

Following the Pigeon-Hole Principle further, if we use a wire that conducts magnetism, we could form a helix that would produce the aetheric poles. From the helix structure of the generator we may expect a clear and controllable distinction of the photonic and the gravitonic poles in a dipole form. It may be amplified by wrapping the helix around a pipe or rod of optic fiber. We may also expect a measurable electric charge on the outer cylinder of this helix.

> **Note:** In Catechism Physics, each helix is said to require a current. The term 'current' facilitates a popular way of talking about this phenomenon, but carried polarization may present an illusion of a physical traveling as waves do on the ocean.
> **Note:** In an anti-material World it seems plausible that all of these poles and charges would be reversed, which completes a symmetry.

CB does not characterize gravitonic convergence as traveling. Gravitonic convergence fields have shape, intensity, and represent a polarity of the photon-graviton field. If gravitonic convergence does not travel, then for CB neither does its opposite polarity, photonic divergence. As here proposed, the EMA field holds three polarity modes perpendicular to each other, therefore this stipulation holds for all of them. The photon-graviton field is directionally polarized in accordance with the generating system (matter or antimatter). Relative to the generating system, this polarization does not travel. However, there are two other notions of movement to consider. If the system moves, its photon-graviton field moves with it. This field is an *ex parte* regard of the EMA-field of which it is one of the three perpendiculars. This EMA-field may also act as a carrier of disturbances to it. For example, it can carry radio signals and though the field itself may not be traveling, the radio signals that it carries can be regarded as traveling through it or on it. It is a variation of energy level that is traveling and not the EMA-field that carries it.

We know that waves on the ocean surface produce the illusion of traveling even though careful tests show that the water is following a swell and a spread alternation in a rhythm that moves it up and down in place. This up and down alternation can produce the illusion that the wave is traveling and this in turn can create the illusion that the water is traveling. If we ask the naïve observer if the water is traveling the usual answer is 'yes'. If we ask those who carefully study the phenomenon the answer is 'no'. But a tsunami seems to travel, yet the water that carries it to the shoreline does not travel. At the shoreline an event that appears different in kind takes place though it is not. First the water moves from the shoreline toward the tsunami swell leaving a wide lowering spread of water and this is followed by the swell filling in the spread and the water floods inland. What actually occurs is that the swell is so large that when it fills in the spread it carries water inland. From this state the untrapped water will flow back out again to complete the down and up cycle of the carrier. In the vernacular it makes sense for us to say that the tsunami is traveling and it is easier for us to talk about it this way. Yet all that is happening is that ocean water is swelling then spreading, up and down in place.

In the laboratory we can synchronize the water waves of constant amplitude such that our perception is put into a state of immanent reversibility. Facing normal to the wave machine, the wave can be followed left to right or right to left as if the direction chosen is arbitrary. If we decide on one direction or the other, it is a bias of interpretation rather than statement of fact. This puts our interpretation of any speed we associate with light in a different perspective.

Three EMA Rider Modes. As with the tsunami we can send a disturbance, a kind of overlay phenomenon that rides on the EMA. We currently use three EMA rider modes characterized as modulations. These riders were first developed for communications. The three communication riders are: AM, FM, PM. These are the Amplitude, Frequency, and Phase modulations. They can be applied to any of the three polarity sets of the EMA field, and like the tsunami they can travel on the EMA field at a variety of speeds that depend on the energy level used at their inception. At night, what happens when we turn the lights on? Again what we have is an overlay travel effect that does not entail physical traveling. When we send radio signals to control satellites over great distances these overlay signals increase latency in both sending and receiving that correspond to the distances. It is said that light takes eight minutes to get from the Sun to Earth. Even so, this timing would only concern the overlay phenomena that ride on the carrier field within the photon-graviton polarities. It may be difficult to decide which part is the carrier and which part is the carried. As with the synchronized water waves it makes little sense to assign a direction of movement to the carrier, though it makes sense to assign a speed and direction to the carried modulations. Any directional movements assigned to the pg-field itself are illusory. It is in this sense that the pg-field does not travel. However, we may conceive of a dampening or an amplification of overlay modulations. In general, the bodies (matter or anti-matter) generating EMA establish polarizations and it is these polarizations that we can modulate.

It may be true that a disturbance to an established light polarization can in the manner of the tsunami be regarded as traveling and like the tsunami we can assign velocities to this disturbance. Using a laser pencil to polarize the carrier field to the Moon and back, interpreting this as a traveling unnatural disturbance to the EMA photon-graviton field polarity allows us to use it for estimating and comparing variations in distance from the Earth to the Moon. This form of measure remains parametric for as long as the illusory aspects of this mode of travel in neighborhoods are not accounted for.

Star Cycles. The driving force in this story of the Universe in our locale is the convergence of material systems. Material systems converge to a sorting out to sink, neutral, or float buoyancy. This form of convergence is a function of size and compactness. In general, the more compactness a material object has the stronger the convergence power per unit of surface area. A sufficiently small object under neighborhood pressure can accrete to another small object or accrete to systems already large enough to generate float buoyancy with other large objects. The accretion cycle begins with the available materials in the occurrence region. In our region, we can surmise that this cycle goes from gas and dust, comets and asteroids, to planets, to fire-stars, and the current growth cycle for our locale ends when our Sun becomes a nova. The materials left after the nova may then again converge into globes of material systems. This does not imply that the nova brings this process in full circle. It is a transition link for a limited series of cycles that progressively increase the quantity of heavier elements per unit volume.

We have observed stars that we deduce are made from the lighter elements. We conceive that the star cycle starts as plasma. This accretes and compacts into convergent material systems under regulations dominated by charged particles. These regulations are different from a CB system. The Swedish physicist, Hannes Alfvén (1908-1995), developed the general description of this stage of condensation. At first he pictured a universal co-extensive phenomenon happening at some earlier date. He later allowed that this condensation could occur periodically at various times in various locales, rather than as a universal co-extensive phenomenon happening at one time in the past. This implies that the Big Bang may never have occurred.

These plasma states are designated here as the first generation of material systems in an evolution cycle several times prior to the one our Sun is in. They continue to accrete and grow into fire-stars. Pulsing facilitates the cooling rate. The pulsing magnitude is related inversely to the convergence power and the neighborhood pressure. Accretion continues until the CIS tension can no longer sustain structural integrity. One result may be the nova type that can generate the low temperatures required for fusion into some elements that are heavier than those in the star before its physical integrity bursts into a nova. These clouds of gas and dust, in turn, converge into globes of material forming the second generation of material systems and these systems contain a proportion of heavier elements that did not exist in the previous generation. These cycles continue until we reach the distribution of elements found in our region of space. Our Sun is likely not a first or second generation star. It may not even be a third generation star. We do not know the time required for these generations of stars. This time factor varies according to accretion rates and the supply of material debris.

The current belief is that far distant objects are homomorphic with early forms. We need to have clear notions about the possible distortions acquired with the great distances, especially the interference with the data carrier propagation (see the work of Halton C. Alp). With respect to visual telescopes, many of the most distant galaxies look just like what we presume our own looks like. Based on radio wave interpretations the radio astronomers claim they can detect at great distances the primordial universe, a conclusion consistent with a variety of Creationists and denied by CB.

These cycles of creating heavier elements and greater proportions of heavier elements finally create a star that has such powerful convergence that the energy released in its nova destroys the atomicity of its components. This is a kind of *supernova* and a likely amplified source of gamma radiation. What is pictured here is that the power of the explosion blows apart all material leaving virtually nothing within the required proximity to effect a fusion under the required temperature environment. The energy level of the explosion is too high for sustaining structures in the form of material atoms. The debris either takes the form of first generation plasma or it achieves a sufficiently negative kelvin temperature to generate antimatter. Theoretically, antimatter objects can cluster to form convergence centers for attracting material stars into forming galaxies or can converge to clusters of material objects and form anti-galaxies. It could also be that matter-antimatter annihilation at the galaxy core becomes a source of plasma. This segment of the story requires more information. The plasma state brings the spiral of development back onto itself; the period closes to begin again. These cycles suggest that present estimates for the age of the Universe (for those who believe it has an age) may be nonsense or underestimated.

Contrary to the one Big Bang theory, if our Universe continues from here to eternity, this cyclic theory characterizes a temporal infinity of smaller bangs. As they are not synchronized to occur at the same moment, collecting them into more comprehensive groups allows a form of statistical averaging that permits treating these groups as steady-states. The *age* of a steady-state Universe has no dereference. However, it is interesting to estimate the cycle times. The Big Bang theory was conceived to have a one-time beginning and with it is conceived a contracting end, a view consistent with a Creationist theory embalmed with universal entropy.

> **Note:** The Big Bang physicists say that when the Universe is expanding there is no center yet we must accept one day it may shrink back to the singularity. They argue that there is no space (or time) until the Big Bang, and no space implies no location for the singularity (baptized 'singularity' to imbue it with the power of sole Creator). And so something has been created from nothing. Stephan W. Hawking accepts this as scientific evidence consistent with Creation and so we may class Big Bang as a Creationist theory. Does he share with Newton the need of religion to make the system (that he believes in) work.

The Little Bang theory has no notion of a beginning or an end for the Universe and entropy has no universal dereference in it. In large part this is consistent with the steady-state, or continuous creation theory of Sir Fred Hoyle (who with irony coined the term 'Big Bang' to reference a theory he opposed).

Without the nova the elements required for the human body would not exist. We are made from heavier elements that are created by the self-destruction of stars. All life forms (that we know of) require materials created by the self-destruction of stars that grew beyond the limits of structural integrity. In effect, all forms of life derive from star-dust.

The next concern is how the revolving systems of our solar neighborhood were formed. This includes the types of systems we find in it and what influences govern their movements.

Fire-Stars, Planets, Moons, and Satellites. Satellites run the gamut of possible structures. These include gas, dust, comets, meteors, planetoids (asteroids), moons, planets, and fire-stars. The term 'satellite' takes its precision from its intended dereference. We say that the Moon is a satellite to the Earth, but there is a sense in which it is just as true that the Earth is a satellite to the Moon. In effect the Earth and the Moon form a binary system. Without pushing this to extreme precision we can use the notion that the material object that has the largest movement about the other is the satellite to the object that has the least movement in material object systems. Thus it is acceptable to say that the Moon is a satellite to the Earth, but not so accurate to say that the Earth is a satellite to the Moon. It is conventional to say that the Earth is a satellite to the Sun. The Earth/Moon system is a system satellite to the Sun but the Moon is not usually spoken of as a satellite to the Sun. We could call it a subsatellite if this distinction has some use. In cases where it is not so clear which system is the satellite, we shall move, without strictness, to the terms 'binary system', 'trinary system', etc.

There are three possible characterizations of satellite paths relative to a given neighborhood pressure. Either the path tends to a circle or the path tends to spiral. If the path tends to spiral, it either spirals outward or it spirals inward. If it spirals inward, CB calls this a sink path until it sorts, etc.

All planets are pristine fire-stars and grow from accreting space debris. Our Earth grows in this way. The accreting planets that survive will finally develop into fire-stars. In our own system, Jupiter is the planet nearest to becoming a fire-star. In fact, all the Jovian planets of our solar system generate more radiation than they receive from the Sun. We do not yet know how many nova cycles are required to create a star like our Sun but for CB it is certain that one-day our Sun will either lose sufficient energy for it to collapse or gain so much energy that it will explode. The Sun energy level depends on accretion rates in the neighborhood, its CIS, and its own interior material breakdown rates.

Characterizing material-systems accretion events as ongoing suggests that circle orbits have only a virtual existence in a momentary instant for homeostatic systems that accrete to surpass the boundary convergence. Boundary convergence is that bulk measure which separates accretion bulk (small objects, irregular in shape) from floating bulk (large objects, well-rounded or spherical in shape). Accretion bulk has its repelling overpowered by a host bulk such that it sinks to the host surface. A floating object has sufficient bulk and distance such that its convergence is not squelched by a neighborhood convergence, i.e., its contribution to the repelling relation has sufficient magnitude (for a given neighborhood pressure) to prevent its accretion to another host, i.e., it floats. The Earth floats about the Moon, the Moon floats about the Earth. As the objects accrete material this increases the magnitude of mutual convergences which concurrently increases the repelling power. This can drive the hosts that converged to their isodynamic limits to spiral apart. The Earth-Moon system spirals apart. The current rate of about three centimeters per solar year has been derived from orbital considerations and measurements by laser (accuracy subject to the limits of a carried signal).

A more precise distinction between moons and satellites is useful. Herein a moon is a satellite to a planet but a satellite to a planet is not necessarily a moon. Moons have near circular orbits or they have orbits that spiral outward. Phobos is a satellite sinking to the surface of Mars; this sinking denies it the status of a moon. Deimos is further out and smaller than Phobos and therefore likely should also be denied the status of moon. Under satellites, CB includes material objects that sink to accretion following spiral trajectories. In CB we define a moon as a material object that sustains a repelling with its planet sufficient for float buoyancy (buoyance). The moon goes around a planet, the planet goes around a fire-star. The planet is a satellite to its fire-star. For it to maintain the status of planet it must spiral outward at a rate proportional to the system homeostasis (mutual accretions). These distinctions have borderline cases that can baffle classification. Jupiter seems to be a pristine star that already produces measurably more radiation than it receives, and some of its moons have characteristics we associate with planets. In a sense, the planets that are spiraling out from the Sun are the moons of the Sun. If we do not lose sight of these possibilities, and we do not need to be more precise, these ambiguities need not bother us. The danger lies in allowing a particular interface to determine how we think about these celestial systems rather than using the interface as a tool. In the CB science-system, accretions of matter are fundamentally similar in their series across formation, development, and dissolution.

In a manner of speaking we could say that it concerns the birth, growth, and death of stars. The first main differences to look for in a given region are the proportions of distinct elements found there. With the debris and orphan planets coming from outside the solar system, care is required not to confuse external material with local material. We need methods for 'finger-printing' the material under scrutiny.

ADOPTED ORPHANS. CB allows for the possibility that the solar satellites are orphans picked up by the solar system. It is possible that the adopted planets are older than the Sun. In view of their heavy element concentration it is possible that their material was formed by older exploding stars. They spiral inward until they are isodynamic. This suggests another possible origin of catastrophic events on the Earth. By bulk, if we assume equal opportunity for accretion, the Earth is the oldest of the rocky Sun satellites. If we follow this logic, then Venus and Mercury may (though not necessarily) have entered the system after the Earth did. This implies they spiraled across the Earth's orbit to position themselves.

We cannot be sure where this crossing could occur but the mutual repelling effects could finally position them, each in turn, diametrically opposed during the crossing of the Earth orbit. The possible variations allow a variety of possible levels of catastrophe. Determining the level of catastrophe for these crossings is in part still guesswork. In the case of the Moon there is circumstantial evidence that suggests that the Moon is an orphan adopted by the Earth. At the point of crossing, the vector component tangent to the Earth orbit would match the Earth's, approximately following Kepler's Third Law. Their speeds being slightly different, they could finally move to a proximity that would allow a capture.

Huron Glaciation. If the Earth were an orphan adopted by the Sun, then there should be evidence that prior to its current proximity to the Sun that it was once far removed. This implies that we might be able to find evidence of an extremely long ice age prior to its positioning in the solar system. In reviewing geologic evidence, there are indications that the Earth did have such a long ice age. It is known as the Huron Glaciation. Present estimates suggest it lasted almost a billion years (at least 900 million years). There are three layers, dated between 1.8 and 2.7 billion years ago. The two interglacial layers appear to be deposits from intervening milder spells, but the evidence is still too sparse for a final judgment. The actual surface extent of the glaciations is not yet well determined. The Snow-Ball theory claimed that the Earth was completely covered with ice and snow some 600 million years ago. In recent years this has been adjusted to 700 million years ago. Either case implies that evolution of life on Earth could not have gotten very far before that time. This suggests that the evidence of earlier life found in Australia would require a re-evaluation. Dating methods need improvement. This suggests an even smaller window for evolution. This is on-going research for a theory that has not been disproved.

The Orphan Theory derived from CB considerations suggests that planetary systems are likely a common phenomenon in our Universe. It also suggests that *finding a star without an orbiting planet is likely a rare find.*

> **Note:** It is likely that all of the debris, moons, and planets of our solar system are orphans caught in the maelstrom of our Sun. That they display important differences even though sustaining millions of years of accretion tells us that our solar system in geological time is still young.

Mars, Venus, Mercury. Though currently regarded as born in our solar system, CB analysis suggests that these planets are mavericks formed outside the solar system and then came through the asteroid belt. This would help explain their variety. For planets all forming at the same time as our Sun we would expect more similarities then we find. Furthermore, four billion years is likely enough time so that if a group of adopted mavericks were early adoptions, they would accrete enough similar materials in the new locale to appear more similar than they do now. Mars either captured Phobos and Deimos somewhere en route, or near its present position, etc. Whatever its origin, the present location of Mars is out of order for the sorting of solar planets according to E-power. This suggests that it is still sinking toward its isodynamic convergence with the Sun. In this argument, Mars is also greater than the solar boundary bulk and therefore will not sink to accretion. It will sink in a spiral until it is isodynamic with solar convergence.

For its present bulk we can guess that this would lie inside the Venus orbit. But if Mars is spiraling toward the Sun, it likely will take many millions of years to spiral inward even one million kilometers. In the meantime it is collecting debris, many metric tons per day and this, by increments, diminishes the sink acceleration.

We could ask whether the Earth is yet isodynamic with the Sun. Venus (radius 6052 km) is not much smaller than the Earth (radius 6407 km), and it is over 40 million kilometers closer to the Sun. We first consider the geometric fact that most of the volume of a sphere is toward the surface of the sphere. This is simple geometry. In terms of the radius one half of the sphere volume is in the outer fifth of the radius. In a locale with a constant volume accreting rate the spheres grow slower and slower in diameter through time. What is suggested here is that in similar-material systems small differences in radii can produce large differences for isodynamic distances.

The Earth is likely moving slowly away from the Sun (Earth-Sun distance is variable and difficult to measure). The Earth-Moon relation can also contribute to this movement and to this belief. As the Earth gains over 200 metric tons per day, the Earth convergence becomes stronger. This increases its repelling magnitude. Though Sun and Earth accretions increase the power for buoyancy, moving us away from the Sun, the increase in Earth size makes the Earth generate more of its own heat (a much slower process than the so-called green-house effect). All the Jovian planets generate more radiation then they receive. The Ice-planet surface conditions likely highly reduce the radiation readings for them. All the rock planets generate less radiation then they receive.

Splooch Craters. Two crater types are well-known, one from volcanism and the other from surface to surface impact. CB suggests a third type from buoyance collisions, here named *splooch craters*. For this we conceive that two large objects under gravitational convergence bob below their isodynamic distance, generating great surface pressure on semi-rigid surfaces. This pressure may be sufficient to form impact-like craters. The CIS introduces differences in how the Roche Limit affects hosts in such collisions. The stronger the CIS tension the more stable the host. It is a form of impact, but the two objects converging to near parallel paths never materially touch surface to surface and they do not bob deeply enough to lose their structural integrity. Furthermore, the power of this CB bounce has an equal and opposite reaction (isostasy) on the opposite side of the participating celestial objects.

The Moon exhibits a number of craters that seem to correspond to this definition of splooch crater. That is to say, on one side of the Moon we have a crater or set of craters with no apparent impact material. On the other side we have *maria* that resemble the Deccan Traps of India. On the side of the Moon unseen from Earth there is a triangle of three craters, the largest, Mare Orientale, is over 900 km across, dated 3.8 billion years old. These are interpreted as impact craters. Opposed to this on the face we see from Earth is a triangle of three maria, large lava plains. These plains, Mare Tranquillitatis (Sea of Tranquility), Mare Crisium (Sea of Crises), Mare Fecunditatis (Sea of Fecundity), are likely from the same epoch though their dust covering is not.

If the Earth-Moon capture took place, there could be bobbing that might generate more than one splooch crater. The largest of the lava plains group is with Mare Imbrium (Sea of Rains) and Oceanus Procellarum (Ocean of Storms) and others nearby. This group of maria may represent isostatic reactions to a strong bobbing buoyance collision. The surface rupture in the Moon's southern hemisphere, not visible from Earth, is as far south on the far side (in opposition) as this group of maria is north on the side seen from Earth. In a Moon map color-coded for relative altitudes, this surface rupture is seen as the largest and the deepest of all the Moon's surface structures.

If the craters were surface to surface impact craters or if the craters were two convergence neighborhoods pressuring against each other that form splooch craters, lava plains could form in opposition to them. This represents the equal and opposed reaction to the crater formation that is modified in its transmission. Evidence circumstantial for denying surface impact derives from the lack of debris from an impact object. If there is a lack of impact debris, then we decide between volcanic or a splooch crater, or some fourth way we do not yet know about.

If the foregoing argument holds, then we may have a way to find a tentative date for the Earth acquiring the Moon. Exploration of the craters that are most likely from the acquisition epoch should reveal no 'foreign' debris that date from the formation event.

This implies that there is a corresponding splooch crater on the Earth, with a corresponding equal and opposite reaction. If we accept the estimated date for the Moon event, it could be that the center of the Pacific Ocean has the corresponding splooch crater and that the breakup of Pangea may be associated with this event. There seems to be geologic evidence that the Earth surface was already solid at this epoch, but this does not preclude the possibility that part of the Earth surface material was liquefied and reformed.

The Moon capture would have wreaked havoc on the Earth and generated severe and rapid weather changes. It might also have been this event that formed the Earth tectonic plates. If it occurred 3.8 billion years ago, then on Earth there may be no acceptable trace of the event. It has not been proven that dating methods using radioactivity half-life would yield same long-range results under different convergence power. The dating methods are not always so sure, but as long as we used similar testing we might at least get a comparable sequence of events.

Even so, this does not yet dismiss the possibility that the Moon is a fragment formed from part of the Earth and that a close encounter with another large celestial object could have formed the tectonic plates and loosened a plate that was sufficiently large to generate on separation a repelling effect. Whether in consequence the plate broke up or liquefied are further possible speculations. We note however that the dissimilarities in Moon rock discovered in the Apollo Moon landings collection suggest more credence for the adoption thesis.

If we pursue the large passing object conjecture we might consider:

Mercury Passing the Earth. The estimated age of the surface of Mercury has the right date, c3.85 billion years old. If Mercury entered the solar system as an 'adopted orphan' it could have passed the Earth orbit on its way to its isodynamic buoyancy with the Sun. This is about the right timing for Pangea as well.

> **Note:** Geologists are naming and renaming continental tables in their research to establish an evolutionary sequence for these tables. This does not vitiate the thesis that the break-up of continental tables is likely related to the formation of tectonic plates no matter what they decide to name the tables of that time.

Mercury also has a candidate splooch crater. Though dating estimates put the event at 3.85 billion years ago, there is no evidence of a corresponding lava plain. But what might be the equal and opposite reaction translated into a crust deformation on the opposite side. This may relate to the distance of the buoyance collision or the nature of Mercury's interior structure, etc.

As yet we cannot determine whether the Moon was captured after Mercury passed by, or already captured, or (though less likely) formed from the passage. The estimated age of the lava plains of the Moon suggest that the current Moon surface was not well formed at the time of this proposed passage of Mercury. But the age estimate for Mercury's possible splooch crater is not yet reliable.

Venus Passing the Earth. Judging by the CB implied planet characteristics (orbital eccentricity, revolution and rotation conformity, ecliptic conformity, bulk sorting), Mercury and the Moon have been in place longer than Venus or Pluto. Venus still rotates slightly in the 'wrong' direction, implying that it may be significantly newer to its position then Mercury or Earth or Mars. Mercury has a higher eccentricity (0.2056), but this may be due to its proximity to the Sun. Its orbital plane is also the closest to the solar equatorial plane. If Venus is a newer 'adopted orphan', this implies that it had to pass the Earth orbit to attain an isodynamic relation with the solar neighborhood. The surface of Venus seems totally renewed at an estimated age of 500 million years ago. If Venus made a splooch crater with the Earth (not a necessary event) the crater on Venus might have disappeared in the renewal of its surface. We could seek on Earth evidence of a possible 500 million year ago disturbance. There is in fact recorded a severe mass extinction on Earth about as old as the estimated age of the Venus surface. No matter where Venus crossed the Earth's orbit, it would likely have caused sufficient disturbance to cause severe weather effects on Earth. There is evidence that 700 million years ago the Earth was covered with ice and snow and this may correspond to a time when Venus shared the Earth's orbit, possibly in opposition, which would place the Earth farther out than now. This still leaves the how and why as open questions for the Venus surface reformation.

During the Cambrian (543-510 million years ago) there were at least four major extinctions. The event at the Early Cambrian epoch boundary wiped out the olnellids (oldest group of trilobites), as well as the archaeocyathids (primary reef-building organisms). The later three extinctions distribute irregularly around the Late Cambrian epoch boundary. These four extinctions wreaked havoc on trilobites, brachiopods, and conodonts and produced an important distribution change in the life forms on Earth.

Mars Splooch. A Mars surface description sounds like the Moon with a polar axis passing through the densely cratered side (southerly) and through the center of the lava plains on the other side (northerly). This suggests that splooch craters may be discovered at the Martian South Pole. The Martian North Pole resembles the Deccan Traps. A few years ago a near South Pole crater was discovered of adequate size to correspond to this prediction.

An older hypothesis, still entertained, is that Mars is the former moon of an exploded planet that now forms the asteroid belt. These considerations have many possible different scenarios. If the Earth is splooch bumped, what effect does this have on the Moon, etc. These suggestions are presented as hypotheses to consider. Splooch craters may explain a number of unusually large craters on celestial rocky objects and may answer why the objects with these craters were not broken up by the event.

Mimas Splooch. The largest crater on Mimas (satellite of Saturn) should have blown Mimas apart if it were a surface to surface impact crater. Here it is suggested that it is the result of a powerful buoyance collision. This might also allow a recoil that would soften the effect enough to maintain Mimas integrity. A more speculative view is that Mimas may have been a part of a larger body that was blown apart by the buoyance collision before adoption by our solar system, etc.

Chixulub Crater. At Chixulub on the Yucatan peninsula there is a crater estimated 180 km in diameter that was discovered during a widespread search for oil in the area. The age of this crater is now estimated at 64.98 ± 0.05 million years. This places the event in the late Cretaceous/Tertiary (K/T) boundary. At this epoch this places the Yucatan peninsula in a close opposition to India where we find the Indian lava plains known as the Deccan Traps. The Deccan Traps are formed from a huge outpouring of lava that covers much of western India and are also dated at 65 million years old. This huge eruption of lava could be a reaction to the Chixulub crater formation. *This Indian event likely had violent affects on the Earth's climate* in addition to whatever effects might have occurred from ground zero.

For the Chixulub crater, size estimates for a surface-to-surface impact object that exceed six kilometers are likely exaggerated. Early in the find some geologists estimated the size at 20 kilometers, which for CB hampers the possibility of a direct impact. This Chixulub crater has been hard to detect and has increased in estimated diameters from its first discovery. It may yet turn out larger than the current estimate of 180 kilometers. Recent exploration shows evidence of debris, which implies the Chixulub Crater is an impact crater. There is also evidence of a layer of metallic debris, which suggests that the meteor went into melt-down at impact.

> **Note:** This impact event is currently associated with the great extinction of dinosaurs. A prior theory for this great extinction is that a proliferation of rodent like creatures, from which humans may have descended, ate the dinosaur eggs. As chickens are believed related descendents of dinosaurs, it seems the 'rodents' are still eating 'dinosaur' eggs.

Pluto and Charon. In looking at the history, Pluto and Charon form a two-object system that was discovered outside the orbit of Neptune as only Pluto, 18 February 1930 by Clyde W. Tombaugh (1906-1997) at the Lowell observatory in Arizona.

The discovery image of Charon with Pluto was taken on 2 July 1978 using the US Naval Observatory 1.54 meter telescope. US Naval observatory scientist, James W. Christy was examining these plates to refine the orbit of Pluto when he discovered Pluto's companion, Charon. Recently these objects have been inside the orbit of Neptune, this was so since 28 November 1978 and they passed outside again 12 May 2000.

There are some important anomalies that deny the status of planet to Pluto. The first is that Pluto forms too great an angle with the ecliptic, 17.17 degrees. The other planets are all within 3.5 degrees of each other. Pluto has crossed the orbit of Neptune and this observation defies all previous estimates for its path. It is generally assumed that all orbits around the Sun are conic sections. This assumption allows orbit determination with knowing only the distance and velocity of the object assumed in orbit. Without these orbit assumptions the method fails. This method was used on Pluto by assuming that Pluto's trajectory was an orbit around the Sun. If we assume that an orbit for Pluto is false, then we need to go back to the drawing board to find out what Pluto is really doing. To find out what Pluto is really doing we have to do it the hard way, careful observations and a laborious search for its locus. We do not know how many times Pluto has gone around the Sun. We do not even know if it has ever gone around the Sun. With sufficient initial conditions it may be just grazing the solar system.

Following the laws of buoyancy, it seems that Pluto and Charon came from outside the solar system. If Pluto and Charon are not just grazing the solar system and are not captured by one of the larger planets, they must eventually traverse the orbits of all the outer planets and the inner planets. The Pluto-Charon system does not have sufficient buoyancy to sustain its present distance from the Sun.

Either the Pluto-Charon system comes from a non-circular highly eccentric path about the Sun or our solar system has collided with a stray celestial object. In either case, there is a possibility that it will reach deeper into the solar system and it is large enough to wreak havoc in the regularities of the rock planets. It is the largest comet-like structure that has ever moved toward our human territory under our observation.

The orbit eccentricity of the Pluto-Charon system is (1981) calculated as 0.250, the largest eccentricity of the so-called planets. It may be that this calculation is falsified by assuming its trajectory is an orbit. Mercury is second worst with 0.206, in this case due to the required path of least impedance in proximity to the Sun, it follows the closest to the solar equatorial plane. Venus and Neptune are the least eccentric, 0.007 and 0.009.

That Charon and Pluto are paired is ambiguous for interpretation. Charon is either in an accreting spiral or they are both above boundary convergence. That they are so close to each other suggests that if they are in buoyance to each other that Charon is not much above boundary convergence. It may also suggest that they recently broke apart.

It is feasible that a celestial object exceeding boundary convergence has an extraneous influence (initial condition) that sends it toward the Sun such that it plunges below the reversal point so far that Sun convergence exceeds the object convergence. The bubble of convergence may be compressed but it is not readily squelched—the object would bob up or else its distorted CIS breaks it up into pieces that may be small enough to get squelched.

If Pluto and Charon ever get closer to the Sun we can expect sublimation and melt-down for both of them. Whether it is sufficient to prevent a danger to our biosphere we have yet to determine. They may be just grazing the solar system.

CRITICAL TO NEWTONIAN GRAVITATION

1. Earth-Moon Tides. The Moon tidal effect on the Earth is about three degrees in the front of a direct Earth-Moon line. This was not known in Newton's time when it was thought that the Moon's tidal effect was directly under the Moon. To make a more accurate fit to the Newton system, the current *ad hoc* assumption is that the Moon draws the tides and the Earth rotation pushes them forward, a manner of making the cart pull the horse. This curious sequence is not required in Convergence Buoyancy because it is expected that the overhead Moon has two major synchronous effects, one is the relational bulk distribution that elongates the CIS in both bodies and induces tides, the other is the obverse surface pressure on both bodies. With respect to the Earth surface, the Moon pressure pushes the tides rather than pulls the tides. Therefore, we expect the tides to be slightly ahead of the Moon overhead position. Similar arguments can be used for the gravitational Sun tides. However, the Sun maelstrom tides, such as the difference in height of the Pacific and Atlantic Oceans at the Panama isthmus, are driven by the maelstrom pressures from solar rotation. And note that no atmospheric jet-stream moves from east to west. They are likely solar maelstrom driven.

2. Moon and Biorhythms. The influence of the human body on its own accretion weight on the Earth surface is only after the 22nd decimal place and this is many orders less than the fifth decimal place influence of the Moon. This implies that in the three-body problem of Moon, Earth, and human on Earth surface, that the Moon has a measurable tidal effect on the Earth measure of weights (likely a factor for the no precision agreement on big-G measures).

Critical Experiment: Using weighing devices accurate to at least six decimal places (cgs) measure the weight of the same object following the Moon cycle (29 days).

In the Newtonian system the overhead Moon implies that the human will weigh less because of the Moon's attraction. In the CB framework the overhead Moon implies that the human will be under a higher gravitational pressure and metabolism will be higher rendering sleep more difficult, because of this increased convergence pressure. Following CB Law III, isostasy produces the same effects on the opposite side of the Earth. With Full Moon overhead, or with New Moon on the opposite side of the Earth, people do not sleep as well (research by author 1987 Heidelberg).

The effect of the Sun on the Earth gravitational pressure is also many orders greater than our own invariant weight (*ex parte* weight). This is still an open area for research that we associate with biorhythms. One experiment conclusion from sleeping in caves is that in the cave the human body follows the lunar day (25 hrs) rather than the solar day (24 hrs). Many blind people also show a Moon day rhythm. Today there are a number of well-documented associations between human metabolism and lunar cycles without understanding how the associations actually work. CB provides a mechanism for it.

3. Moon Anomalies. Dr. Robert Jastrow, first chairperson of the NASA Lunar Exploration Committee, prior to the Moon landings said "The Moon is the Rosetta Stone of the planets." For conventional science, this has proven to be a false statement. The six Moon landings retrieved data that conflicted with most expectations. The conflicting data have been classed as anomalies. That is to say they do not seem to follow any known laws. Jastrow's statement seems prophetic for CB. The anomalies are as follows:

There is hardly any iron on the Moon. This and other mineral differences suggest that the Moon was *not* formed near the Earth. If the Moon was not formed near the Earth then we are faced with the problem of how it could get so exactly placed as a satellite to the Earth. Newtonian analysis yields a very low probability for this occurrence.

Response: CB allows a high probability that the rocky planets could be orphans picked up by the solar system. Similarly, the Earth could pick up the Moon.

The Moon is dated older than expected. In fact it dates over a billion years older than the estimates for the age of the Earth. The oldest Earth rock is dated 3.5 billion years old. The same dating method dates many Moon rocks at 4.5 billion years old. One Moon rock was dated 5.3 billion years old. The dust the rocks were found in was dated one billion years older than the rocks. This makes the Moon dust older than the current estimates for the age of the solar system (whether we are dating Moon rocks as if from some fixed formation time or debris that sank to the Moon surface more recently is an additional decision problem).

Response: If periodic rates are slower on the Moon (as CB claims), there is a possibility that the age of the rocks and dust are even older than the Earth based calculations indicate. Thus CB theory is more consistent with the Earth adoption of an orphan Moon. The Moon may be even much older since the lower convergence power likely directly slows the rate of radioactive decay.

There is no indication that the Moon was ever hot enough to produce volcanic eruptions and yet the maria are seas of solidified molten rock. Almost 4/5ths of all the maria face the Earth. The opposite side of the Moon has many more craters and mountains.

Response: This structured relationship of craters and maria in opposition is consistent with CB Law III as an isostatic reaction modified in its transmission. Newton and CB are both consistent in principle for surface-to-surface impacts. However, CB also allows for splooch cratering. We would next need to decide between surface-to-surface impact or splooch craters. This requires a search for extralunestrial debris, etc.

30 to 60 kilometers below the center of the maria are mascons. These are large, dense, circular bulks placed like bulls-eyes in each of the maria. The maria surface is very hard. Astronauts, using specially designed drills, could only penetrate a few centimeters.

Response: These circular bulks likely plug the blowholes that released the magma that formed the maria. The blow-holes are likely isostatic reactions to the pressure from a possible splooching process or the surface to surface collisions on the opposite side of the Moon. This deep source for the maria may also account for the hardness of the surfaced material. We found no volcanic cones associated with these mascons.

The interior of the Moon was found lighter than expected. It was expected that the Moon structure studies would reveal that the interior of the Moon would have heavier rock and mineral material. It is just the opposite. Newtonian measurements indicate that the heaviest material is on the surface and so heavy that it is calculated that the interior is either hollow or of very light material. The Moon surface mean density is 3.34 g/cm^3 (Newtonian). The Earth surface mean density is about 2.8 g/cm^3 (Newtonian), which corresponds to about 5.6 g/cm^3 (CB mean compactness).

Response: CB theory suggests that all celestial objects with a well-formed CIS have a high probability of being hollow. However, there may be some increased compactness due to slower periodicity rates, which in turn affects corpuscular vibration levels. The sunlight, virtually without atmospheric filtering, would render a high surface material agitation but the Moon's low convergence power, **E**, would sustain only a low affect on periodicity rates. It would be interesting to see if Moon rocks that have been on Earth for more than thirty years show any measurable change in compactness.

The Moon seems to lack a dipole magnetic field yet the Moon rocks were magnetized. Their magnetism is weak but measurable.

Response: Some manifestation of magnetism is expected from every celestial object. Why it has not consolidated into a holistic lunar dipole field is not yet known, but even the Earth is said to have a cyclic change of its polarity. If polarity can change then it may be that the Moon is in an in-between stage. It may also relate to the fact that relative to the Earth, the Moon does not rotate. This lack of rotation may also allow the magnetic poles to form however weak at a placement different from expectation.

4. shuttle-craft Flight. The shuttle-craft flights do not produce true microconvergence environments. They are more accurately described as mechanically neutralized buoyancy environments. In Newtonian terms the shuttle flights are in 'free-fall'. This neutralized buoyancy state is not natural, it is sustained by producing measured constraints on the Earth convergence effects. The natural movement of the shuttle-craft is to sink to the Earth surface. For longterm stays the constraints must be applied to the shuttle-craft under the control of its own engines to maintain its trajectory. *The Earth convergence polarity is still well defined at these near Earth trajectories.* It is a misconception to say that the shuttle-craft can orbit the Earth. They all follow accretion trajectories. In CB, accretion trajectories are not orbits.

When an airplane takes a dive at a fixed, well-defined rate of change and the passengers experience neutral buoyancy, we do not say that the airplane is in orbit. Neither does a roller coaster take us into moments of orbit. Neither does jumping out of an airplane take us into orbit even though we approximate neutral buoyancy in the early part of the 'fall', etc.

However, in reference to convergence, experiments can be done to separate phenomenal effects into those that are sensitive to polarity without 'weight' (neutralized buoyancy) and those that require weight polarity, e.g., plants orient according to weight. At first, neutral buoyancy experiments seemed to indicate that vertebrate Calcium metabolism required gravitation for proper functioning. The proper metabolism of Calcium depends in part on vitamin D_3. NASA people have tested vitamin D_2 (vegetarian D), which they took as equivalent, but this is established under the pressure of normal surface gravitation and they ignored work done with the Eskimos that denies equivalence. It should *not* be regarded as evident that the two vitamin forms have the same power of action in neutral buoyancy or 'microgravitation' neighborhoods (micro-convergence effects). Not even liquid water can have the same molecular clustering in neutralized buoyancy. This may affect assimilation of water soluble nutrients (the D vitamins are oil soluble and store in the liver).

> **Note:** Polar bear hair works like optic fibers conducting light (including ultraviolet) to the skin level while at the same time maintaining insulation against heat loss, a fact Eric Laithewaite (1921—1997) would like to have known. Polar bear liver has such a high quantity of vitamin D_3 that humans can OD on it with one serving.

Note that three months in neutral buoyancy environments is sufficient to account for a 5% bone loss. The longest stays in space were almost one year and 8% bone loss was recorded. The situation is likely much worse in microconvergence zones where photon-graviton polarity is too fine to distinguish (nearly parallel lines), etc.

Metabolism rates are proportional to gravitational convergence pressure. We have yet to determine all the effects of a slower metabolism in microconvergence or neutralized buoyancy, but coincidently there are correspondences with the degenerative effects we associate with aging. Astronauts in microconvergence or neutralized buoyancy and aging people both suffer bone loss. The major common factor is the slower metabolism. *It is likely the slower metabolism that inhibits the homeostasis of bone structures by diminishing the proper metabolism of calcium.*

In order to allow long stays for astronauts in microconvergence or neutralized buoyancy environments, we must find a means for them to sustain their metabolic rates above the minimum threshold required for the proper metabolism of calcium to sustain bone homeostasis. It is also true that every thought and movement depends on calcium and therefore on the proper metabolism of calcium.

Should we discover an Earth-like planet with a higher gravitational convergence we can anticipate that life forms would be smaller. The increased convergence raises agitation levels and metabolism rates would increase. Within certain limits, smallness facilitates a tolerable level of heat dampening with respect to a human body under higher than normal gravitational convergence. On Earth, children require a higher metabolic rate to maintain temperatures equivalent to those of an adult. This suggests that it could be more difficult for them to adapt to microconvergence environments. Their metabolism might have an overtaxing decline.

People age at different rates and, in general, physical aging on Mars would likely be chronologically slower, etc. Bone loss on Mars would likely occur at an earlier physical age (as distinct from chronological age). The children of the first human denizens of Mars will likely grow larger and have a slower metabolism. Coming back to Earth they would feel uncomfortably warm, weak, and shuffle walk.

Observation: One of the temporary changes in neutral buoyancy is an increase in heart rate. This is induced by the sudden release from the convergence pressure. John Glenn, at age 77, averaged 95 heart beats per minute on his last trip. Average heart-rate for the rest of the crew in the early exposure to neutralized buoyancy was 115 beats per minute. These rates are supposedly not related to CO_2 levels. Further data are required to know the adaptation rate and where it levels off. On the Earth surface the heart rates are used for estimating metabolic rates since heart beat rates adjust to CO_2 levels in the blood.

CB Hypothesis: *As the heart rate adapts to the release from convergence pressure, the metabolism slows down below the threshold required for bone homeostasis.*

Old persons and astronauts can suffer from short term memory loss and disorientation. A dysfunctional Calcium metabolism likely underpins these problems. In general, the increase of metabolic rate through exercise is short-lived, especially in microgravity.

What astronauts need are drugs that accelerate the metabolism without wreaking havoc on the body. Until then, astronauts and old people might jury-rig a bone-loss slow-down by imbibing calcium and vitamin D_3 just before raising the metabolism through exercises.

5. Experiments in Neutral Buoyancy. We need a test that determines the effects of spin convergence on the surface of a spinning object.

The following experiments are possible.

Plants orient themselves according to the polarity of convergence. For most the roots grow toward g-convergence and the leaves grow away from this convergence.

The plant growth follows the shape of the convergence neighborhood it participates in. This result has already been seen in neutral buoyancy environments, but what occurred was not understood.

> **Note:** In recent years tomatoes have been grown at what appears as upside down. The roots are in a hanging container and the tomato branches grow hanging out from the bottom. On closer examination the hidden roots actually turn down and the leaves turn up. It is interesting to compare this with growing potatoes where the tubers are above and the potato branches and leaves hang down. The tomato and the potato are biologically closely related. Potatoes in hanging water pots grow branches and leaves and the branches hang down, whereas the leaves turn up, etc.

In neutralized buoyancy (e.g., space-shuttle when its sink acceleration is equal to Earth E at a fixed altitude), plants were spun at the edge of wheels with the intent of simulating acceleration due to gravity. The expectation was that the plant stems and leaves would grow toward the center and the roots would grow outward. This supposition is based on the mass equivalence principle, used tacitly by Newton, and expressly stated by the Einsteins (see Chapter 1).

In fact, the roots grew toward the center and the leaves grew away from the center, as if the center were like a gravitation sink. This is the reverse of expectation for Newtonian physics. It is precisely what is expected for CB physics.

Make a wheel that can hold plants on its rim. In any convergence neighborhood, the state of no-spin allows the plant to follow the neighborhood convergence.

Spin the wheel. If the wheel is already squelched in a strong neighborhood then the leaves grow toward the center and the roots grow away from the center and this seems to confirm the Newtonian equivalence principle (gravitational mass, inertial mass).

If the wheel is in neutralized buoyancy, then with no-spin the plants display a confused structure. The roots and leaves show no particular orientation.

If we spin the wheel in neutralized buoyancy, then the roots grow toward the center and the leaves grow outward and the mass equivalence principle seems violated.

Some NASA scientists suggested that the roots were attracted by water, but this is as logical as asserting that the stem and leaves were repelled by water. It also leaves us to wonder why no-spin in neutralized buoyancy results in a confused structure in the presence of water.

6. Centrifuge in Space. The CB expectation is that the centrifuge does not work as well in a shuttle-craft convergence neutralizing trajectory as it does on the Earth surface. With even greater distances from a convergence center, where we have microconvergence (near parallel convergence lines, etc.), it must fail Newtonian expectations altogether. In a microconvergence neighborhood, convergence polarity is very weak (too weak to measure). On a wheel space-station, with spin convergence greater than the neighborhood microconvergence, a human on the inner rim would be launched toward the center rather than feeling weighted against the rim.

A CB analysis advocates that the neighborhood convergence polarity provides a physical reference-frame for indicating the change in direction to the object-and-its-field. Therefore, when neighborhood convergence goes below the convergence magnitude of a debris object, the polarity of the neighborhood only tenuously indicates the change of direction. In the shuttle-craft, the neutralized buoyancy 'weight' is higher than the *ex parte* weight but still millions of times smaller than its Earth surface weight.

In the shuttle-craft, under neutralized buoyancy, astronauts standing in the shuttle-craft bay can move multi-ton satellites with the push of one finger. The multi-ton satellites are still small objects (debris) and their bulk relative to the Earth is represented as a fraction so small that the measure is beyond the precision of present instruments, but one finger can move them. This is observed. Using Newtonian physics, scientists highly over estimated the efforts that would be required for manipulating these satellites, i.e., they did not anticipate what in fact occurred.

Objects 'freed' from neighborhood convergence can turn at instant right angles or instant reverses, with no detectable effects. When this is observed it exhibits the full dereference for an inertia-free physics. If the space-station wheel is close enough to the Earth, then it will 'work' at least fractionally (as in the MIR spacelab), but in a large scale deep-space wheel, the spinning must fail altogether to exhibit Newtonian centrifugal force. Centrifugal force is an inertial reaction and there is no inertia. Without a convergence neighborhood no changes in velocity can be conveyed to a moving object.

Thought Experiment. Consider the possible phases in field deformations of a spinning plastic system sinking into a stronger convergence reference-frame. A spinning plastic system starts in a microconvergence and is sinking toward the Earth surface; the field deformation passes from near sphere to oblate sphere to flat to negative sphere. In terms of convergence fields this goes from Riemann (positive curve) to Euclid (flat) to Lobachevski (negative curve) shaped fields. As long as the system convergence is greater than that of its neighborhood its field will tend to be convex. Surface tension also contributes to convexity. When its convergence power is equal to its immediate neighborhood it will tend to be flat. When it is less than the neighborhood it will tend to be concave. Convex, flat, and concave are relative to the neighborhood field shape, e.g., a radially flat line in a curved field follows the curve of the field (the 'geodesic').

These field shapes help predict what will happen to plants on a constraint driven spinning wheel rim. In the Riemannian field (the sphere is a special case of Riemann space shapes) the roots grow toward the center and the stems and leaves away from the center. In the Euclidean field the directions are confused. In a Lobachevskian field the stem and leaves grow toward the center and the roots away from the center.

As mentioned before, one of the interesting consequences of this spin convergence effect is that the wheel conception for a space-station cannot work the way it was originally intended. Even though the kinematics motion implies acceleration toward the center as a push from the revolving floor, experiment will likely show that this acceleration, rather than effecting an inertia that holds us to the floor, launches us toward the center. Our invariant (*ex parte*) weights are too small. It is this effect that suggests why the spinning roots grow toward the center in neutral buoyancy.

Experiment. In a wheel with radial tubes containing a mixture of water and oil in contrasting colors, spin at various rates in the three field shapes that are modeled after {Riemann, Euclid, Lobachevski} spaces. In a Lobachevski field (host surface, or strong polarity) the water should move out toward the rim and the oil should move toward the hub. In a Euclidean field the two should form a confusion mixture. In Riemann field (microconvergence) the water should move toward the hub and the oil should move toward the rim. With respect to a well-defined polarity, the water goes toward the rim and the oil is displaced toward the hub. In near zero polarity power, any water at the rim will be launched toward the hub, the transition should finally leave the oil toward the rim and the water toward the hub.

7. Atomic Periodicities. Is the effect of convergence on atomic periodicities, directly or inversely proportional?

Newton Theory. This is a non-Newtonian consideration. Using the Einstein theory, in which the Einsteins claim includes Newton physics as a special case, they state that the relation is *inverse*; the stronger the gravitation the slower the periodicities. With this, the Einsteins claim there would be a corresponding redshift to prove it.

CB Theory. Asserts from observations that periodicity rates of corpuscular vibrations vary directly with convergence pressure, the stronger the convergence, the higher the periodicity rates (within the limits of physical integrity).

Material Implications: Dating Moon rocks (long-term age) cannot be done without a transformation system. If the Earth and the Moon were the same age then the surface of the Earth should appear "older" than the surface of the Moon since rates of change are faster on the Earth surface. This seems not the case. Under present modes of analysis, all evidence available from the Moon indicates that the Moon is older than the Earth. This also is consistent with the CB conclusion that the Moon is an adopted orphan.

Even though rates of periodicity are faster on the Earth than on the Moon, we do not know what effect the higher convergence pressure may have on the half-life of radioactive decay. The results of surface analysis are suspect for two main reasons:

1. The non-terminating accretion of debris from space. (How do we determine the origin of a Moon rock?)
2. The half-life of radioactive materials is likely different on Earth from that on the Moon.

CB predicts that higher convergence power accelerates the half-life decay rate, and this may be why Plutonium in a natural state is not found on the Earth surface, whereas it is found in a natural state on the Moon. This suggests that for heavy elements that the breakdown may be significantly slower on the Moon's surface than on the Earth's surface. Even so, Moon rocks appear older than Earth rocks and this leaves us in a contradiction from previous beliefs. Part of the solution to this may come from a more careful sorting out of whether the Moon rocks are long term residents or new arrivals.

On Earth we have breaching stones that test half a billion years younger than long-term surface stones. When we go below the surface of the Earth we know that according to CB the tension between the bulk above and bulk below this depth lowers the convergence pressure, and concomitantly a lowering of convergence power. This could allow for a lengthening of half-life of radioactive materials at these interior depths. The affect is that these interior materials appear to be younger since the half-life is longer.

NASA suggests that cosmic ray bombardments may account for the presence of Plutonium on the Moon.

We need more information. We need experiments.

Periodicity Experiment: Place an atomic clock on the Moon surface. Atomic clocks have been miniaturized (2004), which renders this experiment more feasible. It is likely that the Moon surface clock will function more slowly. Again, this may be why we find Plutonium in its natural state on the Moon surface. Since time is measured by periodicity, for CB it can only be uniform in a constant photon-graviton pressure zone.

Even so, it has been reported that GIS clocks ignore periodicity problems implied by theoretical sources (the Einsteins, et. al.) and so far have found no such problems.

Furthermore, what we might expect is that the spectrum of periodicities for a given pressure zone relative to what they would be in another pressure zone, do not vary in a Eudoxian manner. This can be surmised from the fact that the number of dimensions is greater than two. If we had two distinct periodicity rates in one pressure zone, their ratio to each other likely changes in a different pressure zone. This follows since the volume of their periodicities vary as the cube, etc.

Whether our measure of time, in fact, is uniform or non-uniform may be undecidable, but this measure of time requires an operational methodology for quantifying time in its phenomenal representations (comparison of periodicity rates or counting of periods).

Note again that periodicities on the Sun are thought to generate the visible light, x-rays (which may relate to fissioning of heavier elements), gamma rays. Earth periodicities are mostly in the infrared region. In this comparison there is no subtlety. The Einsteins are wrong. *Stronger convergence power increases periodicity rates.*

Critical Test. Biological implications: One of the problems observed in near zero weight environments is the slowing down of metabolism. An astronaut, in space, experiences a slowing of metabolism. This is consistent with the supposition that periodicities slow down with distance from a center of convergence. This slowing of metabolism would tend to make time appear to the astronaut as passing more quickly. In older people there is a slowing of metabolism which gives them a similar sense of time passing more quickly. Thus for human experience, the dilation of metabolic rate makes time appear contracted, though it allows them to live longer. Even so, an immediate effect for human exposure to neutral buoyancy is a faster heartbeat. This is not from a faster metabolism, it is from the sudden reduction of convergence pressure. Over time this faster rate is not sustained if the vehicle atmosphere can be kept in normal proportions. Too much carbon dioxide can raise the heartbeat and give the illusion of a higher metabolism. We know that too much carbon dioxide can be deadly even if there is a normal level of oxygen. There would be irony if controlling the carbon dioxide in artificial biospheres would be adequate to the requirements of proper calcium metabolism, but likely it is not.

Reasoning by analogy it may seem that going toward microconvergence would release the pressure on periodicities and therefore lower resistance to the periodicity rates. This interpretation fails to signal the effect of pressure on dimensions. Under pressure the material dimensions are more compressed and the periodic rates have shorter distances. The effect is likely to produce more periods per unit time and therefore faster periodicities.

Time cycles vary inversely with their altitude from a convergence center, the higher the altitude, the slower the cycles. The atomic change of rate is likely small but is rendered more visible with long-term observation.

8. Weather Effects. There is a long tradition that the 19-year Moon cycle coordinates with Earth weather conditions. This belief is the basis for almanac long-term weather prediction. The following conjectures suggest another possible test for deciding between the Newtonian and CB reference-frames: With respect to the Earth-Moon relation to Earth weather, there are two major view-points of interest, (1) with respect to the Sun surface, (2) with respect to the Earth surface.

(1) With respect to the Sun surface, at Quarter Moons when the Earth-Moon line is parallel to a tangent on the Sun surface the Earth-Moon fulcrum moves inward toward the Sun (the Earth is rendered closer to the Sun). At the Quarter Moons this fulcrum is at its nearest position to the Sun for a given cycle. The movement toward the Sun raises the surface temperature on the Earth and the Moon. As the Earth-Moon line moves toward a perpendicular to the Sun surface the Earth-Moon fulcrum moves away from the Sun. The initial movement from the parallel to perpendicular position augments the pressure between the Earth and Moon. Therefore, on the average we should expect the Quarter Moons to be synchronic with a lower gravitational convergence pressure. Though a nearing to the Sun implies a raise in temperature and heat, the lower pressure suggests a more efficient dampening. Barring local noise, quarter Moons tend to be the warmer times of the month. Full and New Moons are synchronic with a following drop in temperature and heat. The coldest time of the month should be near New Moon. The Earth is furthest from the Sun for the given cycle and as it moves from New Moon toward Quarter Moon the convergence pressure between the Earth and Moon diminishes. Thus the New Moon associates with both heat dampening and a lowering of temperature. The most significant temperature drops would occur on the average just after Full and after New Moon. Because of the way that the Earth-Moon fulcrum varies in its distance from the Sun for isodynamic movement, there is also a dampening effect on the temperature changes, making the changes less extreme.

(2) With respect to the Earth surface, overhead Moon (3 degree lag) marks the highest pressure, horizon Moon (perpendicular to overhead) marks the least pressure. Furthermore, overhead Full Moon and overhead New Moon associate with maximum pressure. This maximum is achieved through two components; one the alignment with the Sun, two the overhead position. Overhead Quarter Moons mark less pressure than overhead Full or New Moon. Quarter Moons at the horizon mark the least pressure with respect to an observer on the Earth surface. These various pressures have a direct influence on the temperature and radiation gradients. One of the main difficulties for measures of these characteristics is that the required accuracy may be embedded in the surface environmental noise. The prediction is that higher pressures raise the heat, which in turn may support higher temperatures. Also, when the Earth-Moon alignment is parallel to a Sun surface tangent, the Earth-Moon fulcrum is closest to the Sun and the Earth and Moon are under the least pressure. Similar considerations can account for the variations we find in the tides.

There is nothing in Newton that directly predicts this state of weather dependency. The derivation of long-term weather cycles by the Serbian astrophysicist Milutin Milankovitch (1879—1958), from variations in the Earth orbit shows agreement with other modes of climate change investigation.

On a shorter term consideration, for the Earth northern Winters all the planets were in one of their close alignments and remained in this close opposition to the Earth for at least two consecutive Winters (1997-1999). The Earth was alone on one side of the Sun with every major convergence object in its locale in opposition to it. This solar system configuration repels the Earth further away from the Sun than any other configuration. It can be regarded as a kind of isodynamic reaction of the Earth to its neighborhood to maintain isostasy. In December 1997 Venus and Mercury came up 'behind' the Earth and add to the repelling in addition to the isodynamic push out. These events may have kept the Earth sufficiently more distant from the Sun and the Northern Hemisphere temperatures went down unusually low for December/January. The Northern Hemisphere was the hardest hit with a severe winter and a hot summer. The Southern Hemisphere had a cool summer and a mild winter. One of the peculiar side effects is that there is less restraint on Earth radiation. This implies that there is an increase in potential difference, an increased Earth contribution toward raising the surface temperature of the Earth. This suggests the source of an increased El Niño effect since water is highly compliant to these changes. The El Niño effect can render the Winters uneven.

CB Entropy. The usual interpretation for the measure of entropy in a system is the loss of useable energy in that system. For CB stellar evolution, there are periods of growth and periods of destruction. The growth period for material objects is usually many times longer than the destruction period. The growth period is the period of accretion, the period when accretion dominates in the formation of material systems. This period is anti-entropic until through accretion a material object obtains to a size and pressure that drives the CIS to an energy level that can no longer sustain structural integrity. The object goes to nova, the dynamic process that destroys the anti-entropic homeostasis.

Homeostasis can take on an immanently reversible interpretation in that the presence of debris contributes to a change in the isodynamic state. In this sense it is a form of system entropy in that the anti-entropic driver energy can no longer be used for maintaining the original isodynamic state. On the other hand, the isodynamic restraints drive the system through new isodynamic states and this building of new states may again be characterized as anti-entropic.

This dual conceptualization is quantifiable in terms of the accretions and their effects. For example, the Earth accretes measurably over 200 metric tons of debris every day (it is harder to estimate the quantities of cosmic dust, etc.). The Moon is also accreting material. The Earth and the Moon are separating close to three centimeters per year, etc. This can be used for measuring the entropy relative to previous states, while at the same time it measures the anti-entropy of the new states. This is what is indicated by an 'immanently reversible' interpretation.

A constraint is a disturbance to isodynamic movement. The reaction to the constraint is a regression pressure driven by the neighborhood restraints. Here it is important to note that all regression is toward an isodynamic state though not necessarily toward the original isodynamic state. If it regresses toward the original isodynamic state we say that this regression is anti-entropic. If it regresses toward a new isodynamic state, then relative to the original state it is entropic but relative to the new state it is anti-entropic.

In this sense entropy only has a limited dereference for indicating a difference from the original state whatever that difference might be. Even though there may be a great difference, the isodynamic processes prevail. After a nova the debris may condense into a star, though the new star may have differences from the star that lost its physical integrity.

Homeostasis is when we apply gradual accretion to an isodynamic system. It allows for a kind of growing change to a system that otherwise would be purely isodynamic. The isodynamic aspect is restraint driven and restraints are anti-entropic. CB entropy can only occur within a limited scope of time and location within the evolutionary cycles of stars. Otherwise entropy is always confined to smaller-scale limited and local occurrences, such as breaking a glass, burning wood, dying, or some other Lilliputian scale loss of organization (that loses usable energy for a given system or a capacity to do work for that system).

CHAOS THEORY. The Pythagoreans divided the World into Cosmos and Chaos. Cosmos is Apollonian and follows laws. Chaos is Dionysian and follows no laws. Today we associate Chaos with randomness and we search for methods that put order into what at first appears as unintelligible randomness. Probability and statistics are brought into the description. It seems that human thought has limits to the complexity it can handle and shifts methods according to the complexity of the problems. One of Schrödinger's favorite examples for order out of Chaos is to take a divided tank, one side with purple dyed water, the other side with no dye. He states that if the molecular movement is random, then it is certain that when the divider is removed that the entire quantity of water will become purple. Of course he is wrong in this manner of statement. It is more accurate to say that if the molecular movement is random, then given enough time, every possible distribution may occur. Note that I say 'may' rather than 'will' because, for the given conditions, no particular distribution is a necessary one. Of course, most molecular distributions would show a purple look and experiments would likely seem to confirm his conclusion. However, the likelihood of perfect homogeneity is just as rare as the original separation. Furthermore, we could emphasize the distribution of the dilution rather than the purple, etc.

One of the newer subject appellations for approaches to data modeling is Chaos Theory. It is still in the exploratory stage. Its architectonic is not yet fully defined. Whether it remains as a separate species of subject matter or melds into modes for statistical analysis remains to be seen. It started out as offering new data distribution modes that reveal previously unseen recursions. This area of Chaos Theory easily melds into advanced statistical methods for data reduction and leads to further exploration of how the recursions generate. In hydrodynamics, for example, there has been some success in modeling transition zones between smooth flowing and turbulence, etc.

Another recursion model is fractal geometry. Mandelbrot's work represents very complex appearances by very simple formulas. His methods are undergoing continual extensions in Chaos Theory. One of its attempted applications is in data compaction that reduces the storage space used in computer memory. However, this is at the expense of computation time to produce it.

Chaos Theory techniques have proven useful for modeling certain modes of complexity. The predominant activity is to find simpler representations for apparently complex phenomena, a motive that underpins the construction of any science-system. Whether it is a new paradigm or an extension of some older ones remains to be seen.

BUOYANCY CATEGORY SUMMARIES

In isodynamic relations we consider three state types {host-host, debris-host, debris-debris}.

- **Host-Host (Float).** For the convergence of two host objects, by definition they converge to float buoyancy (buoyance). In this isodynamic state there is always a distance between the two objects. From the geometry, the calculation places the Convergence Reversal Node (CRN) exterior to and between the two surfaces. Their revolution around each other sustains an isodynamic convergence that correlates to the neighborhood pressure.

- **Debris-Host (Accretion).** For the convergence of debris to a host, by definition they converge to sink buoyancy (accretion). In this isodynamic state the distance between the two objects is regarded as nil. It may seem that the CRN must be at their tangent or interior to one of the two objects. The actual CRN is under the host surface.
- **Debris-Debris**. If two debris accrete to one another we can assume that there are effects on the composite convergence power that vitiate the generated repelling, e.g., the neighborhood pressure can drive debris toward compaction. It may be that a neighborhood convergence passing between two debris objects could also draw them toward each other with a kind of Bernoulli effect. The CIS in smaller debris objects is too weak to produce the opacity that amplifies the repelling effect of host objects.

SINK BUOYANCY (Principle of Accretion). We distinguish the variable CRN that can occur in non-isodynamic states and the *isodynamic* CRN. Prior to an isodynamic state, as two gravitational convergence objects approach each other their CRN varies in its position for two converging objects, using their surface convergence, **e** for debris, **E** for hosts, we calculate from their respective surfaces to find a position where their mutual convergences exhibit equal convergence power.

When we calculate the distances between debris moving toward accretion and a target host, we find that the target host **E** effect is greater than the debris **e** at every position between them. This is a definatory characteristic of a squelching in process. This state has no Convergence Reversal

Zone (CRZ) between the debris and the target host surfaces. It also establishes a demarcation rule to decide if an object spiraling around a host is debris or another host. For example, if we could determine the convergence power of Phobos we would likely find that Phobos is already at an altitude from Mars (currently about 6,000 km) where the Martian **E** effect on the Phobos surface is greater than that of the Phobos **e** (Phobos, $e < 1$ ms^{-2}). This would clearly establish Phobos as debris that will accrete to the surface of Mars and we expect that the CRN is below the Martian surface. Following Newtonian reasoning, Phobos is accreting because it orbits Mars below the synchronous orbit radius. For CB, Phobos is debris, etc. Deimos, both smaller and further out, is also debris.

Several hundred metric tons of objects sink to the Earth every day. About 200 tons per day are visible pieces of stone or metal. These objects sinking to the Earth surface are 'small' objects. Even 20 km in diameter is rather small next to the Earth size. An inventory of objects that are known to have sunk to the Earth's surface shows no direct evidence of anything larger than 6 kilometers average diameter actually striking in one place.

There also seems to be a close statistical correspondence of massive extinctions with the accretion of larger debris.

If an object near 20 kilometers in diameter went into accretion course with the Earth, it would likely resemble the course of Phobos. Phobos is 22.2 km (27 x 21.6 x 18.8) in diameter and about 6000 km above the Martian surface spiraling down in a collision course. As with Phobos sinking to the surface of Mars it would take millions of years of spiraling to finally strike the Earth. It would not be a straight down plunge as popular science magazines are wont to show. The repelling reaction renders improbable such a direct plunge of objects this large. Other initial conditions must act for a direct plunge to take place and the setting for such conditions is difficult to obtain. Furthermore, with objects large enough to be stressed by tidal effects (stress associated with CIS distortion and a breakdown of surface tension) after approaching to a distance that distorts the object to a critical shape, the sinking object breaks into pieces too small relative to the milieu for further destructive tidal effects in these pieces. The tidal effects become too small for further breakdown of surface tension. The smaller objects fall faster to the Earth than the original object. We can expect that when the surface tension of Phobos is weakened by sufficient proximity to the Martian surface that Phobos will likely break into smaller pieces (assuming it is stone).

Considering Phobos and Deimos, it seems that objects in our solar neighborhood with a near 22 kilometer diameter may border a rough limit for accretion bulk with respect to host object convergence. It may be that objects larger than 22 kilometers diameter are sufficiently large to define a (CRZ) that is positioned sufficiently above its surface to allow bobbing back into its isodynamic position after being subjected to some aberrational influence to its orbit. Otherwise it is an accretion trajectory.

Of all the meteorites found on Earth the biggest and heaviest are made of iron. The largest recorded intact meteorite is the Hoba meteorite found in Namibia in 1920. It weighed 60,000 kg. The largest recorded stony meteorite is the Jilin meteorite found in China in 1976. It weighed 1,770 kg. This suggests that iron meteors entering the Earth atmosphere are more resistant to tidal breakdown as well as to impact breakdown. We also expect that if a metallic meteor is affected by magnetism that its sink trajectory will reflect this.

One of the current arguments for explaining why host objects do not sink to the Earth's surface is that objects of such size are exceedingly rare. It seems that the larger an object is, the more rare it is (objects of any size are rare). It is estimated that we can expect about three craters that exceed 10 kilometers in diameter, every million years. But rare is a matter of interpretation. Actually, we have around 13,000 planetoids mapped in the solar system (2004 A.D.); most of them lie between Mars and Jupiter. The largest number of these together forms a flattened torus shape with the outer edge about half way to Jupiter from Mars. Careful examination of this torus reveals harmonic partitions (Kirkwood gaps) similar to those exhibited by Saturn's rings. At least a thousand other planetoids are known to cross the Earth's path and there is speculation now about the probability of one ever crashing onto the Earth's surface. Interest in this question is heightened by the observed accretion of the Shoemaker-Levi 9 comet with Jupiter (1994). Ida and Dactyl are two asteroids in the Sun rings; Dactyl is the smaller and is characterized as 'revolving' around Ida. Neither object is spherical. They are likely spiraling to collision, an accretion under neighborhood pressure if it exceeds their repelling power.

This dampening restraint from neighborhood host convergence explains why no human-made satellite can stay in an orbit. They are too small. All of them are in trajectories and will sink to accretion unless they are sustained by constraints to the neighborhood convergence. Delayed sinking can be obtained by placement in CB Lagrange points.

Near-Earth Asteroid 3753 Cruithne (croo-een-ya). Cruithne, an asteroid about 5 kilometers in diameter, currently seems to be sharing the Earth's orbit. Its trajectory has been studied by a Canadian team led by Paul Wiegert of Queen's University, Canada. It has been proposed that Cruithne follows a horse-shoe 'orbit' running a close parallel to the Earth orbit. Its nearest Earth pass is 0.1 AU, about 15 million kilometers. When it gets near the Earth it reverses its course until it comes near the Earth from the other direction and reverses again. With CB, this complicated trajectory is plausible through the natural repelling generated by convergence proximity. However, we do not yet know if this asteroid is metallic or stony or a combination of the two. We need to know this to estimate influences that might distort the CB relationship. The CB requirements for a host bulk places Cruithne in the debris category (whatever the case). This implies that it is in a trajectory and not in an orbit. Therefore Cruithne must either be overpowered by an increasing proximity to the Earth, which could lead to a massive extinction, or it must continue its descent toward the Sun.

Boundary bulk for an object **A** relative to its host **H** is fixed for a given neighborhood pressure. The **E** power of **H** is a required contributor to that pressure as well as the determinant for the direction along the trajectory (barring other material considerations such as magnetism or electrical charges). These boundary bulk considerations do not preclude the possibility of the CIS rupturing the structure when it approaches too close to a larger object to sustain the CIS integrity limit. Neither does it preclude that surface tension can rupture under the stress of proximity to a convergence neighborhood.

We can assume for isodynamic buoyance between two host objects that the effects of the composite convergence power are equal to the generated repelling (CB Law III). If debris objects accrete to each other or to host objects, then we acknowledge that the generated repelling is simply weaker than their composite convergence power in the neighborhood. For the case of debris accreting to a host, the *isodynamic* CRN is always at a point within the host. The debris can obtain a proximity where the host overpowers the debris. This drives the debris-host accretion.

At first, colliding galaxies seem to contradict these statements. However, if we allow the use of current speculations that the galactic centers are antimatter, the antimatter center could have a strong attraction interaction with the material of another galaxy or with the material centers of an antigalaxy (center material, peripheral antimaterial), etc.

NEUTRAL BUOYANCY (Magnitude of Boundary Bulk). Boundary bulk is the demarcation between debris and host objects. We next explore some questions concerning the clarity of this demarcation.

Question: Is boundary bulk functionally independent of a variety of neighborhoods?

Comment: What is connoted by the question is that if an object exhibits a boundary bulk magnitude that is functionally independent of any particular neighborhood, then the boundary bulk can be associated with a fixed magnitude. The influences that define a bulk as a boundary bulk include the convergence power of its host, the neighborhood pressure, and the material characteristics. Furthermore, the opacity that generates from even a weak CIS can increase the repelling power of the boundary bulk. With respect to the material characteristics we can have influences, such as magnetism or electrical charges that under certain conditions can function as constraints on the gravitational convergence considerations. The boundary bulk is never functionally independent of these factors and therefore varies according to a composite of these influences. Though this prohibits associating a fixed magnitude for boundary bulks it does not preclude a narrow scope for the variation within certain groupings such as the rocky planets.

Question: How do we determine a bulk boundary magnitude?

Comment: As implied by the previous comment, the determination of boundary bulks is complicated by a number of factors. For now we shall avoid the details of this determination and look to finding an estimate among the rocky planets. Phobos may put boundary bulk within the scope of a good guess for the rocky planets. It is the inner satellite that is spiraling slowly to an accretion to Mars. The sink estimates from 3 to 50 million years suggest that the bulk magnitude of Phobos is near the boundary bulk with respect to Mars. However it still collects debris, therefore even though its speed augments toward accretion, its descent acceleration rate can diminish by increments. If Phobos could collect enough debris to pass the boundary bulk point, **k** (Exhibit 2.4), this would compel a regress from Mars. However, 3 to 50 million years is likely not enough time to pass the boundary bulk point. Phobos is rounded but not a sphere and this implies that its CIS has a weaker influence than the agglutination of its material (surface tension). This is the expected state of an object below the boundary bulk.

FLOAT BUOYANCY (Principle of Buoyance). For our solar system, all objects sinking into it are subject to restraints. When a host object revolves around a larger host object, such as the Earth around the Sun, it is impelled by the Sun with three main restraints. The first restraint is that the plane of Earth orbit must pass through the convergence center of the Sun, the direction of greatest convergence power. The second restraint is that the Earth must revolve with the solar maelstrom. The third restraint comes from the variable angular velocity greatest at the Sun equator and least at its poles. All three act at once, therefore their order of presentation is arbitrary.

Because the solar equator region rotates faster than the solar axis, it seems that this would induce less effort toward the poles, but, as on Earth, the solar equator convergence is likely weaker than its polar convergence. This stronger polar convergence power has a stronger repelling effect that would tend to push the planet revolutions toward the solar equatorial plane. Thus though the Sun maelstrom determines the direction of revolution for the Earth, the Earth is impelled toward solar equatorial plane. The closer to the Sun, the closer the orbital tilts are to the solar equatorial plane. The pressure that drives this drift varies inversely for distance from the Sun surface. It can be shown that no *orbiting* body can get far enough from the Sun to have its revolution plane contain the solar axis and still have a solar orbit. It is interesting to note that, with the exception of Mercury, all of the planets are closer to the revolution plane of Jupiter than to the equatorial plane of the Sun.

The planets in isodynamic orbits with orbital tilt have another impelling that is proportional to the tilt from the solar equatorial plane. This impelling drives a rotation of the orbital plane. Pluto is 17.17 degrees off the ecliptic, which for CB with its size and distance from the Sun denies that it is isodynamic with the Sun.

Small objects make open spirals toward the Sun. On the way to the Sun they may be captured by the outer planets. This in part may account for larger sizes in the outer planets; they have the opportunity to accrete spiraling debris first. It should be found that the asteroid belt has many asteroids sinking to accretion toward the Sun, but like Phobos the spiraling inward (sinking) can take millions of years for the ones visible from Earth (2004 AD). From this scenario it is not difficult to imagine that Phobos and Deimos were captured by Mars (in their spiral descent from the asteroid belt, or when Mars passed through the asteroid belt). Mars is now in an ideal position for accretion from the asteroid belt. Furthermore, Mars can function as a shepherd to the asteroid belt and thus slow the increase of asteroids crossing the Earth orbit.

It seems contradictory that Mars can be so small and yet further from the Sun than the Earth. Mars has been characterized as a dead planet and has been thought to be an older planet than the Earth, a planet that died. CB suggests a different interpretation.

Mars is likely newer to the solar system than the Earth, a satellite for the Sun, in one of the best inner planet positions for hosting accretion. New and young are not synonyms here. *A planet new to our solar system could be older than the Sun.* All the rubble spiraling in from the asteroid belt below boundary convergence will pass through the orbit of Mars first. Being close to the asteroid belt, the meteor activity on Mars should be significantly higher than on the Earth. Photographs of the Martian surface seem to confirm that Mars is a collector of rubble. However, Mars also likely has stones breaching through its surface just as the Earth has.

One of the key ideas derived from CB principles is the possibility of fire-stars adopting planets. This mode of forming star systems of planets is distinctly different from Swedenborg's condensation theory, popularized by Kant in his summer courses on cosmology. The condensation theory implies that there should be much more similarity among the planets than is currently observed.

SUMMARY CONCLUSIONS. One of the implications of CB is that the evolution of material systems is remarkably uniform. The probability that a star does not have a complete system similar in its parts to our own solar system is highly diminished. This does not say that there is a life-bearing planet around every star. It declares that every star likely has a system of planets, an asteroid belt, and a collection of comets and meteors (does the Oort cloud exist?) or is evolving toward that structure. It suggests that stars that do not conform to this evolution are quite rare or do not exist. It further suggests that during the life span of a main sequence star like the Sun that it could host at least one life-bearing planet at some time in the course of its history.

It is Venus that is the likely candidate for the next planet to become Earth-like in a natural way in our system. The fact that we are capable of creating colonies on Mars sets the imagination soaring in regard to the possibilities of civilizations in older star systems that may have evolved far beyond us. However, we have yet to show that evolution necessitates producing a civilization.

. At least 90 percent of the main sequence stars have companion stars that have been detected. The Sirius binary system is deemed about 2.65 parsecs away (8.7 light years). Though we may regard visiting there as a distant dream, we no longer regard this proposition as an impossible dream. In light of CB theory there is a stronger probability that we will either find a life-bearing planet there, or the remains of one, or one that is forming.

Bone homeostasis as well as every thought and movement depends on the proper metabolism of calcium. Elderly people experience a diminished metabolic rate. A diminishing of gravitational convergence pressure also lowers the metabolic rate. Both of these cases are associated with bone loss and this suggests that *it is the slowing of the metabolic rate that inhibits the homeostasis of calcium metabolism.*

All regression is toward an isodynamic state though not necessarily toward the original isodynamic state. If it regresses toward the original isodynamic state we say that this regression is anti-entropic. If it regresses toward a new isodynamic state, then relative to the original state it is entropic but relative to the new state it is anti-entropic. This shows an immanent reversibility in the interpretation of CB entropy. On the Lilliputian scale, entropy remains the measure of energy that is unusable in a closed system.

The EMA field may be characterized in terms that are immanently reversible between bootstrap and non-bootstrap models. In the bootstrap model the EMA is the field. If we could prove that the aether referenced by Maxwell and Michelson and others existed, then for CB we would say that the EMA polarities are carried on this field. Thus the speed of light becomes the speed of a particular polarization of

EMA carried by the aether. In the case of the bootstrap model, the speed of light is again a polarization that travels, but in this case it is the EMA itself that is the carrier. For EMA this leaves us with an open question. Does it follow the bootstrap or nonbootstrap mode? At the present time, immanent reversibility precludes any necessity to decide.

END Chapter 6

Chapter 7: DIALOGUE

"If,..., we frame the hypothesis that the configuration, motion, or action of the material system is of a certain definite kind, and if the results of this hypothesis agree with the phenomena, then, unless we can prove that no other hypothesis would account for the phenomena, we must still admit the possibility of our hypothesis being a wrong one."
J. Clerk Maxwell, *Matter and Motion*, sec. 148.

Introduction. *When we believe too much in an apodeictic reference-frame we get caught in a form of catechism thinking. Catechism thinking generates expectations in us that can override our observations of matters-of-fact and can drive us to ignore observations that controvert our beliefs. It can also become an obstacle to developing new frameworks for intelligibility. We can help prevent this internalized blindness by training ourselves through interrogation to test reference-frames as tools. It is through the attitude of interrogation that we can prevent our confidence in a given reference-frame from becoming a mode of belief. This does not preclude the possibility that through investigations we come to believe more strongly in a given reference-frame rather than another.*

There are two general modes for science procedures; one is exploratory, one is confirmatory, and either intent can lead to the other. Exploratory is the search for hypotheses to test and confirmatory is the testing of hypotheses.

All foundation principles for a reference-frame result from the use of limit reasoning for their definitions. It is important, when using limit reasoning, to maintain an awareness of the actual vagueness of what we observe. The use of limits applied to observations leads to a crispness that transcends the observations. This transcendental reasoning may lead to effective results or it may lead to a false clarity, a misleading clarity. The use of limits, when used with intent, is for constructing apodeictic reference-frames that surpass the clarity of the original observations. These reference-frames are then subject to testing and debugging, directly or indirectly.

The Newtonian and the CB reference-frames are mutually exclusive. To decide on one over the other is to change the foundations for judgment of physical phenomena. If we show that CB includes the physical phenomena that Newtonian theory accounts for and furthermore predicts or accounts for physical phenomena that Newtonian theory cannot predict or account for, then we may claim to have critical observations that favor CB theory over Newtonian theory.

DIALOGUE. Some of the celestial mechanics problems the Newtonian system cannot account for include the following:

1. **With the exception of Mercury, why do the planets follow close the orbit plane of Jupiter?**
2. **Why do the planets all go around the Sun in the same direction?**
3. **Why do all the planets follow near circular orbits?**

4. Why do all the planets but Venus rotate from West to East?

(Newton had no data access to the rotation of Venus, etc.)

Response: The Newton model can give no rationale for these facts. For Newton, such movements are 'accidental' or due to God. There is no particular reason in his theory why the planets cannot go in conic sections around the Sun every which way, including the rotations. In a letter to Bentley dated 17 January 1692/3 Newton said, "So then gravity may put the planets into motion but without the divine power it could never put them into such a Circulating motion as they have about the Sun, and therefore for this as well as other reasons I am compelled to ascribe the frame of this System to an intelligent agent."

In CB, the rotation of the Sun is the rotation of a convergence field that physically influences the revolutions and rotations of the planets. In this sense it may remind one of the vortex theory of Descartes, the first field theory applied to celestial mechanics.

- The ***first*** *influence* is the sink-float buoyancy relation, the planets are in float buoyancy; small objects are in sink buoyancy. Objects in float buoyancy tend toward circular orbits, objects in sink buoyancy spiral to accretion to a host surface or, under the pressure of a more comprehensive neighborhood, to each other.
- The ***second*** *influence* is that all planets in isodynamic movement go around the Sun in the same direction as the solar convergence, which spins with the Sun. They are caught in the Sun's maelstrom.
- The ***third*** *influence* is that all solar planets in isodynamic movement tend toward a great circle plane of least impedance for their revolution.

- The ***fourth*** *influence* is that isodynamic planet rotation responds to the Sun's maelstrom by rotating from West to East. Venus has yet to adjust fully to this rule.
- The ***fifth*** *influence* that is possible comes from the pressure of the galactic maelstrom on the Sun. Its exact influence is not yet determined though we may surmise that it influences the Sun rotation, which in turn influences the revolution of the planets.

Number 8 below suggests a rationale for why, with the exception of Mercury, planetary orbit planes are closer to the orbit plane of Jupiter than to the equatorial plane of the Sun.

That Venus has yet to achieve the expected rotation could be regarded as circumstantial evidence consistent with the conclusion that Venus is a newer arrival among the rocky orphans. One Earth event candidate for this passage is the second greatest extinction event on the Earth, the Ordovician mass extinction (about 438 million years ago). Another Earth event candidate (perhaps more likely) is the Cambrian (543-510 million years ago) period with at least four major extinctions. This time duration and multiplicity make this one seem more likely, considering the rebounding effects that were likely endured at the occurrence.

There is also a peculiar syncopation in the Venus rotation so that every time it is nearest the Earth the same side faces the Earth. Further study is required to find the requirements that drive this effect. Venus seems to have Sun tidal impedance to rotation, but the East to West rotation is likely more slowed by its counter rotation to the Sun's convergence field. It must finally become spin-locked or evolve into the same West to East direction as all the other planets.

5. If gravitational systems converge to one another, how is it that the large planets are furthest away

from the Sun and the smallest planets are the closest?

Response: It seems that the big planets should have gotten much closer to the Sun before they stabilized in their orbits. For Newton the sizes as well as the various regularities in orbits established by Kepler have not been entirely accounted for. There is no rule for the planets to sort themselves out according to their bulks. CB requires that there is a general sorting that continues to take place according to bulk and its convergence buoyancy relations. Any satellite that is out of natural order is moving into natural order unless otherwise interfered with. Those with the largest convergence power are the farthest removed, etc. Jupiter and Mars seem largely to violate this rule. This sorting and the apparent exceptions to this sorting are further discussed below.

For CB, Newtonian calculations for planetary **g** values can only be rough estimates for **E** values *first* because the **g** values depend on the assumption that the planetary movements are isodynamic (stable). It has not been proven that all designated planets of our solar system are isodynamic with the Sun, etc. *Second* reason is that the Jovian planets have a more significant effect on the float distance. Direct measures will likely reveal that the **g** values are smaller than the **E** values. The CRN accounts for the required correction. It also accounts for the differences from the Kepler determinations.

6. Why has the face of the Moon locked in a one-face position? How can any satellite get locked in such a position? (Using inertia there is one chance in infinity.)

Response: Note that this locked feature is a characteristic of all of the known moons in the solar system. That so many of the solar system moons are spin-locked relative to their planets gives wonder to the fact that none of the planets have locked in with the Sun. This is related to relative sizes affecting the subtended angle, which also affects the shape of the satellite CIS. The solar tidal effects on a planet CIS can have a braking effect on the planet rotation. Distant planets would have a weaker and flatter bulge, which would have a weaker braking effect, allowing for a faster rotation.

Chapter 2 discusses the Convergence Inversion Sphere (CIS) that can influence the shape of convergent objects. The Moon's CIS is drawn out of shape in response to the Earth's convergence. It forms an elongated shape in line to the Earth. The elongation of the CIS influences the physical shape of the Moon which itself exhibits elongation. The tidal effect that the Earth has on the Moon distorts the Moon CIS into an oval shape along the line to the Earth that is just extreme enough to physically restrain the Moon to librations of the same face (wobble effects). Relative to the Earth, the Moon is spin-locked; relative to the Earth, it cannot rotate.

The deepest CIS depth is likely on the obverse and the reverse side. It is the extremity of this CIS shape that prevents the Moon from spin relative to the Earth.

Note that the shape of the CIS is determined by material distribution adjusted for distance. Tidal effects have two components, one is the CIS and the other is the effects on material distribution. These physical effects on the object shape derive from different convergence relations.

7. How do gravitational systems capture satellites without degrading the system they already have?

Response: Kepler's Third Law states:

In terms of distance and time, the squares of the periodic times of two planets are to each other as the cubes of their mean distances from the Sun.

This Law works fairly well with the inner planets but the larger outer planets have a stronger participation in distancing, which Kepler could not consider. This makes the Law in error proportionate to how bulky the converging host object is. That is why I say that for host size objects that are 'capturing' smaller host size objects, the capture sequence *approximately* tends toward this law. CB can account for the error and adjust the Law to fit this phenomenon. The adjusted Law implies that when an object spirals into the vicinity of a prospective host, its velocity closely matches the host velocity (i.e., speed and direction). Under isodynamic movement an accretion course that generates more violence does not occur without extraordinary (rare) initial conditions. However, it provides a mechanism that can produce splooch craters.

8. Why do the planet elliptic orbits rotate about the Sun?

***Response*:** The planetary orbits are restrained toward the great circle plane of least variation. Insofar as they are not precisely in this plane they are subject to an additional variation in their revolutions. Relative to the Sun's rotation the planets are moving up and down the Sun surface latitudes. Each latitude has its corresponding convergence rotation speed, maximum at equator, minimum at heliographic poles, etc.

If the angular velocity on the Sun's surface was everywhere equal, the planets would revolve closer to the solar equatorial plane. The solar equator convergence is likely weaker than its polar convergence. This stronger polar convergence power has a stronger repelling effect that would tend to push the planet revolutions toward the solar equatorial plane. Thus though the Sun maelstrom determines the direction of revolution for the Earth, the Earth is impelled toward solar equatorial plane. The closer to the Sun, the closer the orbital tilts are to the solar equatorial plane. The pressure that drives this drift varies inversely for distance from the Sun surface. It can be shown that no *orbiting* body can get far enough from the Sun to have its revolution plane contain the solar axis and still have a solar orbit. It is interesting to note that, with the exception of Mercury, all of the planets are closer to the revolution plane of Jupiter than to the equatorial plane of the Sun.

That calculations show that the orbit planes of the planets, with the exception of Mercury, are closer to the orbit plane of Jupiter than the solar equatorial plane is a late discovery. The ecliptic is still the major reference plane. Like the geoid it was invented as a reference plane.

9. Why is there no known comet that has formed a near circular orbit about the Sun?

Response: To date the Oort cloud remains a hypothesis. It is consistent with CB that there may be a ring of comets that form near circular orbits around the Sun out where their convergence power has not been overcome by the Sun's convergence. This suggests why, if there is an Oort Cloud, it likely exists very far out. Knowing the e of a comet, we can estimate how far out the Oort Cloud could be. Note also that due to the weak CIS in the comets that very little aberration would be required to move them out of the Oort Cloud.

Comets are likely a major source that supplies the asteroid belt. Their paths grow toward near-circular spirals that spiral in toward the Sun. This reduction of eccentricity becomes stronger as the tilt to the ecliptic lessens and this may place many of them in the asteroid belt where they finally lose their gaseous content. If they are sufficiently below the boundary bulk level, then they spiral in to accretion more rapidly. These statements do not preclude maverick objects picked up every which way by the solar system, though all the levels of natural spiraling would tend to reduce their number.

10. What sustains the Earth's spin?

Response: In Newtonian physics it is assumed that the Earth spin is sustained by inertia and that tidal effects slow the spin making the Earth turn slower and slower. CB denies this.

The tidal effects of the Moon and the Sun complicate determining the convergence pressure of the Earth. In particular, the tidal effects change the shape of the Earth by generating isostatic bulges. CB presumes that the Earth spin is mainly driven by the solar maelstrom. The isostatic tidal bulges are CIS distortions that resist the Earth rotation, a kind of friction. This tidal friction implies that the equatorial convergence rotation speed is lower than for the full convergence pressure of the Earth surface. In other words, since the tidal friction slows the Earth spin, the Earth spins with a drag.

Again we find two conflicting drivers that affect the spin rate; one is the spin driven by the Sun maelstrom, the other is the friction on this spin driven by changes in the Earth shape through tidal effects. A more rigid Earth (less distortion) could spin faster, etc.

PROBLEMS OUTSIDE NEWTON'S UNIVERSE OF DISCOURSE

11. Why did Pluto come inside the orbit of Neptune?

Response: Pluto is too small to sustain the float buoyancy level required for such a great distance from the Sun. Its high eccentricity, large angle of 'orbit' inclination, and its passing inside the orbit of Neptune suggest that our calculations are wrong or it has collided with our solar system and is either grazing the solar system or has a collision trajectory. The solar system pressures that drive Pluto could diminish the collision status until Pluto is isodynamic with the solar system.

12. Why does Pluto have such a steep angle to the ecliptic?

Response: Pluto and Charon do not have the important traits that help us determine that they are in a stable (float) buoyancy. For such low powered convergence objects, the distance is too far for float buoyancy with the Sun. And this combined with the steep angle implies that they have not related to the Sun for as long as the other satellites that have less inclination to the solar plane of equatorial rotation. This steep incline also denies the speculation that Pluto is an escaped moon from Neptune, etc. They seem to be comet-like objects. It would be interesting to know when this entrance to our solar system took place. It was James W. Christy, an astronomer from the U.S. Naval Observatory, working to obtain more precision in Pluto's path that led to his discovery of Charon in 1978. Whatever the case, they are not yet in an isodynamic relation with the Sun system. This implies that a determination of Pluto's path cannot use the shortcuts provided by the assumption of elliptic orbits. We can therefore expect in the near future a need for further corrections to Pluto's trajectory for its determination.

13. Is Pluto a planet?

Response: According to CB, planets must have near circular orbits on planes with low inclination from the solar plane of equatorial rotation. These orbits are in the form of near circular spirals with inclinations close to the ecliptic. Even with this one restriction, Pluto cannot yet be regarded as a planet of the Sun.

14. Why do the large planets seem to decrease in size from the ring of asteroids out to Neptune?

Response: Some care needs to be taken in judging the bulk of these large planets as they are likely hollow and they are thickly covered with gases. Jupiter, Saturn, Uranus, and Neptune are called the gas planets or Jovian planets. This may account for a curious reversal of order according to apparent size, but this apparent size relates to surface gases. On Uranus and Neptune much of the surface gases are frozen, which could give them a more compacted bulk per unit volume then Saturn and Jupiter. As for the sorting out of planets by bulk, there is a predicted order with the laws of convergence, large bulk planets should be the farthest from the Sun, etc. The Newton 'mass' calculations show a better order then the calculation for equatorial **g**. However, this **g** is regarded as a point mass **g**. No regard is given to the fact that the **g**'s of these larger planets are operating over a much larger surface. The role of convergence pressure on this sorting is affected by surface area, an effect that is usually ignored. With this adjustment, only Jupiter and Mars seem out of the general order of planets.

15. Why is Mars, a planet smaller than Earth, further away from the Sun than the Earth? This seems to contradict the CB pattern of inner planets.

Response: Mars is likely a new candidate planet, possibly an adopted maverick. It may be slowly spiraling toward the Sun. This could take millions of years to sink a million kilometers. As it accretes surface material, this slows the spiraling. If the accretion level and buoyance level finally match, then the next accretion begins a regress from the Sun. Its current isodynamic float buoyancy position, if judged by Newton mass would place it below Venus.

16. Why are the inner planets much smaller than the outer planets? How can we account for their physical differences?

Response: A possible conjecture here is that these inner (rocky) planets are orphans adopted by the Sun. The asteroid belt is a possible birthplace of planets, but it is more likely that our rocky planets are mavericks from outer space. These pristine planets (stars) spiral in toward the Sun until they reach their isodynamic level with the Sun's convergence. If sufficient bulk is obtained, these new planets begin a slow regress away from the Sun proportional to their accretions and the Sun's accretions.

As adopted orphans it is plausible to assume that their otherwise unaccounted for differences are due to their variety of origins. That they still sustain important differences suggests that geologically our solar system is still young.

17. Why are the inner planets, excluding Mars, arranged large to small going toward the Sun?

Response: It may be that the inner planets have already reached their isodynamic levels or nearly so. This would naturally sort them out according to E_n and surface area. Minimal eccentricity suggests maximum stabilization. This would suggest that Mercury, with its somewhat larger eccentricity, is either still not in its proper place or there are other factors not yet considered. Its smallness and proximity to the Sun allows the Sun spin a more powerful influence on Mercury's path. This drives Mercury to orbit nearest to the solar equatorial plane. It also gives the ellipse of its orbit a measurable rotation rate. It also shows that 'debris' is a relative term.

18. Why are Uranus and Neptune less in a gaseous state then Saturn and Jupiter?

Response: The first suggestion is that much of the gases are frozen and therefore these planets appear less voluminous. However, comets are an import source for gases. When comets enter the solar system, their angles to the great circle of the Sun's rotation can be steep. This implies that the Oort cloud, if it exists, has some variations in tilts. The tilt variations could place many of the comet trajectories out of range for capture by the outermost planets. By the time they get to Saturn and Jupiter, they are much closer to the plane, a greater probability for accretion or capture. This implies the possibility of a resorting of the outer planets that from their current state would require millions of years.

19. Why do Saturn and Jupiter have many more moons than Neptune and Uranus?

Response: This was partly answered above. These are the last great planets to precede the Sun's rings of asteroids. The observation of so many moons and satellites suggests that these giants are well placed for capturing incoming asteroids and comets that are close to isodynamic movement. The 1994 collision of the Shoemaker-Levi 9 comet with Jupiter is the largest example of a planetary capture to accretion with a photographic record.

20. Why do Prometheus and Pandora, two shepherd moons of Saturn, repel the F-ring?

Response: Convergences in proximity compete, driven by a repelling. If the repelling is not squelched to accretion, then we see what appears to be a physical repelling that drives a floating effect. The repelling effect that the shepherd moons have on the rings gives the appearance of shepherding the ring material into a well-formed ring. Other Jovian planets also have shepherd moons that appear to repel their ring material.

21. Why is Phoebe, a satellite of Saturn, in retrograde to Saturn's spin?

Response: Phoebe is the outermost satellite of Saturn, almost thirteen million km. It is about 220 km in diameter and has the largest eccentricity of all of the known large objects orbiting Saturn. In retrograde we judge its inclination as 177.00 degrees. This is about $150°$ relative to the plane of Saturn's orbit. These data suggest that Phoebe is a fairly recent object captured by the Saturn convergence. The size of Phoebe suggests that it has not yet reached its float position relative to Saturn. We can expect that it will spiral inward until it is isodynamic within the Saturnian system. Because it is in retrograde we can expect some dramatic changes before isodynamic movement is achieved. For it to stabilize at its present size would place it as one of the innermost satellites. It must also reverse its revolution. The non-spherical shape of Phoebe suggests that material surface tension has not yet been over-powered by the CIS.

22. Why do large gravitational objects have rings around them?

Response: All the Jovian planets have some form of rings (Neptune has only partial rings in the form of crescents). All rotating convergence objects above boundary bulk likely have rings of debris around them. The rotation implies a tidal effect, which in turn implies that the hollow planets are thinner at their equators than at their axis poles. This further implies that their equatorial region is thinner than their polar regions, which makes the polar regions have a stronger repelling power.

The mean size of the objects in these rings corresponds to the size of the host and the available rubble. These rings represent objects that are spiraling to accretion with the host and to date the bulk of the rings circle their host in a space interior to the host moon orbits.

The rings of Saturn are here thought to contain objects that are sinking toward Saturn. However, the rings are supposed here as on-the-average evenly distributed objects that converge and mutually repel. But for such small objects this repelling is slight and fragile to any influence. Objects escape the rings and sink to the surface of Saturn. The rings in turn are replenished from incoming debris.

Dense rings could be convergence traps that interface with buoyance-accretion isodynamic interaction on an individual object with the ring host. As an individual object, the motion toward isodynamic interaction would directly lead to accretion with the host, but getting trapped in the rings can resist or delay this accretion interaction. If the interior material for Saturn were evenly distributed, rather than rings we would have bubbles of debris 'enclosing' Saturn. For CB the ring formation of Saturn is driven by an interior material distribution that is thicker at the poles than at its equator. For CB this is a general assumption for celestial bodies that have rings of debris. Spoking in rings is a kind of convergence clustering. A NASA scientist (cf Carolyn C. Porco) found evidence that spoking is synchronized with Saturn's electromagnetic field.

23. Related to the previous question, why are there rings of asteroids between Mars and Jupiter?

Response: The asteroids catalogued are just less than 13,000 (2004 A.D.) and most of them are spread in a torus of hundreds of millions of kilometers. There are likely billions of smaller objects in the Sun rings and mostly invisible to our Earth view. These objects are all spiraling toward the Sun to accretion or to float buoyance. This Sun concentric ring set is not regarded as dense but it exhibits distribution harmonics (in this case the Kirkwood gaps) just as other ring systems do. From the CB perspective, the Sun rotation indicates a galactic tidal effect. For a hollow Sun this implies that the equatorial region is thinner than the polar regions. Therefore the polar regions of the Sun have a stronger repelling then its equatorial region and this in large part helps to keep the rings like a torus in a revolution plane. The angle of this plane is close to the Jupiter orbital plane. The inner circumference and outer circumference from the Sun are in large part due to Jupiter and Mars functioning there as shepherd planets.

24. Why are there so many rocks strewn about the Martian surface?

Response: Mars continues to collect debris. In general, small debris are collected in greater number than large debris. Many millions of tons of stones will strike the surface of Mars before Phobos strikes. The Martian surface may provide a hint as to the quantity of smaller debris in the Sun rings unseen by our telescopes. However, Mars also likely has stones breaching through its surface just as the Earth has.

25. Why is the Moon receding from the Earth?

Response: The Earth and the Moon are both accreting debris. This accreted debris increases the convergence powers of both objects. This increases the repelling pressures between them and raises the isodynamic level of buoyancy. Currently we estimate a separation of nearly three centimeters per year. The two main drivers for this distance are the convergence powers and the more comprehensive neighborhood pressure.

26. Why is the Moon three degrees behind the tides rather than over them or ahead of them?

Response: Earth tides due to the Moon have at least two major drivers. One component is the CIS, the shape of which is determined by material distribution adjusted for distance. The other component is the convergence between them. The Earth will tend to a spherical shape unless otherwise interfered with. The repelling between the Earth and the Moon is expressed as a pressure on the surfaces. This pressure is expressed in part as an isostatic flattening. This flattening produces isostatic bulges along a line normal to the pressure. These bulges are called tides. On Earth, these tides are land tides and water tides (the atmosphere also has tides but more influenced by radiation gradients). The CB perspective indicates that the lunar tide is pushed by the Moon rather than pulled by the Moon. Since the lunar tide is only three degrees ahead of the overhead Moon position this implies a strong lag. The lag would be considerably smaller if the power link between the Earth and the Moon were Eudoxian rather than an inverse square rule embedded in a more comprehensive pressure zone.

27. How old is a planet? (This question is more difficult than popular science would have us believe.)

Response: Since planet growth is through accretion there is little sense in choosing arbitrary rocks and doing elaborate age tests to assign age to the planet. Growth by accretion denies this conclusion made from the cited evidence. Judging the age of the Earth by the age of stones that sink to the Earth is definitely not a well thought out method. Furthermore, it is a *non sequitur* extrapolation to assign these results to the age of the solar system. All stones on Earth are of celestial origin no matter what changes may have been incurred in Earth processes.

Curiously, under present test modes, most surface stones are concluded older than stone material breaching from below the mantle. The difference is about half a billion years. These stones breaching from within the Earth are either aging more slowly than the surface stones, or their source (a nova) is younger than the sources of present debris. Surface stones are exposed to maximum Earth convergence, which suggests that periodicity rates due to convergence are higher on the surface than in certain depths of the Earth interior (not too proximate to the CIS). This could accelerate surface 'aging' (not yet proven). Even so, the interior is hotter and generates higher levels of radiance as we approach the CIS, a logical consequence from the Earth mantle that offers impedance to heat dampening.

28. Why are human bodies not spheres?

Response: This is not a strange question to those who have followed the argument. Human corporal existence lies in the zone of transition between gravitation and surface tension. If we were dominated by our own gravitational convergence (rather than Earth's, say) we would tend toward a spherical shape, if we were dominated by surface tension convergence we would be spherical. It is our existence as determined by our bodies in the Earth convergence that allows us to have a more complex shape.

In space where surface tension is the more dominant convergence for small objects, the body swells somewhat, a small approach toward spherical. A more plastic material more easily deforms. Bones have very low plasticity and high surface tension, but the body fluids are highly plastic and redistribute free of our gravity pressure controls.

29. Is there an assignable causal order in regard to rotation in a neighborhood?

Response: In a manner of speaking, revolving (as distinct from rotation) is a continuation of convergence. Two objects revolving around each other have a calculable centripetal convergence rate. For circular revolution it is expressed as, v^2/r. Without a third object this revolution is invisible for spin-locked revolutions. However, if at least one object rotates relative to the other this could be taken as an indication of a revolution driven rotation. In convergence buoyancy there is no such thing as rotation for an isolated object. Any evidence of rotation would indicate that the object is in a neighborhood. Continuing in *ex parte* terms, we could say that it is the interaction of convergence fields that generates revolution and rotation can be generated by revolutions.

There are also tidal effects on rotation. The larger the tidal effects the slower the rotation relative to a host object. With respect to Moon-Earth, relative to the Earth, the Moon has its slowest possible rotation, once per revolution. If it had a slower rotation it would have to be due to influences other than the Earth. Under present conditions, these other influences would likely show in the Moon librations.

This synchronic interaction also helps us understand where the heaviest influences are applied. They are most heavily applied to the equatorial regions even for Uranus with its tilt. This explains why the Jovian planets and the Sun have axes that rotate more slowly than their equatorial regions. For the Jovian planets it is their unfrozen gases that follow this pattern.

The surface convergences calculated for Saturn, Uranus, and Neptune have them following a predicted order of more to less. Jupiter remains problematic. Jupiter is problematic for the Newtonian system as well. Its equatorial **g** power is 2.528 Earth **g**. Jupiter has an oblateness just as Saturn does, but it is not sufficient for an inertial interpretation. From the Newtonian perspective, Jupiter should be more flattened by its rotation. This should certainly be the case if inertia were a true physical principle with a determinable effect on the shape of Jupiter.

A more careful study of Uranus moons suggests in CB terms that the Uranian poles are thicker in material than its equatorial region. In large part this is shown with the fact that these moons tend to revolve around the equatorial plane of Uranus. This also implies that the Uranian CIS is deepest at its polar regions.

30. How does CB affect the light grazing the Sun?

Response: For light grazing the limb of the Sun we calculate the Newtonian aberration: $a(R_o) = 2GM_o$

Putting in numbers:

$M_o = 1.99 \times 10^{23}$ gm   Sun mass
$R_o = 6.96 \times 10^{10}$ cm   Sun radius

$G = 6.67 \times 10^{-8}$ cgs units   gravitational constant
$a(R_o) = 4.245 \times 10^{-8}$ rad
radians = 0.875"

Experimental observations are more closely consistent with double this result, i.e., **1.75"**.

After examining the Eddington solar eclipse study it is found that only a few of the stars followed within reasonable accuracy the Einstein prediction. The variation seems consistent with an optical interpretation (geometric optics). From the discussion of the first critical test for the CIS we conclude that these Earth weight measures are taken in a CIS reference-frame that would yield results that differ by a factor of 2. Thus in Convergence Buoyancy the factor of 2 removes the erroneous result, though now we have to explain why the value for M_o remains valid. The calculation of big-G (gravitational constant) has been performed using an apparatus that requires two bulks ('masses') that have their respective magnitudes determined by Earth weight measures (Michell-Cavendish experiments). This explanation involves the mode used for obtaining the value, a method using ratios (discussed in another context).

If this reasoning is correct, then the supporters of Einstein Relativity now have to debug their reasoning where they could not include the CIS effect on measures of Earth 'mass'.

31. What are the major known possible causes for global warming?

Response: In the past 120 years there have been four scares with a promise of disaster; global cooling (1895), global warming (1920's), global cooling (1975), and now we are back to global warming again. 28 April 1975, Time magazine announced that the world faced disaster from global cooling. From about 1940 until around 1979 world temperatures were falling (39 years). It was about 1979 that they reversed direction into a general rise begun in the 1980s. In the rough this suggests that 2009-2019 may be the beginning of a transition stage back to a general global cooling. At present the most feasible theory offered for these changes are the Milankovitch Cycles calculated by the Serbian engineer and scientist, Milutin Milankovitch (1879-1958). He calculated that variations in eccentricity, axial tilt, and recession of the Earth orbit are determinants of longterm climate cycles on Earth. These scares are relatively short-term, but the reversals may again be indications that the influence for change is more from orbital considerations than from human driven events.

END Chapter 7

EPILOGUE

This exposition presents theory and speculations contrary to the historical observations of Pierre Duhem (1859—1916) who echoes the beliefs and warnings of Galileo, Newton, and others against using physics theory for explanations, "la théorie physique n'était pas une explication" (physics theory was not an explanation). My approach represents this anti-Newtonian hypothesis for convergence as a likely story insofar as for the moment I consider it one of the more plausible models that might show how the interior of the closed watch could work. In this I rest in the tradition that such models may aid in understanding. For me, there is irony in how we can impose order on our observations, whether it is visible or not.

Even if errors are found in this presentation, I do not believe that a fatal error will be found for the foundational notion that for host bulks a repelling generates between them when in proximity. Furthermore, the repelling power of debris is squelched when in the proximity of a host.

A threshold is crossed and I hope others will join the work to carry it toward a fuller development. There is much left to do. My current project is the construction of the mathematical summaries for CB theory. Much of this mathematical modeling has affinities to Ludwig Boltzmann's mathematical development of his gas theory. Boltzmann, in fact, uses an underpinning model that strongly resembles a form of CB on the molecular level. In volume one of his *Lectures on Gas Theory* (1896) he states in the Introduction that for a molecule:

> "If it approaches a neighboring molecule it is repelled by it, but if it moves farther away there is an attraction."

His description of this foundational principle of movement is defended on the basis of the *a posteriori* accountability that his derived theory provides for the phenomena under scrutiny. CB provides a more detailed qualitative development of this model that invokes principles of physical description that provide an *a priori* accountability for this form of movement. Furthermore, the *a posteriori* accountability remains intact.

Star Laws: An Introduction to Convergence Buoyancy has a companion volume entitled, *Star Laws Addenda: Duality Theory*. This is a study concerning the presuppositions in Science-Systems and Design that underpin the structuring of CB Theory.

End of Book I
Star Laws: An Introduction to Convergence Buoyancy

Star Laws

ADDENDA: Duality Theory

Book II
A Treatise on Metaphysics

by

Ross Lee Graham, Ph.D.

The reasonable man adapts himself to his world; the unreasonable one persists in trying to adapt the world to himself. Therefore all progress depends on the unreasonable man.

-- George Bernard Shaw, *Man and Superman*, act 3.

Time-Module Books
Brașov, Romania

Star Laws Addenda: Duality Theory

All Rights Reserved © 2004, 2009 by Ross Lee Graham, PhD

No part of this book may be reproduced or transmitted in any form or by any means, graphic, electronic, or mechanical, including photocopying, recording, taping, or by any information storage retrieval system, without the permission in writing from the copyright holder.

Time-Module Books
Published in Braşov, Romania

Printed in the USA by lulu.com

Two Volumes in One ISBN: on page 2

Distributed by Barnes & Noble

TABLE OF CONTENTS

BOOK II: Duality Theory

    Abstract 375

    Foreword 377

ADDENDA: Science-Systems Analysis

    Addendum 1: Science-Systems 385

    Addendum 2: Status of Logic in Science-Systems 419

    Addendum 3: Formal Duality in Finite Logics 449

    Addendum 4: Data-Field Duality 483

    Addendum 5: Reference-Frame Duality 507

    Addendum 6: Ptolemy, Copernicus, Kepler 543

    Addendum 7: Fermat's Last Theorem 569

Appendix: Number References for Calculations 587
Bibliography 589

    ...everything that comes into being comes into being from its contrary...
                            --- Aristotle, *On the Heavens*

ABSTRACT

Star Laws, Addenda: Duality Theory, is a study concerning some of the key presuppositions in Science-Systems Analysis and Design that underpin the structuring of the Convergence Buoyancy Theory, a non-inertial Physics.

This study unshrouds some of the instrumental powers of using dual concepts overtly in modeling physical theories. The underlying principle is the formulation of dual concepts and dual systems of thought. Therefore, these Addenda expositions address the subject of duality with special emphasis on clarifying how to dereference the term. It is the use of dual analysis that led to the anti-Newtonian concept of Convergence Buoyancy, a concept that provides a different interpretation for two foundational concepts of Newtonian Physics, mass and inertia.

The Addenda also include notes on other aspects of the presuppositions underpinning this Convergence Buoyancy Theory.

Ross Lee Graham
Email: drrosslg@gmail.com
Mid Sweden University
Sundsvall, Sweden

Ides of March 2004

FOREWORD

Introduction. *When we speak of the Universe, integrity binds us to speak from a basis formed from what we claim to know. But because what we want to know is beyond any powers of ordinary knowing, we have not escaped the inherent metaphysics of cosmology. We have never escaped the role that belief plays in our comprehension of the Universe. Our beliefs intertwine in everything that we claim to know. These beliefs help us cover up great gaps of ignorance and help us sustain pretensions of comprehension that far exceed ordinary human abilities.*

Educating Scientists. Training scientists has become a form of catechism. Re-evaluating or endeavoring to speculate in a form different from the imposed catechism is subjected to an in-house Inquisition. Anyone indulging in such out-house speculations can expect rejection from the self-accredited scientific community.

Creative teaching can be strangled by the constraints of accreditation, the syllabus, the licensed teacher. The original intention of these constraints was to guarantee the 'quality' and 'transferability' of course content. The end result is that an in-house Inquisition takes the vim and vigor out of schools. This defect is especially evident in the American education system where, except for isolated instances, learning equals submediocre textbook training courses.

In universities and colleges, the first year syllabus for Physics is so full of material that no student can hope to critique the fundamentals on the first time around. There is no time given to critique these 'fundamentals'. Teachers program the students to cast problems into a certain model set. Even though they may show several ways to approach a problem, the instruction never gets beyond the guided tour.

No paths are tread that were not tread before. Such teachers do not even believe relevant new ideas can come up in a class discussion and they are so predisposed that they neither recognize nor respect new ideas if they do come up.

Many teachers, with an answer in mind, pose questions they believe make students think. It is a performance in class that resembles the game of Charades. The students try to guess what the teacher is thinking. The one who guesses it first is the winner. Other student guesses, including those with more originality, may be regarded as wrong or not quite there yet. With such teachers no exploration is possible. It is like taking a guided tour through a prefabricated house. The students are not taught to question why it was built that way.

Seeking the presuppositions that underlie the problems is not a part of this training. It is a case where the students have their lights on but they are treated as if nobody is home. It is important for students who want to develop creative thought to educate themselves beyond such classroom activities. In the words of Mark Twain, "Don't let your schooling interfere with your education."

In effect, the American school system, in most of its instances, does not address itself to the problems involved in developing science theory. There are very few people trained or involved in this sort of work. Science is taught more like rites of initiation, a form of catechism.

Today, most of the people with the label 'scientist' are more accurately regarded as highly trained technologists or technicians. Their work is not so much in the study of science foundations as it is in the implementation of science.

Scientific Activity. We have two general categories for science activity; the theoretical, the practical. The theoretical is usually concerned with analysis and design of solutions for specific problems. Usually this work is in the context of accepted science-systems. On rare occasions this may lead to new foundations, a new paradigm, a new science-system. Practical science centers on the development of implementations that derive from theoretical work as well as from trial and error. These categories are not mutually exclusive, but each has its dominant activity.

Thomas Alva Edison (1847-1931) was not a scientist but he was a great technician and he was assisted by some of the world's finest technicians. His genius is a paradigm for technician creativity. Each mode of science has its mode of genius. History shows a few individuals who crossed over the lines of demarcation. Michael Faraday (1791-1867) was a technician who became one of the greatest of all experimental scientists and theorists (electromagnetic induction, the electric motor, the foundations of field theory, and the laws of electrolysis, etc.). Nicolas Leonard Sadi Carnot (1796-1832) was a technologist who contributed to physics theory (Carnot Cycle, 1824), etc. He is one of the few engineers who discovered a foundation principle for the Physics of his time.

When we look at scientific activity, the technicians and technologists are about the only people that we find visible. Their work is creative and in it there is room for genius, but to regard them as theorists distorts their primary intent. They are technologists, the engineers of science, the ones who design the implementations; or they are technicians, the ones who build the engineer designs. In general, their central intent is to implement science, not to create it. However, today their great numbers in the realm of science give them the political clout to direct funds and publicity away from theoretical science. Furthermore, profits and returns are more feasible and immediate in implementation, a strong argument for investor support.

Creative Science, *The Third World of Science*. I have observed many Physics classes around the world. I found only one that teaches that the concept of mass is entwined in a circular definition. This was at the University of Paris and my teacher was d'Espagnat. However, for most schools, no principle is put in question to stimulate original thought on fundamental questions. Alternatives to established methods are neither encouraged nor sought.

If you want to do creative thinking in the foundations of science you have your best chances in the pristine areas of science that have not yet completed their architectonic, have not yet turned into a licensing catechism. For the developed areas of science there are firmly entrenched referees who can close the door on your work. If you persist in an unwanted direction of research, those same referees may form the Inquisition that excommunicates you, that damns you as crackpot. Questioning fundamental principles is encouraged only in areas that are still under-developed, the Third World of Science.

In spite of these warnings I endeavor in my exposition of Convergence Buoyancy to show why two foundation concepts of Physics, mass and inertia, should be abandoned. I relegate these concepts to apodeictic fictions that have become obstacles to further progress in Physics. Furthermore, I redefine the concept of heat and I refine the distinction between heat and temperature. All of these revisions are based on or generate from an anti-Newtonian model, Convergence Buoyancy (CB).

Outline of Addenda. These Addenda delineate many of the presuppositions that guided the structuring of CB theory in Book I.

From experience I have learned to expect that a system dual to another system is internally consistent if that other system is consistent. This can be true even in the cases where they may be mutually exclusive (contradict each other) when applied to a data-field. This text presents a general proof for this expectation in mathematical logic. It also offers many implementations that attest its plausibility in physical interpretations.

Addendum 1 gives some general features of Science-Systems. It is a relatively new way of treating sciences, a kind of science of sciences. At this point in time it is neither desirable nor possible to give Science-Systems an exact definition. There is still much exploratory work to be done before a possible architectonic can be well-defined.

Addendum 2 continues discussing Science-Systems. In this addendum it is in terms of three general theses that may be used to indicate specific constraints in the use of logical principles in science investigations. It also introduces useful vocabulary for talking about reference-frames of any kind.

Addendum 3 uses logic as a model for duality and develops a more precise vocabulary for understanding dual systems. It also introduces a novel basis for a logic where values become the variables and formulas become the variable instantiations. In effect, it introduces a logic that is dual to the usual algebraic expression for Boolean Logics. It works in the order required by computers, but this is a first presentation of its basis.

Addendum 4 applies the concept of duality to data-field mode and introduces the terms *heteronymic* and *homonymic* dualities and a few more terms associated with this analysis. The difficulties in describing *heteronymic* duality are illustrated with geometric examples. *homonymic* duality is illustrated with Boolean logic examples. These examples are used paradigmatically. This renders the difficulties more apodeictic. It also shows that development in this direction requires concomitant language development.

Addendum 5 characterizes features of duality that apply to reference-frames for intelligibility and reference-frames for measurement. Galileo's remarkable treatment of speed is analyzed for its rigor and its presuppositions. These forms of duality also lead to a novel critique of Einstein Relativity theories.

Addendum 6 applies the terms of the previous addenda to an analysis of one of the largest scale dualities in the history of science, the duality between the Ptolemy and Copernicus World (Universe) Systems. This Addendum ends with an outline of Kepler's contributions to celestial mechanics. His three laws required the remarkably precise observations catalogued by Tycho Brahe. How Kepler discovered his three laws also involves a dual approach to scientific experiment.

Addendum 7 shows my general proof for Fermat's Last Theorem using the mathematical constraints Fermat was subject to.

End Foreword

ADDENDA
Science-Systems Analysis

Addendum 1
SCIENCE-SYSTEMS

INTRODUCTION. *Many scientists claim induction is proper to sciences, then they insist that experiments are not well-defined unless they are defined in terms of a theory. It is a contradiction to claim induction as the preferred method of science and then insist that all experiments should be deduced from theory. This represents a confusion between exploratory and confirmatory science.*

Francis Bacon (1561-1626), a pioneer in Science-Systems Analysis and Design, believed himself to be the first to define and advocate induction methods as a system for science investigations. He advocated reasoning from particular well-defined experiments to intermediate principles to more abstract principles. Even today many academics would question the claim that an experiment can be well-defined outside of a theoretical basis. Bacon would meet with academic resistance to his methodology of beginning with well-defined experiments that are not based on a theory.

> **Note:** Bacon formulated the hypothesis that ice can preserve chicken meat. He did experiments to prove it (and subsequently died of pneumonia). He had no theory. Theories, hunches, and guesswork can generate hypotheses, but hypothesis and theory are not synonyms.

Bacon knew that his methodology was counter to the academic beliefs of his time. He was trained for three years at the University of Cambridge and knew well the academic preferences. Syllogism deductions were preferred, starting from a general principle, to deduce the particulars. Experiment was still not in vogue for academic 'scientists'. In fact, doing things with one's own hands was regarded with disdain. (Lord Kelvin introduced *student* laboratory experiments in the 19th Century.) Bacon presented a critique of what science should be. He outlines a methodology dual to the methodology of his contemporaries in academia.

Newton and many others were inspired by the work of Francis Bacon. His work signals the advent of a new era in science, which antedates Descartes (1596-1650) and finally had a more powerful influence in the developing methodology of experimental science. In particular, Bacon emphasized the prejudices that interfere with science investigations. He discussed how these prejudices, the **Idols of the Mind**, interfere with the methods proper to reasoning by induction. The aim of science for Bacon was to gain power, power over nature, to bend nature and its actions to human purposes. This power is expressed in new inventions and the riches they gain for us. He put experiments into two classes, enlightening and profitable. The enlightening were of no use in themselves though they led to the formulation of axioms and principles. The ultimate value for these axioms and principles is in the utility of what can be produced from them. Thus pursuing knowledge for its own sake is turned to the knowledge that brings about technological advancement. Science is thus to be judged not merely by the truth of its statements about the world but whether these statements are productive to human advantage in the material world. This is the fundamental dictum of the *utilitarians* who thrived in the 19th century. Bacon is perhaps the greatest intellectual advocate for seeking *useful* knowledge, a view that gained much force in time. Today we witness a renewed debate between the advocates of knowledge for its own sake and the advocates of useful knowledge. Whatever the motive for research, no age has seen more people involved in the search for knowledge as are so active today (though most of them eliminate themselves from foundational discoveries because they do not understand what Bacon was getting at).

Francis Bacon published *The Great Instauration* in 1620, or rather what he had finished of it. The plan of the work was vastly ambitious, more ambitious than the accomplishment. The work was to encompass all sciences, a prospectus that today might be responded to with ironic smiles or outright scoffing. The work was to have six parts as follows:

1. The Divisions of the Sciences.
2. The New Organon, or Directions concerning the Interpretation of Nature.
3. The Phenomena of the Universe, or a Natural and Experimental History for the Foundations of Philosophy.
4. The Ladder of the Intellect.
5. The Forerunner, or Anticipations of the New Philosophy.
6. The New Philosophy, or Active Science.

From a current perspective this menu resembles the work of a Systems Analyst. However, his menu remained largely a projection. The first part, he said, was to be found partly expressed in the Second Book of the *Proficience and Advancement of Learning, Divine and Human* (1605). The second part is found in the two books of the *Novum Organum*. The third part was to have been a "Natural and Experimental History for the Foundation of Philosophy." Instead, he left us a short work entitled *Parasceve* (preparation) which, rather than the history itself, is a description and delineation of what this history should be. He excuses this by telling us that it is a thing of very great size and would require great labor and many people to help. With this we cannot disagree. He is one of the first to come to sense the enormity of such a project. Parts 4, 5, and 6 were never written. Thus part one was replaced in apparent *ad hoc* fashion by the *Advancement of Learning* of 1605, part three was reduced to an outline for the history requirements rather than the history itself, and we find only part two a new and complete work, *Novum Organum*. Part two and three appeared in 1620 under the title of *The Great Restauration*. It remains today one of the great works of human thought, an inspired work of genius.

Bacon left us a menu and a map but like Moses, he guides us to the Promised Land which few credit him with entering. Even so, some of the world's finest minds, Newton among them, were inspired by Bacon and have saluted his genius. Today, many space scientists agree with Bacon's methodology (for reasons I argue in Book I: Chapter 4).

GALILEO'S TELESCOPIC DISCOVERIES. Galileo reinvented the telescope and applied it to astronomy. He had difficulties getting academics of reputation to look through his telescopes. One rationale given for this refusal was that there is no proof that the telescope merely magnifies the objects that it is pointed at. Today people have gone to the opposite extreme. There is much trust that magnification is all that the telescope is doing. For example, astronomers need to know more precisely what effects lenses and mirrors have on the spectra of observed objects. Studying the effects that the instruments produce for observation is not emphasized in the astronomy curriculum.

In addition to his academic and science interests Galileo was trained as an artist, this included drawing in perspective. He belonged to the Florentine Guild of Artists. He had learned to use *chiaro oscuro*, the use of light and darkness for perspective. This training provided an excellent background for interpreting his telescopic observations.

1. **The Moon is like the Earth.** When Galileo saw the Moon he saw shades of light and darkness, a corrugated effect that his artist eye could interpret as a three-dimensional landscape illuminated by the Sun. He saw dark gray areas that he thought at first were seas and even today we keep names like Mare Imbrium reminding us of this first interpretation. Later he said they were not seas and he pictured the Moon as a dead Earth. A remarkable deduction centers on his interpretation of white points in the dark regions of the Moon. He deduced that they were mountaintops illuminated by the Sun. For mountains in the light regions he applied geometry to calculate their altitudes. One such measure he calculated translates near 6 km high which today with more exact measuring instruments we calculate as 5.7 km, a remarkable agreement showing how sophisticated he was with his far less sophisticated devices. Seeing the Moon as another Earth was a shattering experience. It is the first strong argument against the uniqueness of the Earth. If the Earth is like the Moon, since the Moon moves, why not the Earth. By analogy, the Copernican requirement that the Earth moves is made more plausible. However, it does not directly address the physical objections to this movement that Ptolemy discussed (see Addendum 6 for details on Ptolemy).

2. **The Earth shines.** It was current to accept that the Moon is opaque. This opaqueness was easy to observe during the eclipses of the Sun. Therefore, when he saw a dim shining from the shaded area of the Moon he

knew it was not from lunar transparency. He deduced that the Earth shines and this shine reflects from the Moon in like manner as the bright areas of the Moon reflect the light of the Sun. This was one more step toward seeing a likeness between the Moon and the Earth. The Earth is not unique.

3. **The number of stars is greatly increased.** From this it was evident to Galileo that no one had ever really seen the heavens, so why trust the conclusions of previous observers. This conclusion undermines the validity of reasoning by authority, it shows that authorities are not infallible. It does not directly address accepting the Copernican heliocentric system, but it does promote the use of observations to accredit or discredit any authority, any system.

4. **The Milky Way is a collection of stars.** This contradicted all previous interpretations, reduced them to nonsense. This became an argument for regarding certain sacrosanct authorities as authors of nonsense. This observation provides a clear argument for denying the sacrosanctness of these authorities if the evidence derived from the telescope is believed. Though not specifically favorable or unfavorable to the Copernican system, such observations as this and the previous one (#3) clear the way for seriously considering new points of view, new systems.

5. **Jupiter has moons.** This was to Galileo's thought his greatest discovery. He named the four moons he saw the Medician planets. He was astonished by his set of observations of these little planets going about Jupiter. Jupiter moved and did not lose these moons, therefore why can't the Earth move and not lose the Moon. The uniqueness of the Earth-Moon relation was hereby lost. The Jupiterian system was also likened to a mini-Copernican system. Copernicus is

justified by analogy. Unfortunate for Galileo's preference, this argument could also be shifted and used in favor of the Tycho Brahe (1546-1601) system. In the Brahe system the Earth remains the center of the Universe with the Sun and the Moon (both regarded as planets, 'wanderers') revolve around it and the other planets revolve around the Sun.

6. **Each planet is a disk and the stars remain stars.** From this Galileo reasoned as additional evidence that the planets were like the Earth and the stars were indeed far away. These observations add more weight to the argument that the Earth is like the other planets, so why not allow it to move like the other planets. These observations also provide evidence that the greater distance Copernicus required for the sphere of fixed stars is plausible.

7. **The Sun has spots and rotates.** These spots sustained stability in shape over long periods of time and by their movement made possible the interpretation that the Sun rotates. Though this does not answer to the problems in the consideration of rotation, it does accredit by analogy. If the Sun rotates, why not the Earth also.

8. **Venus has phases just like the Moon.** Galileo concluded from this that, like the Moon, Venus shines by reflected light. Furthermore he reasoned this as conclusive proof that Venus moves around the Sun. The phases he observed were consistent with this conclusion, which contradicts Ptolemy. It is consistent with Copernicus though it is not a proof for his system because it is also consistent with the Tycho Brahe system. Copernicus himself uses the Venus variation in brightness to argue against Ptolemy.

Galileo believed that by observations and reasoning he had shown clearly that the heliocentric system is correct and that the Ptolemaic system is contradicted. We agree that he showed the Ptolemaic system is wrong. We do not agree that he showed the heliocentric system is correct. It is true that the heliocentric system was not contradicted by his observations, but non-contradiction is not a sufficient criterion for stating that it is proven. The system of Tycho Brahe was geocentric and Galileo had not presented any arguments that would reject the Tycho Brahe system.

Even though the work of Descartes, Kepler, and Newton greatly increased the plausibility of a heliocentric system, it was not until 1833 when Bessel first observed parallax that the case was put to rest. This 'observation' of parallax was not just a matter of simple observation. Note that it required compensatory calculations that made the proof more complex than this mention of it suggests. This 'observed' parallax proved that the Earth revolved around the Sun. This form of parallax was named **heliocentric parallax**. Until Bessel's rigorous arguments, the heliocentric system remained a hypothesis.

At the time of Bessel most scientists believed that the parallax would finally be observed. In this regard we see that Bessel's arguments were offered to a welcoming audience. A less friendly audience might not have been so open to all the compensatory calculations required for the proof.

SYSTEMS THEORY. Systems Theory is a meta-science. It is concerned with the possible formal relations within any subject. As such, the primary analysis is qualitative rather than quantitative. The quantitative aspects cannot exist until the qualitative aspects of the system are well-defined. There is nothing to measure until magnitude characteristics are defined.

A Science-System is a formal system used as an interface between its user and matters-of-fact. The Science-System dereferences to matters-of-fact. The matters-of-fact are understood in terms of the Science-System. This is not a definition of Science-System but rather is a statement of its main use. It may vary from a single useful concept or formula, etc., to a large scope theory. The problem of defining matters-of-fact is discussed in Addendum 2.

Science-System Life-Cycle. Following current systems theory, in a science-system life-cycle we may categorize four main phases:

1. analysis,
2. design,
3. implementation,
4. support (maintenance).

These four phases can interplay and overlap, but the general order is sustained. Design makes no sense without analysis, implementation makes no sense without design, and there is nothing to support without implementation.

1. **Analysis** begins with a study phase. This includes the study of currently accepted systems and how they came to be (genesis). These studies are done to identify the problems in the use of a science-system as an interface between its user and matters-of-fact. History is especially useful in coming to terms with conceptual modes that block our vision from the way things might be. By studying conceptual blockage in the history of science we become more wary of the presuppositions in the present modes of conceptualization.
2. **Design** is concerned with creating comprehensive logical systems that might answer to the problems identified in the study phase. Maps, models, and logic are the

main themes of design. These map-model-logic systems are then subjected to selection criteria that direct how we come to prefer one system over another. Experimental results may be required, a form of implementation that interplays with design.

Note: These first two phases define the emphasis of the theoretical scientist.

3. **Implementation** is concerned with putting a Science-System design to use. This phase can give useful feedback for design choices and adjustments. Implementation is applications oriented. This is the phase that defines the highest realm of the technologists, the engineers. The maps, models, and logic of the design phase are turned into user applications and may use trial and error as well as theory. Applications are physical extensions of the design. They are sought through experiments.

Though each phase has its genre of break-throughs, it is usually those found in the implementation phase that get stressed by the communications media. Implementation break-throughs usually have something tangible about them that can be displayed and thus are well-formed for show-and-tell. Even if the breakthrough is obscured somewhat, such as the pigs engineered to have human hearts, or human skin, or human blood, or human insulin, the media can at least show moving images of these pigs and tell in general terms what we have in common with them.

The range of implementations development is largely delimited by financial support. This financial support is also a major selection factor. In this sense, industrial or government interests or any other money sources are the major influences in defining the targets for development.

4. **Support** or maintenance of science-systems is the pursuit to sustain or refine the science-system implementations. Control, adjustment, and prediction are the main activities of support or maintenance. Support includes handling exceptions.

   Under the guidance of technologists, the technicians produce improvements or enhancements that are part of the maintenance and support activities in science. The 1995-1996 French nuclear tests belong to this phase of science-systems. As the implementations of genetic engineering get to be more and more routine they will shift from mostly break-through activity to mostly maintenance and support.

Note: These last two phases define the emphasis of the practical scientist.

Newton participated in both forms of science, theoretical and practical (experimental). (Even Bacon's chicken experiment fits this Life-cycle.)

A PRINCIPLE OF CHOICE. A major principle of choice for Science-Systems design selection is Ockham's Razor.

Ockham's Razor. William of Ockham (c1295-1349) was one of the finest scholars of his time. He is most remembered for the famous rule known as Ockham's Razor: *Entia non sunt multiplicanda praeter necessitatem* (Entities are not multiplied unless necessary). Curiously, we learn this rule only by tradition, as it has never been found in any of his extant writings. One of his statements consistent with this rule is: *Nunquam ponenda est pluralitas sine necessitate* (Never assume more without necessity). This is found in his theological work on the Sentences of Peter Lombard entitled, *Super Quattuor Libros Sententiarum* (ed. Lugd., 1495, i, dist. 27, qu. 2, K).

In his *Summa Totius Logicae*, i, 12, Ockham cites the principle of economy (parsimony): *Frusta fit per plura quod potest fieri per pauciera* (It is vain to do with more what can be done with less). The principle use made by Ockham of the principle of economy was in the elimination of pseudo-explanatory entities. He expresses his criterion in his statement that nothing is to be assumed as necessary in accounting for any fact unless it is established by evident experience or evident reasoning or is required by the articles of faith (see Kneale, W. & M., *The Development of Logic*).

Galileo quotes Ockham's Principle of Economy as Aristotle's (*Two Chief World Systems*) but I have not yet found it in Aristotle's extant works. In the cited context, Galileo praises this principle and uses it against the Aristotelians who claim it among their worthy principles. Galileo says under the name of Salviati:

> "Giving our attention, then, to the movable bodies, and not questioning that it is a shorter and readier operation to move the Earth than the universe, and paying attention to the many other simplifications and conveniences that follow from merely this one, it is much more probable that the diurnal motion belongs to the Earth alone than to the rest of the universe excepting the Earth. This is supported by a very true maxim of Aristotle's which teaches that *frusta fit per plura quod postest fieri per pauciera.*" (*Great Books of the Western World*, ed. 1952,, vol. 28, p123)

We see, then, that Galileo used this maxim in a statement that appears pro-Aristotle. However, the context renders this ironic because the statement is used against the Aristotelian framework.

Newton's Rules of Reasoning. In Newton we see the maxim take on two aspects expressed in Newton's Rules of Reasoning. These Rules are placed in the beginning of Book III of the *Principia*, entitled 'The System of the World'. This is the book on practical applications of the mathematical theory developed in the first two books.

Newton's Four Rules of Reasoning are:

> "**RULE I:** We are to admit no more causes of natural things than such as are both true and sufficient to explain their appearances.
> **RULE II:** Therefore to the same natural effects we must, as far as possible, assign the same causes.
> **RULE III:** The qualities of bodies, which admit neither intensification nor remission of degrees, and which are found to belong to all bodies within the reach of our experiments, are to be esteemed the universal qualities of all bodies whatsoever.
> **RULE IV:** In experimental philosophy we are to look upon propositions inferred by general induction from phenomena as accurately or very nearly true, notwithstanding any contrary hypotheses that may be imagined, till such time as other phenomena occur, by which they may either be made more accurate, or liable to exceptions." (*Great Books of the Western World*, ed. 1952, vol. 34)

It is in the first two Rules we see the razor has two edges. On the one edge stated in Rule I he tells us not to multiply the causes for particular effects. On the other edge stated in Rule II we are to include as many effects in nature as we can under the same causes. He treats these reductions aspectually rather than as independent of each other. Rule II begins with the word 'Therefore' showing that he regards it as a consequence of or corollary to Rule I.

RULE IV at once gives a dictum against inventing hypotheses (causal or otherwise) which could impede the general inductions of experimental philosophers and tells us another aspect of the Ockham principle of economy. Pseudo-contrarieties are not to be invented for counter arguments. Pseudo-explanatory entities are not to be invented either for or against an argument. The rejection of such hypotheses for counter arguments relieves the inductive procedure of a strangling impediment. When Newton said he does not make hypotheses, this is the kind of making that he was referring to. In *Star Laws: An Introduction to Convergence Buoyancy* I argue that Newton's notions of mass and inertia are both pseudo-explanatory entities (apodeictic fictions).

INTRODUCTION TO THE NOTION OF DUALITY. A central theme in this exposition concerns the recognition and use of dual concepts and dual systems of thought in physical theory and in particular in reference to Newtonian theory.

A summary definition of duality is:

> **Duality:** Two systems or subsystems with all corresponding *conceptual* orders opposed in pairs (one-to-one) are said to be dual to one another.

For example, if we have an ordered binary string of 0s inor 1s, such as 0001, then we have two conceptual orders to oppose. Oppose the two values, 0 becomes 1, 1 becomes 0. Therefore, 0001 becomes 1110. The other order we can oppose is the sequence. Therefore, 1110 becomes 0111. Thus we say that 0001 and 0111 are dual to each other, which we see is not just a simple negation. In a later context, this duality is shown as unique. In more complex systems the problem of establishing dual correspondences is richer in possibilities. More than one set of dualities may be possible if the phenomena under scrutiny have a non-dual set of more than one systemic description. Even so, there are also cases where systems duality might be expressible as a simple opposition, such as 1001 and 0110. Above, when Galileo spoke of the Universe turning or the Earth turning we have a simple opposition that can generate the Ptolemaic exor Copernican dual World descriptions.

For now, I state as a thesis that each expressed *system* also has a dual expression, its *anti-system*.

Anti-Systems. For this exposition, the prefix 'anti-' is used only in its logical sense of duality. In current systems literature, the anti-system is a dual form of the original system. This sense is also reflected in the formal use of the terms 'thesis' and 'antithesis'. It is also consistent with current literary use in the contrast of 'hero' with 'anti-hero.'

Question: Can two physical systems dual to each other both have validity?

Response: Science-Systems are reference-frames and physics systems are species of Science-Systems. Therefore, to answer this question we have to understand the role of reference-frames. Studies of movement begin with the idea of locus or envelope, the kinematics description.

LOCUS or **(ENVELOPE** (see Addendum 4 for the duality of locus and envelope). Since mechanics concerns the motion of bodies, then one of the first problems is to describe the locus (or envelope) of this motion. This seems readily done through applying principles of synthetic or analytic geometry, etc.

The ancients used synthetic geometry with selected reference points. These reference points, points-of-view, form the center of the reference-frame. Galileo also used synthetic geometry reference-frames. Even so, his acceleration diagrams in his *Dialogs Concerning the Two New Sciences* resemble the arrangement of Cartesian coordinates. Descartes had a reverence for Galileo and rendered him a visit. Galileo may have inspired Descartes, but Galileo never used algebra. He used synthetic geometry only. Today we consider synthetic geometry less convenient than using formulas expressed in analytic terms. An extension of the geometric reference-frame is expressed in the use of tensors, a subject I will not pursue here.

Using analytic geometry, we choose a set of coordinates (a mathematical reference-frame) and we apply mathematical procedures to develop formulas for the motion as it might be represented in this geometric reference-frame.

At first this seems a rather simple notion but let us examine this procedure to see if any characteristic of our moving body indicates constraints on such a procedure. First case, let us say that our task is only to provide a description of the locus of our observed object. What constraints does this imply? Actually from this point of view in mathematics we can say that this task, as defined, allows the most latitude of choice, because any mathematical system of coordinates with a sufficient number of dimensions may be used to this purpose. Furthermore, once our description formula is established we may readily produce transforms between coordinate systems to provide a means of translating our description into any system we desire among the choices open to us. Euler provides us with a well-defined method (in *Introductio in Analysis Infinitorum*).

We may ask in regard to this procedure what in our object remains invariant, that is, what in our object remains independent of the reference-frame used to describe it. We cannot say, for example, that its observed mobility is independent of the reference-frame used. It is quite possible to choose reference-frames that show our object not moving, and again we can choose a reference-frame that shows our object moving in any manner we wish. Observed mobility is thus relative to the frame of reference. These considerations bring to the fore the central problem of invariance. What stays invariant to reference-frame choice? These are the universal qualities Newton refers to in RULE III. What is independent of reference-frame choice? To understand the foundations of this important point, in terms of measurement, we must understand the concept of Eudoxian magnitudes.

EUDOXIAN MAGNITUDES. This concept is fundamental to quantitative reference-frames. In Physics, it is fundamental to the concept of invariant magnitudes. In modern parlance these magnitudes are called Archimedean, but Archimedes (c.298 BC—212 BC) got it from Eudoxus (c.406 BC—c.355 BC). Euclid (c.300 BC) systemized this concept in its most general form in Book V in his *Elements*. Euclid used a synthetic mode of exposition. The following outline is predominantly analytic.

For a characteristic magnitude to be Eudoxian it must be possible to add any instance of itself to itself enough times such that it will exceed any given greater instance of the same characteristic. Thus:

If $a < b$, then there must be an integer, $n \in N$, such that $na > b$,

where multiplying by n represents repeated addition.

From this we see that for a characteristic magnitude to be Eudoxian the magnitude must be ***additive***. Any instances of the characteristic magnitude can be added together to produce another instance of the same characteristic magnitude. This is also known as closure for addition of a characteristic magnitude. For an algorithmic convenience, the addition operator is usually defined as dyadic. (There is no necessary reason for addition to be dyadic.) For addition to be Eudoxian it must obey the following arithmetic rules:

1. **Rule of ISOMETRY:** If equals be added to equals the results are equal. Thus:

   If $a = b$ and $c = d$, then $a + c = b + d$.

Note that this statement remains analytic because $a + c$ is not represented as a closure. If we said $a + c = k$, where a, c, k, are single values (usually numbers or magnitudes), then k is the instance of closure, by definition. Closure is a form of synthesis. This is consistent with Kant (see *Critique of Pure Reason, Introduction,* V,1, p17f, Great Books of the Western World, ed. 1952, vol. 42).

2. **Rule of NON-ORDER:** Addition is independent of order. This Rule has two unnecessary forms:

   Commutativity: $a + b = b + a$, and
   Associativity: $(a + b) + c = a + (b + c)$.

Note that these forms of non-order are artifacts of an analytic notation system. The notation also requires us to put things in patterns that do not require the imposed pattern. *Associativity* is also expressible as

   $a + b + c = k$,

   where we may resolve k with any order of {a,b,c} by an ordinary-arithmetic addition algorithm, without disturbing the equality. Given *commutativity* and *associativity* for the notation system we can prove that addition is independent of order. The proof is usually regarded by formalists as necessary. However, this should not be regarded as profound or essential to the concept of *non-order*. If we accept *non-*

order as the definatory characteristic, then we can prove *commutativity* and *associativity*. The choice of condition order is arbitrary and remains as a system artifact. I emphasize that this is the case. It is necessitated by a notation system choice; it is not necessitated by what is functionally independent of the notation system. For the latter case the proofs are trivial, as follows:

PROOF for COMMUTATIVITY

When '+' is defined as dyadic and non-ordered, a + b may be resolved to k in any order, by definition, then:

$$a + b = k$$
$$b + a = k, \text{ and}$$

things equal to the same thing are equal to each other, therefore,

$$a + b = b + a$$

Q.E.F.

The dual argument is that k can be partitioned into two complementary non-overlapping values, a, b, in arbitrary order. This sounds analytic, but it is not. For this argument, k remains intact. This argument mode also renders this Rule by proof. It is therefore again a theorem rather than an axiom.

It is easy to prove *associativity* using *commutativity* with the Rule of *non-order*.

PROOF for ASSOCIATIVITY

If '+' is defined as dyadic, and a + b + c is non-ordered, then a + b + c may be resolved to k in any order, by definition, as follows:

$$(a + b) + c = k$$
$$a + (b + c) = k$$
$$a) + b + (c = k,$$

Note: Of course, if any of these equalities is not true, then the definition of non-order is not fulfilled, etc.

These statements with *commutativity* provide every notational permutation, and things equal to the same thing are equal to each other, therefore, etc..
Q.E.F.

We also have trivial proofs using the notion of dereference (described in another context).

3. Rule of MONOTONY: If equals be added to inequalities $\{>, <\}$, then the result keeps the original order of inequality. Thus:

If $a > b$, then $a + c > b + c$, or
if $a < b$, then $a + c < b + c$.

These inequalities are the only two possible cases for denial of equality, therefore we also have:

if $a \neq b$, then $a + c \neq b + c$.

If the Arithmetic Rules, 1, 2, 3, hold general subsistence for a domain that has infinite many distinct values we can assert two Ordinal Rules for this domain:

4. Rule of UNIQUENESS: Instances of the characteristic magnitude must be linearly ordered. Linear ordering implies a mutually exclusive relation between any two instances, i.e., a unique relation, either = or > or <, thus:

Either a = b or a ≠ b, not both,
but if a ≠ b, then either a < b or a > b, not both.

This is why computer languages can render 'a ≠ b' as 'a<>b'. The latter is read, "a is less than or greater than b."

5. Rule of TRANSITIVITY: A given unique relation {=, >, <} is *transitive*. Thus:

If a > b and b > c, then a > c,
If a < b and b < c, then a < c,
If a = b and b = c, then a = c.

The denial of a unique relation is not a unique relation.
Even so, the non-unique relations {⊳, ⊀} are transitive. Thus:

If a⊳b and b⊳c, then a⊳c, OR if a ≤ b and b ≤ c, then a ≤ c.
If a⊀b and b⊀c, then a⊀c, OR if a ≥ b and b ≥ c, then a ≥ c.

but the denial of equality (≠) is not necessarily transitive, e.g., if 3 ≠ 2 and 2 ≠ 3 then transitivity implies 3 ≠ 3, etc.. This denial is equivalent to 3 <> 2 and 2 <> 3. Here, transitivity would imply 3 <> 3, which is false, etc.

Only by experiment can we demonstrate (subject to the limits of experimental exactitude and its scalability) that a characteristic physical magnitude can appear Eudoxian or not. If, within the scope of experimental measurements it can appear Eudoxian, then we can reason that this characteristic magnitude is subject to measurement by fixed units, that is, fixed units are possible within the scope established by those measures.

A fixed unit, by definition, is a form of invariance. It is also a kind of reference-frame. Consistent with this we can say that all Eudoxian magnitudes are invariant in that they are measurable by fixed units.

That a characteristic magnitude is invariant or otherwise subjectable to measurement does not imply that the characteristic magnitude is Eudoxian. Examples of non-Eudoxian measures are; pecking order for chickens, scratching order in the Mohs (1773-1839) hardness scale, volcano scale for eruption violence (Hawaii to Volcano).

The pecking order of chickens obtains ordinality, but it is not invariant and it is not additive. The Mohs hardness scale is invariant in its ordinality (uniqueness, transitivity), but it does not render the magnitude of hardness as additive. The volcano scale sustains only a vague (fuzzy) ordinality, because a particular volcano type has a variety in eruption violence that may overlap with other volcano types.

In general, if we can establish an ordinality, then we can assign a kind of measurement. However, if it is not also additive, it cannot be Eudoxian. Notice that *the only arithmetic operator required for the definition of Eudoxian magnitudes is addition*. To complete the definition we include every possible arithmetic constraint (relation) between these magnitudes that is determined by their ordinality.

Measurement. The concept of *measurement* is fundamental to physical quantities. Science-Systems quantitative equations have no physical meaning without measurement. Since some magnitudes are regarded as Eudoxian and some are not we have a defined restriction for Eudoxian measures.

> **DEFINITION:** A *Eudoxian measure* is the count of how many of a fixed unit the magnitude contains, including fractional parts.

The fixed units are determined by convention. The count of a fixed unit expresses the Eudoxian measure. Even so, there are still problems in regard to actual measurement. In practice we still have margins of error. These problems are discussed in a later context.

Uniform cyclic phenomena are convenient for dividing duration into 'invariant' unit time intervals. We use these intervals to measure time. Though we presume an invariance for these intervals we also know that this invariance is apodeictic. There is no known method for finding an absolute invariance. Using the oscillations of quartz crystals or molecules allows the measurement of time intervals to an accuracy of 10^{-6} to 10^{-12} on Earth.

In Newton's Physics, measures of space and time are Eudoxian, weight is Eudoxian, hardness (surface tension) is not. For Einstein Physics, measurements of space and time in space-time are non-Eudoxian, mass in its full generality is non-Eudoxian; they do not address hardness. Hardness may have no well-defined meaning in the world of $\mathbf{E = mc^2}$ (a formula first worked out and published by the Italian engineer, Olinto De Pretto, in 1903). For me, hardness is a form of surface tension. The study of surface tension includes the study of hardness.

Pseudoforces. In Newtonian Physics, when we give consideration to forces then we introduce systemic constraints. How are we to show these forces in an arbitrary coordinate system? This is not an evident procedure. The fact is, we can show that care in choosing reference-frames becomes essential when we wish to analyze forces. It is easy to find a frame of reference that displays pseudoforces. This notion of pseudoforce becomes an essential notion in any physics admitting force. It also has implications regarding the epistemology of force, the essence of force. We shall show

how 'force' becomes a principle for selecting reference-frames and how arbitrariness becomes highly eliminated.

With 'force' there are certain necessary features that the reference-frame must exhibit. In the Newtonian perspective the only reference-frame without pseudoforces is an ***inertial reference-frame***. If we wanted to reason in full circle we could define an inertial reference-frame as one which does not introduce pseudoforces. This is the only Newtonian reference-frame that does not introduce pseudoforces. However, it is easy to show that in fact we can never have a perfect inertial reference-frame. This implies that in the Newtonian perspective we must always find pseudoforces that are lurking in these approximate inertial reference-frames. This condemns Newtonian physics to a physics of approximation even from its theory.

The inertial reference-frame always breaks down at its limit and it is in this sense that I say Newtonian physics and any physics based on inertia are doomed to approximation. That all physics systems are doomed to approximations might be connoted by the physical process of measuring. However, one can use the theory of limits and achieve results that are more exact than possible observation. The plausibility of these 'exact' results rests in the power of the theory that generates them. The theory of limits allows us to explore hypotheses that otherwise could not be explored. Galileo pushed physics to a new level of exactitude using an intuitive notion of limits to eliminate the effects of air resistance in measuring the acceleration of "falling" bodies. At the same time, it was his reasoning to limits that prevented him from achieving the so-called modern definition of inertia. At Galileo's limit, inertia had to follow the circular movement of the Earth's axis. He needed an inertia that kept objects to the surface of a spherical Earth that turns. In this respect, under Galileo's constraints this definition is more intelligent than the 'modern' definition. Even so, system generated exactitude that transcends observation exactitude remains suspect.

LINEAR SYSTEMS. Systems models may be linear, but experience has shown that linear cases are exceptional, often inadequate, often incomplete. Linear systems are easier to handle so that often attempts are made to impose a linear model on events that are non-linear. This practice is under criticism but has not disappeared. Newton's presentation is highly linear in form (as is his definition of inertia). Non-linear analysis in a linear framework elicits great difficulties. This may account for why the three-body problem is so difficult to solve in the Newtonian model.

SYSTEMS TERMS AND SUMMARY DEFINITIONS
The following are special terms for Science-Systems and relate to any systemic Physics exposition.

Summary Definition: A summary definition is a reference reduction form that stands in place of a more complete or rigorous definition. Many scholars apologize by calling them "crude" or "rough," etc., I do not. I regard them as convenient reductions. They can help the reader see where the argument is going. When the argument is finished and properly dereferenced, the summary definition has its proper dereference and is ready for its function as a mnemonic device for its proper dereference. A summary definition is only as rigorous as its proper dereference.

Duality: Systems with all corresponding orders opposed are said to be dual to one another. A fuller delineation of this summary definition is one of the underpinning subjects of this treatise. For duality, usually the definatory elements that form the basis for a system are used for developing the corresponding anti-system orders. If we took the dual of a consequence of a dual, then logically we should be back to the original system.

Object and Boundary: In cases where a system is conceived as part of a larger system we introduce the notion of boundary. The points that contact the system's environment define its *boundary*. This boundary separates what is inside the system from what is outside the system. For this exposition such a bounded system defines the term *object*. The context of the argument determines usage. Thus the solar system can be regarded as an object in one argument and as a system of objects in another argument. There are possible unbounded systems, therefore not all systems are objects. 'Possible' is constrained according to our ignorance.

An object has a sense of completeness in itself, but when regarded as a part of another object, then in itself relative to the new *scope of observation* where it is only a part, it is incomplete.

Closed System: If for apodeictic purposes we regard a bounded system as separate from its environment, that is, within our scale of observation, the environment has negligible influence on the system, we say that this system has a closed boundary and call it a *closed system*. An object (bounded system) need not be closed.

Monad: A monad is a closed object. This is another way of saying that a monad is functionally independent of its environment, therefore monads are functionally independent of each other. Monads are often used in thought experiments and in general can only be approximated in Nature. The degree of approximation is judged relative to the scale of the observations. Any system that is characterized as being merely the sum of its parts is a system made up of *monads*. It is the monads that form the parts of such a system. Thus by definition we can say that a system made up of monads is a system that is merely the sum of its parts. A system of this kind can itself be regarded as a monad if it too can exist in virtual isolation, that is, independently of anything else. I claim that every object that Newton conceives of in his Physics is a monadic 'object'.

An interaction among monads does not contradict the monadic aspect as long as it does not effect (as distinct from 'affect') internal changes. In this sense it is a 'billiard ball' physics. It is a characteristic of mechanistic analysis that monadic objects must touch each other to interact. In this regard Newton could not make sense of gravitation. He could not account for the assumed 'action-at-a-distance' in his monadic conceptualization. It is this monadic characteristic that allows us to say that his Physics is merely the sum of its parts. As for his World (the Universe), we do not know whether it is open or closed, but those who say this Universe is subject to entropy have decided it is closed. As for the action-at-a-distance, Newton remained confounded to the end of his days. It is one rationale for calling his book Mathematical Principles, rather than Physical Principles. This title absolves Newton from having to explain what gravitation is. Mathematical formulas are quantitative relations, not explanations. In fact, within the text he denies knowing the nature of gravitation and does not even decide whether objects he said were attracting one another are being pulled or being pushed together. This opposition remains either in a state of immanent reversibility or somehow a combination of the two.

Newton's presentation in the *Principia* is mathematical and his use of mathematics allows him to avoid concern with a physical theory. His theory of universal gravitation is a mathematical description. The universality has a basis in observations, but the proof of equivalence between terrestrial gravitational phenomena and celestial phenomena is mathematical. He accounts for the appearances, not the Physics. Other than its phenomenal description, he openly admits that he does not know what gravitation is. In this he follows Galileo.

Note: There is no intention in these remarks to demean the greatness that was Newton. Without his remarkable architectonic, Science as we know it might have suffered many delays.

Emergence: Systems Theory is concerned with principles of intertwining and connectedness, how to use modes of analysis to help build holistic models for understanding. The theory is most difficult in cases where the whole is recognized as something more than merely the sum of its parts. This something more is known as an *emergence*. It is in the works of Henri Bergson (1860-1941) that we see this concept first brought to the fore and discussed at length. Chemistry abounds in examples, such as H_2O, that seem to be emergent combinations; it is interesting to ask whether we have here **non-monadic combinations**. Is it possible for monadic combinations to generate an emergence? I argue it is not.

The maverick genius, Jan Christian Smuts (1870-1950) invented the words 'holism' and 'holistic' in his book *Holism and Evolution* (1926), which stresses the importance of structure and function in holistic reference-frames. He discourses on how losses obtain from analysis and how gains obtain from holistic views. Function is put as an emergent quality, a something more than a mere sum of its parts. His ideas had a great influence on the expositions in biology and medicine and have since extended to all systems studies. His contemporary, Henri Bergson, is brilliant and thought provoking and inspired psychologists, e.g., William James, but his manner of exposition did not excite the vast range of applications inspired by Jan Christian Smuts.

Kinematics: When we consider how to model movement independently of Physical theory, the science of motion is called kinematics. Kinematics can be thought of as geometry in motion. Reference-frames are arbitrary for geometry in motion. The most we might require of them is user convenience.

Dynamics: When we consider how to model movement as dependent on Physical influences, this science of motion is called dynamics. Dynamics can be thought of as physics in motion. Reference-frames are *not* arbitrary for physics in motion. The reference-frame must be chosen so as not to introduce pseudophysics into the movement under scrutiny. This judgment is made in terms of the physical systems used.

Faraday Attitude: Begin at the beginning, question the fundamental assumptions, repeat every experiment, judge from personal experience.

In effect, this Faraday attitude is consistent with the belief held by a growing number of physicists that the intelligibility of mathematical expressions of physical relations must be bound to their origins in human experience, which includes intuitive insight and empirical observations. The first explicit expression following this Faraday attitude is exhibited in the works of Maxwell.

David Bohm, author of *The undivided universe; An ontological interpretation of quantum theory*, 1993, spent the latter years of his life rendering this attempt at an interpretation for quantum mechanics that connects it with human experience.

REVERSIBILITY OF TIME

In Mechanics: To be able to determine the past and the future from present conditions implies a conceptual reversibility of time.

From arguments above we can say that any monadic model for Physics can be characterized as being merely the sum of its parts. This characteristic allows us to derive the conclusion that given the position and forces acting on an object in a closed system, the future or past, for positions and forces, are determinable. The past or future positions and forces for these objects within this closed system are predetermined "knowable" analytically, whatever the scale; predetermined "knowable" because of the monadic definitions given to these objects/systems. The objects referred to in this case are monadic, as is the case for any parts that are regarded as independent of each other, as merely summing to a whole. Furthermore, these monads may group into systems of a larger scale as constituent objects.

That this knowing through reversible time is conceptually possible is more grandly stated by Pierre-Simon Laplace (1749-1827):

> "Give one moment to an intelligence that can comprehend all the forces that animate nature and the respective positions of the objects that compose it, if moreover this intelligence were vast enough to subject these data to analysis, it would hold in the same formula the

movements of both the largest bodies in the universe and those of the smallest atom; to it nothing would be uncertain, and the future as well as the past would be present to its view." (Introduction to *Oeuvres* vol. VII, *Theorie Analytique de Probabilites*)

His conclusion is derived from his analytical extensions of the Newtonian model of the World. Time in this sense is reversible. We see also that this connotes a form of determinism.

Laplace knew full well that this kind of knowing is not within the possibilities of ordinary human experience. Even so, to believe the statement requires that we believe in the mechanicalness of the Universe. To believe in this mechanicalness itself requires a knowledge beyond current human possibilities and therefore begs the question. Accepting exaggerated extrapolations from our Lilliputian existence requires a naiveté that has not been adequately stressed.

In Classic Field Theory: With the re-introduction of a field theory by Faraday (Descartes and his vortices was the first) and mathematized by Maxwell we do not have any real change from the monad conceptualization, though many physicists agree that a fundamental change took place. Note also that the problem of 'action at a distance' dissolves in a field theory where everything touches everything else with force fields. I argue that the only difference in conceptualization in regard to monads is that the smallest monads became infinitesimal in their mathematical treatment. The field is still the sum of its parts, the parts now being the field points, and time is still reversible.

From the Maxwell field equations and their arguments, we can determine the electromagnetic forces acting at any particular point. It is true that these forces are related to the field as a whole, but it is the local forces, the forces at any point that are calculable from the field equations. It is this that keeps this mode of analysis monadic. It is true that the equations defining a field must define every point in the field and in this sense are holistic. This is an important change, but it is not fundamentally different in terms of monads that determine a mode of analysis, a mode of thought. For fields, when we integrate we are taking limits of infinite sums, when we differentiate we are deriving dimensional values, such as point value limits. That we speak of infinite sum limits or particular point value limits, both are monadic modes of conception. Integration does not create an emergent entity, integrated objects are indeed only expressions of them as the limit of the sums of their diminishing parts, in effect, an analytical method. This argument regards limits as analytical arguments and therefore not as emergent entities.

Monadic conceptualizations are readily amenable to reduction methods.

Irreversible Time: A newer thrust in Physics research is into phenomena that seem irreversible. For these phenomena *time* becomes a one way arrow. Entropy is commonly given as a characteristic of these phenomena. This example *per se* is worth re-examining since, in its original context, entropy was conceived in a monadic physics. Nevertheless, it suggests that there may be occurrences which we cannot retroject to see genesis, nor can we project these occurrences into the future to foresee some outcome. Accuracy in extrapolation (e.g., prediction) fades rapidly. In Physics, many examples of this problem have been found in problems concerning changes in state. Our present theories are especially weak in modeling certain changes in state. The study of turbulence yields many examples. When does water boil, etc. This problem is not as simple as elementary textbooks imply. It is one of the important research areas of Chaos Theory.

KEY WORDS
Systems Theory, Science-Systems Life-Cycle, Principle of Choice, Ockham's Razor, Duality, Anti-systems, Locus, Eudoxian magnitudes, Eudoxian measures, Pseudoforces, Critical test, Emergence, Object, Closed System, Monad, Reversibility of Time, Irreversible Time, Entropy, Kinematics, Dynamics, Faraday attitude

END Addendum 1

Addendum 2
STATUS OF LOGIC IN SCIENCE-SYSTEMS

Introduction. *In order to understand specific constraints in the use of logical principles in science investigations we need to clarify the logical status of science-systems, that is, the role that logic can play in regard to science-systems. In this regard this addendum develops three theses*:

Thesis One: All science-system premises must dereference to the data-field through tests, directly or indirectly.

Thesis Two: From a given science-system we can use logic to search for new premises to dereference in data-field tests.

Thesis Three: There is no necessary connection between a particular science-system and the data-field it dereferences to.

'Thesis' here connotes that no formal proof is provided. That they are reasonable principles is judged from the results, predictions, descriptions, explanations that are derived from them and appear to hold. Though more general in scope, this status is similar to Newton's three Laws of Motion or

Maxwell's field equations in the fact that they also have no direct derivations or proofs.

Habitual Thought. The milieu that gives rise to the beliefs one adheres to sinks into a background dimmed by habitual acquaintance. Socrates asserted that the unexamined life is not worth living. It would likely be too strong a claim even for those who reflect on their lives that they found the sources of all the habits of custom and belief that direct their lives. As ingrained patterns these habits impel one to prefer one perspective over another possibility. The knowledge and understanding we have of ourselves and the world around us are couched in these habitual modes of thought.

Conventions. A convention refers to any notion determined by agreement, no matter how fuzzy or crisp the determination might be, no matter how tacit or explicit the agreement might be. Such conventions always have a mediate usage.

By conventions we can use money as a measure of value for goods and services, we can use money for exchange of goods and services. Exchange is the mediate use for money; money is the mediation of exchange. The more goods and services we obtain with a fixed amount of money, the more value we ascribe to the money. Money itself is a commodity for exchange, but its value remains determined by what a user may purchase and this determination is by a moveable convention. The original usage for conventions remains mediate. Any other use opens for the possibility of abuse or a misunderstanding of the basis for conventions.

Language conventions, as with all conventions, must be learned. It is somewhat arbitrary which language system be adopted in a community as long as it is sufficiently rich or developable for the requirements of verbal expression. Thus we may speak of a particular language as an arbitrary convention. It is as arbitrary as the dollar system or the euro system.

Dereferencing. When we identify a reference language with the experience it maps, we ignore the mediacy required for valid language reference. It is like regarding the map as the territory (this problem is a major concern in General Semantics).

With toddler children, when we point to an object with the intention that the child regard that object, the child regards the pointing finger. The sequence of events for fuller comprehension requires two steps:

1. regard the pointing finger,
2. identify the object pointed at.

The first step is the ***reference***.
The second step is the ***dereference***.

Young children who see the pointing finger generally do not perceive it as a reference intended for dereferencing. It is taken unwittingly as its own reference. This precludes the intentional dereferencing of the finger. As the child learns a reference language, the problem of reference and dereference is fundamental to what the child is learning. Every symbolic representation implies these two aspects. This subject still merits more research.

Our language is one of our most elaborate resources for referencing our experience. Our experience, however we hold it, is the field for dereference. The couch for the representation of our experience is memory. This includes our previous experience, our present experience, and our anticipated experience.

In the case of the pointing finger, maturity of judgment requires that we know the sequence of seeing the finger and dereferencing the object pointed at. In the case of language reference, there are clear instances where something other than dereferencing seems to happen. The possibility of this other something is enhanced by the dependence of dereferencing to remembered experience. Fluency in speech requires considerable dereferencing to memories. This helps speed verbal expression, but this speed can degenerate into a trade-off with accuracy. Even during the course of a day, memory may diminish or amplify the intensity of its holding of original experience. For some types of experience, this generates a trouble causing distortion in the accuracy of the dereference.

A child who unwittingly takes the pointing finger as a self-reference can learn to point a finger without yet understanding what the original intent was. This important lesson is applicable to adult language use. Most adults use many language utterances that cannot be dereferenced to memories of direct experience. In effect, they use words that point to nothing from their direct experience, but they use them as if they did. If you have been to Larissa and can tell others how to get there, you are a right and good guide. If you have never been to Larissa and can tell others how to get there, you are also a right and good guide, etc.

Formal Reasoning. Formal reasoning is without dereferencing. This is the form of 'thinking' that we can program computers to do. This is the ultimate in formal reasoning. In effect, general purpose computers can compute with formal rules of syntax without dereferencing.

Computer knowledge-based systems were founded on the recognition of the importance of dereferencing. Only a knowledge-based system has the capacity to simulate dereferencing. I say simulate because the current computers can only dereference to second-hand data and information. Even so, much of human 'knowing' is of this second-hand sort. If you are committed to accurate dereferencing, then you hold second-hand information clearly marked as second-hand information until it is properly dereferenced.

There is much current research in computer hardware to extend sensory devices to imitate human sensors. This widens the horizon for computer dereferencing possibilities. The great stumbling block to holistic grasping, for computers a simulation of perception, is to modify sensory input into a perception, the problem of perceptual closure (cf Bergson, Smuts, and also the Gestalt psychologists). In digital computers, data and information remain monadic and in this sense express important limitations of their creators and their current monadic reasoning methods.

Using our reasoning independent of dereferencing is, in a manner of speaking, a ceasing to adhere to matters-of-fact, a ceasing to adhere to content. This mode of ceasing to adhere can be useful in coming to terms with our memories (interpretation of the past) and anticipations (formulating future possibilities). Formulating expectations also sets us up for disappointment or fulfillment. Only if we are able to dereference as we go can we claim we are adhering to matters-of-fact or our memory of matters-of-fact. Our ability to re-adhere to matters-of-fact provides a context for our attempts to implement or experience our anticipations.

In effect, a reference-frame provides a mode for **addressing** personal events (matters-of-fact in sensation, perception, or memory). To dereference these events is equivalent to perceiving **what is at the address**. Our general problem with reference-frames is how do we discern what is put into the event by the reference-frame and what is not. The what-is-not is a data-field dereference. The what-is are artifacts from the reference-frame. Are 'God', 'Freedom', and 'Immortality' artifacts born from reference-frames, or can they be dereferenced to the data-field, to matters-of-fact. Most people do not bother to ask. Kant made these questions a central issue in his *Critique of Pure Reason*.

Many 'well educated' people use terms that are empty when dereferenced and they talk as if these terms could be dereferenced. In effect, they use terms for which they have no valid address for dereferencing. When we treat such invalid addresses as if valid, we have an address where nobody is home, etc., the attempted dereferencing can end up as much ado about nothing.

Faulty dereferencing is a common occurrence. Many of the problems in human references are directly related to the individual incapacity for accurate dereferencing. This is complicated by the fact that internalized reference-frames may guide too strictly in what one pays attention to and what one neglects. In such a state, the reference-frame controls the person and the person is not controlling the reference-frame. For me, the highest achievement in reference language usage is to recognize the full extent of my own possible dereferences in everything I say as I say it. This requires an attitude of interrogation and much practice. The idea is to keep a reference-frame in the status of tool. It is interrogation that allows me to 'cease to adhere' to a reference-frame so that I can control its use. I can then maintain it as a tool that I control, rather than sustain it as a framework that controls me. This is another way of saying that through interrogation I can learn the difference between what I know and do not know through reference-frames.

When I talk about 'gravity' the word is supposed to be a reference. 'Gravity' is like an address. If I pose the question "Why does the book fall to the floor?" and the answer tendered is "gravity," no answer is received unless I can dereference 'gravity'. We are like toddlers looking at a pointing finger, who are challenged to dereference the word 'gravity'. Other than associating this term with certain superficial sequence events, we find no dereference that derives from maturity of judgment (within the context of experience).

Apodeictic Conventions. Certain conventions are used in argument to exhibit principles or give proof. This use in argument is apodeictic, keeping close to the original Greek, 'ἀποδειξις', 'a setting forth', 'arguments in proof of'. Thus conventions used for proof or for instructional exhibition are *apodeictic conventions*. Because they are conventions, their proper usage is mediate.

Reference-frames for intelligibility that are couched in language, are expressed with all the arbitrariness of language, of words, and of grammar. A language, a particular system of words and grammar, may influence the interpretation of any further experience, may order thoughts in ways that include systemic induced error, may impose a system generated bias. A word like 'sunset' may disposition us to regard the Sun as moving around the Earth rather than opening our thoughts to the dual possibility that the Earth rotates. When this occurs, it is not us who are in control of our language use, it is our language that is in 'control' of how we think (cf Friedrich Nietzsche (1844-1900), Edward Sapir (1884-1939), Benjamin Whorf (1897-1941)).

Accepted social behavior deemed 'constructed', 'artificial', or 'formal', is a matter of convention. When through practice we finally act in a seemingly natural and spontaneous fashion we say that these conventions have been 'internalized' and insofar as the structures are used automatically, through habituation or habits of expression, the degree of this internalization may be estimated. Automatism in conventional usage indicates an interpretation through an internalization of the convention when the automatism is a matter of learned behavior of any kind. It is this internalization that renders us fluent, but it can also render invisible to us the possible identifications. Habit makes efficient, but it can also make blind. Furthermore, when you are used to an environment you can become blind to certain things in it from a kind of saturation. There may be an aroma for the atmosphere, but we cannot detect it. It is an aroma lost to us through a saturated experience.

A convention is an artificial device. Its artificiality implies a kind of arbitrariness. This arbitrariness admits a kind of fictional aspect, something invented or made up (whether intentional or unintentional). When we become aware of this invention aspect we find the possibility of a kind of freedom from it. This freedom is a ceasing to adhere to the immediate dictates imposed on us by our learned conventions, our cultural heritage. We may abstract ourselves from it and think about it. But habit can interfere with this 'ceasing to adhere'.

Cultural Cloning. People who remain under the control of their childhood shadows also maintain the necessity for their personal death. Without death those humans predetermined by their childhood shadows can only create an anthill. New paradigms flourish best in the new generation after the retirement of the old generation.

The scope of cultural evolution is expanded by the degree of our inability to clone our offspring culturally. *In communities where cultural cloning is most successful we find the least changes in conduct.*

The obstacles to cultural cloning may be characterized as exterior and interior to you. They range from the learning of liberal attitudes to the inadequate assimilation of the culture you are born into, to acquaintance contact with foreign cultures, to shocking experiences, to wars or natural disasters.

Our conventions define our civilization. Most civilizations have forceful sanctions for individuals who do not respect the accredited conventions. It is through interrogation that we can put conventions back into their proper place of mediate devices for civilized life. This form of ceasing to adhere can remove our identification with conventions. We might define the level of a civilization by the measure of how explicitly its elders teach conventions as conventions. A predominance of *tacit* conventions would signal a low civilization. A predominance of *explicit* conventions would signal a high civilization. Unfortunately, this harsh demarcation places every known civilization on the face of the Earth into the low category. This liberation from convention can have destabilization effects that are not wanted by the authorities nor wanted by the general public of a given society.

Apodeictic conventions, as conventions, inherit a fictional aspect, a kind of freedom for invention. Apodeictic conventions themselves may become restricting principles in a certain mode of thinking, forming an adherence to a way of reasoning that does not adhere to matters-of-fact. This possible form of freedom with respect to science is pejorative. This is why Aristotle believed that with knowledge (science) one loses freedom. If you 'know' the way things are then you lose the freedom of choosing the way things are. The question is, can we know the way things are?

Likely Stories. The impetus to create likely stories has never been adequately explained. Ingenious minds early set themselves to inquire how experience is to be accounted for. Intellectual history is regarded by some scholars as a history of development from mythology to science.

Some ground of justification or explanation is thought of, i.e., a likely story that seems to account for an experience and seems to reconcile it with the current state of beliefs. Sometimes a description is given of an occurrence but its likely story, the story that accounts for it, changes. With time, different stories evolve that become more effective in their applications, displacing those with less power. A kind of 'cultural' selection occurs here. Then the understanding adapts itself to the new story that has been found for explanation, and the description receives a new life and in time the description gets modified to fit the new story better. This series of events is exemplified in the genesis of celestial mechanics. For example, this kind of change in explanation is shown in part by Kepler's kinematics laws of planetary motion. Kepler's explanation was displaced by Newton's in terms of Hooke's suggested use of the inverse square law. Another example, the Lorentz transformation, a description (empirical); Lorentz's explanation was displaced by the Einstein Relativity theories, etc.

Social Reference-Frame. We like to believe that in science our knowing moves toward invariance with respect to any particular social reference-frame. Just as logicians want to believe that a proposition or logic system is invariant to any particular language it is expressed in, scientists want to believe that a science law or science-system is invariant to any particular social reference-frame it is expressed in. Insofar as it is not, it is not science.

Thus we may say that scientific meaning or intelligibility for a science-system derives from its social reference-frame but it is always expressed in an apodeictic reference-frame that is intended to be independent of any particular society, we want this independence as well for our sensing of the data-field. If this independence is achieved then it would represent the demarcation between what is science and what are social beliefs. This is not concluded from a history of how science came into being but rather a stipulation that no scientist would want to deny.

Culture-Independence. Scientists do not want to believe in cultural relativism for science. There seems to be a concerted effort to accredit a monoculture for science that is somehow culture-independent.

The maintainers of science belong to many different cultures and this makes these maintainers a multicultural community. However, the members of this science community seem to have a sense of membership and unity in science, and this makes it look like a kind of monoculture. Most of its members believe that scientific observations are culture-independent.

How is it possible for representatives from so many different cultures to regard scientific inquiry in much the same way? Historians teach us that through time this claim is not true, scientific attitudes change through time. Beliefs in one epoch are different from those in another epoch. On the other hand, coextensively, it is a fact that many different cultures do regard scientific inquiry in much the same way.

Is it that this apparent unity derives from a form of objectivity and that this makes scientific conception modes seem so alike from culture to culture?

Or is it that we view this community in some simplistic way that leaves out the differences we could find if we looked for them.

Or is it that scientists themselves grew from a small loosely connected international group to a culture in itself with its own evolution.

Whatever the case, there seems to be a science community and we often find some means of regarding a community as a monoculture even when it is not.

A number of recent thinkers believe that our science is *not* culture-independent. This is the case for Alfred North Whitehead (1861—1947), Michel Foucault (1926—1984), Thomas S. Kuhn (1922—1996), and others, including me. Observations are not innocent of culture and upbringing. How we look at the world is largely influenced by artificial constructs that we inherit and internalize through training, growing up in a certain society, in a certain family. These experiences even form the basis from which we generate schematic or paradigmatic protocols for what we regard as known or knowable.

I would like to suggest that because of the monoculture assumption, it has been taught to us that science really is objective, that is, truly culture-independent. As said before, this is suggested by the fact that so many different cultures support scientists that seem to agree on what kind of evidence is to be regarded as scientific, or so goes this current belief. The scientific community at large would balk at the suggestion that the particularities of a culture play an important role in what we regard as scientific.

The first question posed here is--Is the scientific community a monoculture? Scientists themselves want to believe that what they do is culture-independent and this belief denies a community pluralism in regard to scientific conclusions. Therefore, the answer that is taught to us is *yes*, the science community is a monoculture.

The Science Community. Shared experiences form the basis of community. If you want to establish a new viewpoint you must obtain approval from the concerned community or you must form a new support group outside that community. The survival and further development of your ideas depend on it. Whatever the case, every person, as a member or as an outsider, who claims to have a new science viewpoint faces the science community. This community represents the powers that be who examine the proofs that claim the new view is valid. If the referees of this community do not accredit the work, then it gets passed over unpublished by the community. This passover is another case where the science community can take on a monolithic appearance. Some scientists (e.g., James Lovelock) claim that today's scientific community has a self-imposed Inquisition, a cult of accredited research. The science community now has its politics and it does not support mavericks.

I would characterize every major change in science viewpoint as being associated with a break from the monolithic scientific community. This break represents a cultural difference, a difference from whatever the current science culture accredits. Entrenched members of the monolith claim that they are open-minded to new points of view. They can offer as evidence that there are arguments within the scientific community, but the first problem in this community is to get the new viewpoint exposed for criticism. This is a difficult problem involving confrontation with politics, bias, and prejudice.

It is a community of people that generates viewpoints and protocols for what we accredit as 'normal', even for what we regard as known or knowable. Kuhn teaches us that "The commitments that govern normal science specify not only what sorts of entities the universe does contain, but also, by implication, those that it does not." This indicates that certain things are systematically left out of the discussion. The educational initiation that prepares and licenses the student for professional practice is based on these constraining commitments. In spite of Whitehead, Foucault, Kuhn, and others, we would like to believe with justification that our scientific observations are culture-independent. If it were so, this would be a large step toward justifying any claim to objectivity for this human endeavor.

Is Objectivity Possible? Many members of the scientific community reject the thesis of cultural dependence and expressly despise it. Cultural dependence denies objectivity.

For myself, asserting cultural dependence seems a strong step in the direction of accurate perception. Historians make it clear that scientific attitudes change through time, i.e., that science is a changing endeavor. Some of these changes are regarded as revolutions in science-systems. Whether the change is portrayed as gradual or revolutionary, either case displays changes over time. This indicates that either objectivity has *not* been achieved in science, or we need a new definition for objectivity. Renaming it 'intersubjective testability' does not solve the problem. If 'objectivity' has no valid dereference, then neither does 'subjectivity' as our understanding of these concepts is dependent on their opposition.

I maintain that culture-independence has not been fully achieved and that this unachieved status places any claim to objectivity in corresponding doubt.

Scientific Truth. Recall that in logic we have a distinction between validity and truth. *Validity* refers to the correctness in reasoning such as syllogisms are said to display. *Truth* in a conclusion is said to follow mediately from the premises if the premises are true and the reasoning is valid. This truth is mediate (hereditary) if it follows from the premises to a conclusion. If the conclusion is logically true but in fact is false, that is, does not correspond to any matters-of-fact, we may still say that the argument is valid when the protocol for reasoning has been followed. Such a case may bother us enough to re-examine premises and maybe even the protocol.

Truth values are a defining characteristic for certain kinds of sentences.

If a sentence may meaningfully be assigned a truth value we call this sentence a *propositional form* (cf Russell and Whitehead). When we assign a truth value to it we call it a *proposition*.

> **Note:** A binary valued proposition value set is an example that contradicts the abstract set theory supposition that the empty set is a member of every set. Here the empty set value sustains the propositional form only.

George Boole conceived these values as numeric. In his most general consideration these values could include any value in the closed interval $[0,1]$ of Real values. This includes the possibility of representing partial truths (more on this in Addendum 4).

Propositions are intended to be independent of the language they are expressed in, just as science statements are intended to be independent of the culture they are expressed in. If the assigned truth value is accepted without formal proof we call the proposition an *axiom*. If the proposition is accepted from a formal proof we call it a *theorem*. No one has ever proven that axioms are necessary. Systems in which axioms seem necessary may actually harbor hidden constraints, unnecessary constraints, systemic artifacts. I use the phrase "seem necessary" because in a later context I show how a logic can be designed without axioms.

Validity is a defining characteristic of an argument. It is a mediation protocol for arguments and may be used for deciding the truth value of a conclusion. Axioms or theorems used as premises provide statements for the argument accepted as true. Establishing direct truth values that involve perception and judgment entails other modes of dereferencing.

For formal logic, the problem of dereferencing propositions to a data-field is metalogical (cf Boole). To start an argument, all we need are premises with assigned truth values and no concern need be given as to how the truth-values were assigned for us to follow the operator rules to new propositions. Furthermore, in the implementation of formal reasoning we are not usually concerned with how the schematics or paradigms for reasoning were deemed correct. In general, we can say that any question concerning formal logic which does not simply entail its implementation according to predetermined rules is a metalogical question. This does not mean that concerning logic we should not ask metalogical questions. The logic theory that underpins the logic operations consists mostly of this sort of inquiry.

In logic theory, the problem of how we assign truth is an important one if formal logic is to have useful de-referencing. The reader may wonder what tools are available for logic theory proofs. As it turns out, current logic theory turns formal proofs back onto its own work and we find that we turn full circle in using logic to prove that logic is consistent. Avoiding this circle is one of the underlying concerns of Addendum 3 where I define a logic without axioms.

This section could be rewritten in a way that is independent of the old validity argument.

Data-Field. Dereferencing ultimately implies a data-field, the field of concern, the field from which we choose our matters-of-fact. What we pay attention to in the data-field I call matters-of-fact. It is our first phase of semantics. If we make a statement referencing matters-of-fact, the statement is the reference and the matter-of-fact is the dereference. If we also use matters-of-fact paradigmatically, then 'reference' and 'dereference' become immanently reversible in their possible applications. In other words, the term 'tree' may dereference to individual trees (schema or formula usage), or an individual tree may be used as a model to de-reference to other individual trees (paradigmatic usage). If we then assign a truth value to the statement, "Maple is a kind of tree," then it becomes a candidate for a premise in an argument.

How we may decide that a particular item of experience is a dereference is not easy to characterize but we generally include all possible differentiations as possible data. An actual datum is whatever we can pay attention to, it constitutes whatever we experience that we might be able to indicate (including concepts, imagination), it may be regarded as indivisible or as highly complex, as fuzzy or crisp. It is a candidate for making a statement concerning matters-of-fact. Proceeding in this way we may discover foundation statements for new systems of thought, remaining vigilant for system generated artifacts.

What is Truth. In this context we may then ask, What is the dereference for truth with respect to systems of thought and their data-fields? Such a question revises the usual notion of Truth. Here we judge a truth relative to systems of thought. Here it takes on a definition that the naive question, What is Truth, does not suggest. I already said that logic has a demarcation between the notions of validity and truth. It is possible that different systems of thought may spread over the 'same' data-field, or more precisely, two systems of thought may have the same data-field spread, though they have different modes of argument, different grammars, different modes of reference. For example, in a mathematics model, we may trace a trajectory with respect to a Cartesian frame of reference or with respect to a polar frame of reference. The formulas representing the trajectory of a moving point are in different terms even when they are for the simplest representations, and furthermore, the respective formulas are incompatible. What is a true representation (corresponding reference/dereference) in one reference-frame loses its dereference when used in the other. Even if we placed the zero points or any other points in coincidence, it may be true that these points relate to one point in the data-field, but the representation of the location is given by the reference-frames and the reference-frames are different. This does not mean that we cannot establish transforms between the two systems so that we can, in a manner of speaking, translate the representations in one system into corresponding representations in the other. This is a case where we conceive that the dereference is invariant but has a variety of possible reference modes.

That we can judge the reference modes as 'corresponding' suggests that the data-field has somehow been used as a reference-frame. The relation between the Celsius (as revised by Linnaeus (Sweden) and Cristin (France), 1745) and the Fahrenheit temperature scales is a simple example. These scales exhibit two different quantitative conventions for the same data-field spread. In regard to representation there is no loss or gain in the accuracy of the representation between the two scales. There is just as accurate a representation in 212° F as in 100° C. The translation from one to the other is a simple formula; $F=(9/5)C + 32$ or $C=(5/9)(F - 32)$. This is also a case where the formulas work even if temperature is non-Eudoxian.

But apodeictic systems also may differ in qualitative conventions. Different reference-frame modes differ in their qualitative distinctions. Though we may find correspondences and modes of translating for qualitative differences, we find greater difficulty in translating between such systems. Something may be lost, something may be gained. And of course the worst scenario is to lose wanted representation and to gain unwanted representation. The Einsteins claimed that Newton's system is a special case of Relativity, indicating that their Relativity theories account for a larger scope of the data-field. Here the Einsteins claim a gain in wanted representation, a more comprehensive data-field spread.

> **Note:** In general, the more differences between two cultures the more culture artifacts that have no direct correspondences. Translating from one language to another includes the problem of perspective mismatches through cultural differences.

One of the characteristics of the relation between the Ptolemy system and the Copernicus system is that one is a geometric transformation of the other. The two theories have equal scope and equal difficulty. The losses and gains in this transformation are detailed in Addendum 6.

With such a relativity of different systems for the data-field spread, how can we speak of Truth which likewise must be considered in relation to different systems? At least we should distinguish different truths relative to different systems of thought or show that there are no differences in this regard. If there are differences in mode of expression from system to system, or reference-frame to reference-frame, then truth, which we are accustomed to assigning to certain statements, is relegated to a relativity that seems to subvert the possibility of dereferencing an 'invariantly true statement.'

Invariant Truth. An invariant truth is one that is independent of reference-frames or systems of thought. The careful reader may say, Yes but what about the data-field? Doesn't it remain invariant to changes made by reference-frames? In coordinate geometry, in Euclidean space, and in Newtonian mechanics we can, though not without some difficulties, establish a theory of invariance. In a specific system of measurement, measuring the length of a finite straight edge yields the same number-neighborhood no matter what Euclidean, Cartesian, or Newtonian reference-frame is used, as long as the unity is a common measure within the constraints of the specific system used. Or so it seems.

Data-Field Access. In a manner of speaking this length that is independent of other reference-frames and even independent of a standard unit as reference-frame, is a feature of the data-field. In a manner of speaking we could say that the data-field is invariant truth and is itself a kind of reference-frame. But this is only a manner of speaking because *when we consider the data-field we must also consider our access to it*. One limitation is already indicated by calling it a data-field. We presume a certain access to data, data by its definition indicates anything that can be differentiated by us and represented either through concept or by image (concept and image are rigorously different as the concept of a circle and the image of a circle exemplify). Beyond this we cannot see.

Transcendental Knowing. In the past it has proven unwise to presume there is no beyond. This 'beyond' has been characterized in several different ways. One way is like the Kantian notion of noumena to which by definition we can never have direct access. Another way is through instruments for mediate access data, data available to us only through instruments. For these arguments I shall ignore the Articles of Faith (a third notion of 'beyond').

We know there are data easily distinguishable with the unaided senses, but our powers of discrimination are limited. Our senses have limited ranges with different types of limitations for each sense, both qualitative and quantitative. For now it seems that these limitations are expressible in three dimensions; scope, resolution, intensity. All of these dimensions have size (ranges), size admits non-linear scopes. Size is couched in time and space. Space, time, and contrast are conditions for measurement (Wittgenstein named the third condition 'color').

Some examples are given to clarify these terms.

> **Scope:** With respect to seeing we have the spectrum of light, with respect to hearing we have the range of tones, etc..
>
> **Resolution:** A synonym for resolving power, it is measured by the minimum difference at which a difference between two values is perceived, the difference between two colors (shade or hue), the

difference between two sounds (timbres, tones), etc..

Intensity: With respect to seeing we have dimness and brightness, with respect to hearing we have softness and loudness, etc..

Size (*Scale, n-dimension Range/Scope*): Range is quantitative. Scope includes qualities and range. Plotinus first pointed out size with respect to sound and Hemholtz was the first to measure it. Size may impose a variety of constraints. With images at extremes we know for our senses some things are too dim or too bright, too small or too large, too close by or too far away for direct determination, etc.. To view the Sun we use smoked glass, etc., to view dim stars we use long exposures to film or use computer enhanced images. Galileo showed that bigger may not be better, a larger lever may break under its own weight, he also showed how the shape of bones requires modification with size (scale), etc. Chemical engineers know that certain results obtained by mixing a few grams in small laboratory conditions are not gotten when these same chemicals are mixed on an industrial scale. This introduces the problem of *scalability*.

Transcendent Instruments. We develop instruments that we reason extend our distinguishing powers beyond the limitations of our natural senses. We now believe that our natural sense-ranges yield only a small fraction of the possible data-field. Our instruments may bring us to new limits which we attempt to breach with even more refined instruments. In some cases some believe we have reached near ultimate limits that instruments will never have the occasion to measure beyond a calculated physical limit. Many physicists believe that no material temperature is possible below a standard value for absolute zero, that the Kelvin scale cannot have negative temperatures. To sustain this absolute zero there can be no absorption and no dissipation of heat. This implies there is no transfer of heat or cold. This should be 0 kelvins ('kelvin' is the unit, not 'degree'), which indicates absolute zero.

It is difficult to eliminate the participation of the measuring apparatus in the creation of the 'observed' results, that is, the observer instruments may influence the observed. A disregard of this possibility can lead to highly artificial theories. Microparticle physics suffers from this (related in more detail in Book I). This disregard obfuscates the definition of 'empirical'. It is this consideration that defends contemporaries of Galileo who disbelieved in his telescope observation interpretations.

Independent Agreement. We like to suppose that investigators working independently of each other would come to the same determinations, a kind of invariance. Our satisfaction with regard to evidence and arguments presented in favor of this assertion depends much on a notion of standards that we have acquired in our experience. This includes standards in two modes, those relating to reference-frames for intelligibility (quality) and those relating to concomitant reference-frames for measurement (quantity). Reference-frames used for measurement are known to be arbitrary within the latitudes of usability and are decided by convention. Reference-frames used for intelligibility are usually not thought of as arbitrary because they form the very basis for how we see ourselves in our world, we live by them and through them. We do not want them to be arbitrary. It is reassuring when investigators working independently of each other can come to the same determinations. But is this illusion?

Apodeictic Fictions. From within a given mode of thinking one may lose or be deprived of choices. This is still quite abstract so let us take an example, a paradigm to illustrate the possible dereferences. To aid in this example I shall first characterize three artificial categories of persons in a social context. When we use categories that are artificially constructed but useful in argument we are using apodeictic constructs. Apodeictic constructs are known to be artificial, apodeictic conventions may or may not be known to be artificial. If these constructs have no dereference to matters-of-fact, they are apodeictic fictions. The following constructs are partly fictional because no actual person exactly fits any of them.

Participants: Insider members of the society, who internalize the habitual modes of thought through a thorough apprenticeship.
Observers: Outsiders who have little or no apprenticeship for internalizing the habitual modes of thought of the visited society.
Participant-Observers: Are outsider-insider members of the society that live in the hyphen of immanent reversibility, an adherence through apprenticeship, a non-adherence through interrogation.

In general, well-conditioned **insiders** do not question custom or law unless it is to assure that they themselves are conforming within the latitude allowed by custom or law. The arbitrariness of social conventions used in their reasoning is rendered invisible to them through habituation or trained denial. If behavior criticism exists, it targets non-conformity and uses sanctions to render conformation.

A person who 'knows' that if they do not get baptized that they will go to hell when they die may be respecting a long well-defined family tradition. As an insider (s)he does not have the perspective required to test for the arbitrariness of this act. (S)he well 'knows' that testing it could have dire consequences.

As an **outsider** one may readily note the differences from one's own beliefs and practices. At the same time there is a possible blindness to the systems of emotions and values of the foreign society, the very drivers of behavior for that society. The outsider comes from another point of view. Even outsiders willing to adapt to a foreign society do so usually with an accent. As an outsider, one may never fully understand the 'grammar' of behavior in the way required to live it without accent. Whatever the degree of accent, fluency requires internalization.

Perhaps a non-gullible outsider comes to know about the local belief concerning baptism. This outsider may become confused, or may decide that it is a useless act, or may be glad that (s)he found out in time. A recognition of choices can impose interrogation onto the observing outsider whether it is wanted or not.

The **participant-observer** is in a state of immanent reversibility, an adherence developed through apprenticeship and a non-adherence obtained through interrogation. This state is regarded as the most effective for students and travelers who want to understand other cultures.

It is through interrogation that the drivers of behavior are rendered more visible. This discussion implies that habits and customs drive our actions and influence what we pay attention to and what we neglect. These internalized drivers are the architects of our reference-frames that we use for defining our world-view. The discussion also suggests that interrogation can help us cease adherence to these internalized conventional principles and beliefs. We become more open to seeing factors significant to a foreign culture that have a different significance from our own, a useful trait for travelers.

Data-Field Testing. A reference-frame may have apodeictic usage. Granting an apodeictic reference-frame that makes use of a data-field, we now ask, what is the nature of the relationship between the apodeictic reference-frame and its data-field? A careful consideration requires regarding possible feedforward/feedback relations between them. In feedforward the reference-frame conditions how we access the data-field. Regarding how the reference-frame influences our access to the data-field we consider what to look at, what we look for, what we neglect, etc. In feedback the data-field can influence the reference-frame. Regarding how the data-field influences the reference-frame we consider modes of disconfirmation, error, and reference-frame tests for robustness. Therefore these condition orders are seen as immanently reversible.

Field-Free Determinations. An apodeictic reference-frame may lead us to new points in the data-field that might not have been regarded without the indications derived from that reference-frame. A classic example is the periodic table of elements developed by Mendeleyev (1834-1907) in 1869. In 1871 he published an improved version that left gaps for elements that were not yet discovered. This led to a systematic search for new elements. Gallium, Germanium, and Scandium were the first predicted elements discovered. In its refined form, the periodic table also provides a model that tells us when to stop looking.

Reference-frames can teach us new ways or places to observe in the data-field. These determinations are called field-free determinations because they first derive from a reference-frame without an immediate regard for the data-field. In a manner of speaking these are field-free truth values and are formal determinations. Should all field-free determinations require data-field testing? The answer for science is yes. All such premises must ultimately dereference to matters-of-fact (selections from the data-field). The answer for formal logic is no. In this sense formal logic is not a science, but its uses for field-free determinations provide a powerful exploration tool for science.

We might ask more particularly how field-free determinations have any significant dereferences. For science this can be characterized as the Problem of Deduction. As yet undiscovered elements were deduced from Mendeleev's periodic table of elements. The truth of the deduction was contingent on discovering the predicted elements. This Problem of Deduction is as important as the better known Problem of Induction. For Science both are contingent on the data-field. It is in their contingency that we find common ground for these respective problems.

Logical Status Theses. From this mosaic exposition we may pull out arguments favorable to the given three general theses concerning the logical status of science-systems. These theses remain as definatory constraints with no proofs offered.

The truth of a science-system premise, being a premise in an apodeictic reference-frame, entails a double relativity, namely, 'truth' relative to data-field (empirical truth), 'truth' relative to system (formal truth). Science-systems experiments are designed to test 'truth' relative to data-field through carefully defined operations with data-field references. These references may be direct or indirect. Newton's three laws of motion are a case of indirect data-field reference. These laws cannot field reference directly, though they underpin many results that have been data-field tested. These tests confirm the laws as robust within certain testing limits. Though these results are from direct tests, it is the consistency that is sustained in these results that is taken as confirming that the general laws are robust within the limits of the data-field testing. This mode of conclusion indirectly measures the robustness to data-field testing. This requirement of direct or indirect testing is reflected in Thesis One.

> **Thesis One:** All science-system premises must dereference to the data-field through tests, directly or indirectly.

Science uses 'truth' relative to system as a tool, mediate reasoning or reasoning by rule may lead to new premises. These new premises are contingent, directly or indirectly, on data-field references, data-field testing. A science-system is logical insofar as it is a system. As a system we may apply principles of logic to develop premises that can be tested. If a premise cannot be dereferenced to a data-field, then we must, as Newton requires, adjust the system to make it more accurate or state that it is liable to such and such exceptions. What is indicated here has often been used in science as a characteristic of procedure. It is because a science-system also has the features of a logic system that it is open to logical procedures. This is reflected in Thesis Two.

> **Thesis Two:** From a given science-system we can use logic to search for new premises to dereference in data-field tests.

Different apodeictic reference-frames may appear to have the same data-field spread, but can this accredit saying that they spread over the same data-field? With respect to our understanding, the data-field does not remain the 'same' because the system for observation is different. The critique of judgment has changed. We are left with the difficult problem of indicating a data-field that itself should not be implicated in the indication. To be implicated in the indication is to say that each mode of perception changes the way we see the data-field. We see the field with 'different eyes'. We are faced anew with the effect of the observer on the observed, not in the physical sense as in the Heisenberg principle, but in the interface sense where apodeictic reference-frames generate bias in observation.

Since a given system may give a predisposition for looking at some things rather than others, this in no way implies a necessary elimination of observing the fringe and beyond of the predispositioning reference-frame. Such wayward regarding may lead to the destruction of an accepted reference-frame leading to a confusion until another reference-frame for judgment is available. Tacit here is the notion that science-systems are contingent on matters-of-fact for their maintenance. This is another way of saying that though data-fields are necessary to the maintenance of apodeictic reference-frames and though apodeictic reference-frames are necessary for the intelligibility of data-fields, the relation between data-field and reference-frame is a contingent one and therefore not a necessary relation in the logical sense of necessary. For science-systems, as apodeictic reference-frames, this is reflected in Thesis Three.

> **Thesis Three:** There is no necessary connection between a particular science-system and the data-field it dereferences to.

This is a restatement of the fundamental thesis that David Hume (1711-1776) teaches us. Charles Sanders Pierce (1839-1914) understood this problem, William James

(1841-1910) agreed. Sir Karl Raimond Popper (1902-1994) capitalized on it but appears to ignore Pierce.

The problems in using deductive or inductive logic in science-systems is exacerbated by the constraints implied by Thesis Three.

SUMMARY CONCLUSIONS. Catechism teaching can discourage discovery. We are in a position for effective interrogation when we increase our awareness of biological limitations, our socialization biases, and what roles these play in our modes of conceptualization, in our modes of actualization. Continual vigilance and interrogation can help us see the implicit constraints from our habitual thought on our possible actions, i.e., render our biases visible. This visibility can help us toward the design of science-systems that are independent of any particular culture. For now, the design of science-systems that are independent of any particular culture remains an unachieved utopia.

There seems to be a kind of progression for effectiveness and scope of science-systems, but this is no reason to assume that a science-system is approaching some final truth. By the simplistic nature of our models for intelligibility and the data reduction modes these models use, it may be that totally adequate modeling is unachievable, at least within the constraints of our present modes of investigation.

The scope of cultural evolution is expanded by our inability to clone our offspring culturally. *In communities where cultural cloning is most successful we find the least change in conduct.* (American public schools of all levels strive for this cloning. It is only in their failure that we find creative thought from their graduates.)

The obstacles to cultural cloning may be characterized as exterior and interior to the person. They range from the learning of liberal attitudes to the inadequate assimilation of the culture you are born into, to acquaintance contact with foreign cultures, to shocking experiences, to wars or natural disasters. These things drive us to interrogation.

KEY WORDS
Apodeictic conventions, constructs, fictions, culture-independence, social reference-frame independence, culture cloning, scientific truth, matters-of-fact, data-field access, transcendental knowing, transcendent instruments, data-field testing, field-free determinations, objectivity, field spread, science community, identification, truth

END Addendum 2

Addendum 3
FORMAL DUALITY IN FINITE LOGICS

Introduction. *Formal logic can be used as a paradigm for clarifying systemic duality concepts that are data-field free. Data-field free duality is also called formal duality.*

Section One shows formal duality through a binary Value-Computation-System (VCS) logic. This logic is based on definitions and does not require axioms.

Section Two shows formal duality in a binary Formula-Transformation-System (FTS) logic. A presentation of FTS logic independent of VCS logic requires axioms. I have chosen (consistent with Quine) to base FTS logic on VCS logic in order to eliminate all axioms.

Furthermore, VCS and FTS are dual approaches to Boolean logic. In VCS, the value string WFFs (Well Formed Formulas) are operated on to produce value strings and FTS WFFs are the formal instantiations of a value string. In FTS, the formula WFFs are operated on to produce other formula WFFs and the value strings are the formal instantiations of formula WFFs.

An operator reduces, expands or otherwise transforms WFFs.

SECTION ONE: VCS Logics

Though logic systems have been devised for use in computers, they have not been developed independently of the usual algebraic Formula-Transformation-Systems (FTS) that use axioms.

One of the characteristics of a VCS logic is the elimination of axioms. The system herein presented is based on operational definitions only and this eliminates the need for axioms.

Binary computer logics can be based on Boolean arithmetics that have only two values. For digital computers these values are represented as 0 or 1. These two digits are called bits (**bi**nary dig**its**). Note that the answer set is {0,1} and the empty set is *not* a member of this set (contrary to abstract set theory). Although we could change '0' to the empty set, this would eliminate the distinction between a propositional form and a false statement. However, a complete logic can be formulated where a statement has a propositional form and is not a proposition (regarding 'empty' and 'false' as synonym values) or is a proposition (where propositions are regarded only as true), but not both.

Leibnitz (1646-1716) was the first to devise a binary arithmetic. George Boole (1815-1864) was the first to apply 0 and 1 values to logic. Konrad Zuse (1910-1995) was the first to implement binary arithmetic in computing machines (1938).

Value-States. This is a notation system for handling many results at a time. It uses strings of values. This value convention is current in computer architecture analysis and design. Each particular case of a string is a value state. Therefore, if these strings are four bits long, they are 4-bit value states. How strings are generated in computation systems may begin with base-strings.

Base-Strings. Base-strings form an alternance of a power-of-two bits. Following the current convention, each base-string starts with 0. This choice is arbitrary and remains immanently reversible for a thoroughgoing interchange. The number of base-strings correlates to length.

Look at the following examples:

    B(1), 01
    B(2), 0011, 0101
    B(3), 00001111, 00110011, 01010101
    B(4), 0000000011111111, 0000111100001111,
           0011001100110011, 0101010101010101
    etc..

where the powers of 2 are respectively: 1, 2, 3, 4, ... which yield string-lengths of 2, 4, 8, 16, etc.. For B(3), B is for base and 3 is the power of 2 that defines the length of the string. It also directly indicates the number of base-strings. Thus B(3) has three base-strings of length 8. For n-base-strings, B(n), the n strings are of length 2^n.

If we arrange B(4) into four rows we have a table of columns of length 4 that represents every combination and permutation of length-4 strings. FTS logic statements with only two distinct variables use a table of this form. T(4) indicates a 'table' and '(4)' indicates 4 rows of length 2^4, that is, 4 rows of length-16. FTS uses T(4) for displaying in columns all the possible value strings of length-4. Among these columns it also makes sense to speak of base-strings. The table for column strings, T(4), has sixteen different column strings, therefore:

$$T(4) = \begin{vmatrix} 0&0&0&0&0&0&0&0&1&1&1&1&1&1&1&1 \\ 0&0&0&0&1&1&1&1&0&0&0&0&1&1&1&1 \\ 0&0&1&1&0&0&1&1&0&0&1&1&0&0&1&1 \\ 0&1&0&1&0&1&0&1&0&1&0&1&0&1&0&1 \end{vmatrix}$$

A table of the form $T(2^n)$ displays in columns every possible string of length-2^n, which implies it has 2 to the power of 2^n different strings. Thus length-1 columns has 2, length-2 has 4, length-4 has 16, length-8 has 256, length-16 has 65,536, etc. In $T(4)$, the index 4 tells us the columns are of length-4 and since $4 = 2^2$, $n = 2$ and this connotes that there are only *two* column base-strings of length-4.

> **Note:** The base-strings definition is chosen for a convenience. It follows current computer architecture preferences. Other base-strings could have been chosen. By definition, any possible base-string set must form a complete set of every combination and permutation of zeros and ones of the base-string length under a given set of operators. However, different choices may introduce different systemic artifacts (e.g., in regard to their requirements for proof).

Note also that the base-strings with integer lengths not expressible as an integer power of two have as yet no assigned significance as to what they could represent here. The T series that represent only the positive integer powers of 2 are useful for justifying the FTS forms (shown below).

Even though each column of $T(4)$ uses 4-bit value-states we can see that they look like the first sixteen binary numerals, in order left to right. This is a useful mnemonic device. For the numerals less than 1000 we just fill in the front spaces with zeros to maintain four bit states. Thus 1 is written 0001. *Though these value-strings are juxtaposed in fixed permutations, they remain only strings of individual values.* In effect, this is a case of positional notation that is useful for reducing the length of logic formulas.

In general, ordinary Boolean algebra formulas use letters to represent unknown values that we can determine completely or partially through systemic constraints. However, in VCS this is not the case. In VCS value-strings, the letters represent every known possible value, i.e., they are not unknowns. This is one of the dual characteristics of VCS logic to FTS logic. Furthermore, we could write T(4) = ABCD. ABCD is here regarded as shorthand for representing all sixteen possibilities in 4-bit strings and should not be regarded as an algebraic expression that uses unknowns. We are not using this sort of algebra here, no term is unknown.

Similarly, we can construct and represent any $T(2^n)$.

Variable Strings. In VCS, the value-states are used as variables and the instantiations for these variables are FTS formulas. Therefore, VCS variables have FTS formulas as values. The instantiations of VCS value-states are FTS formulas. For example, in T(4), the string 0000 is instantiated by KpNp, or NApNp, or etc., following the constraints of the argument.

> **DEFINITION:** When a WFF of a given system cannot be diminished in length, we say that for this system this WFF is *prime*.

In T(4), each of the 16 possible length-4 strings are WFFs that we can treat as variables. The variables (value strings) cannot be reduced in length for T(4), therefore, they are also prime WFFs. It will be shown below that all $T(2^n)$ formulas (combinations of variables and operators) in normal form (every variable expressed in the length of the longest string) reduce to one variable.

The formulas based on $T(2^n)$ may contain one or more variables, one or more operators. The *minimal prime* string is defined as a single variable with a monadic operator. For example, in VCS: T(4), N0000 and N0001 are minimal primes, using N for negation. For example, in FTS, Np or p are minimal primes where N is negation and p is a position (posited). They are called *minimal primes* because they express the shortest possible strings for the defined system. FTS minimal primes are like p, q, Np, Nq, etc. Minimal primes are used also as minimal length formulas (for producing a specific string of possible values). Using computer convention in an argument with two statements, for string 0011 we could assign the formula, p, for 0101 we could assign the formula, q. These instantiations (p and q) of VCS variables are the two smallest possible for the T(4) column base-strings convention. Here they instantiate formulas for the base-strings.

As George Boole remarked, these names, such as p and q, may represent classes or statements. This makes both FTS and VCS logics reversible in their representation modes. They can be sentential logics or class logics, depending on the chosen representation mode. With appropriate symbols the two modes might be expressed together in various combinations.

> **NEGATION** in binary value logic changes the value to its complement. Therefore, using N for negation, N0 = 1, N1 = 0.

To negate 0 is to posit 1, to negate 1 is to posit 0. For formula brevity we represent the string equation N0, N0, N0, N0 = 1, 1, 1, 1 as N0000 = 1111.

Negation and Reversal. The following two theorems are important but require only simple proofs based on definition of negation.

> **DOUBLE NEGATION THEOREM:** Double negation is equivalent to positing the operand.

EXAMPLE: NN0001 = 0001.
(Since N0001 = 1110 and N1110 = 0001 by definition)

The proof is trivial.

BASE-STRING REVERSAL THEOREM: The order of a base-string is reversed by negation.

PROOF: This follows from the definition of base-strings (i.e., the convention agreed upon). Starting with 0, they have a fixed alternance of substrings of 0s and substrings of 1s, each substring length is the same power of 2. When negated, the alternance is immediately reversed. Therefore the resultant string order is reversed. Therefore, etc., Q.E.F.

COROLLARY: Reversing the base-strings that generate a given result string will reverse the result string.

This proof is also trivial. If the generating strings and columns are upside down (reverse order) then all the results as columns will also be upside down, which is to say in reverse order. Q.E.F., Q.E.D.

Complementary Symmetry. One of the characteristics of this T(4) convention is that you can choose eight columns excluding their negations and the unchosen eight columns are the negations. This is called *complementary symmetry*. Base-strings as well as base-string complements are also said to have this complementary symmetry within the string. If the Universe were made up equally of matter and antimatter, then it could be partitioned into complementary symmetries.

Complementary Equivalence. Since the sixteen states in T(4) are the only possible binary 4-bit states, it should be clear that no matter what order we find them in that if we negate all of them that not only will they all be changed to the opposed value-string, but that we will still have sixteen distinct 4-bit value states. This is called *complementary equivalence*.

These observations lead to the following general THEOREM:

COMPLEMENTARY EQUIVALENCE: A Boolean Table is equivalent to its negation, or
$$T(2^n) = NT(2^n).$$

PROOF: The proof is based on the pigeon-hole principle. This mode of proof is possible from the fact that a given Table has a precise number of columns, no more, no less. Each column is distinct. Each Table also contains every possible value-string of the variable length. Therefore, if we negate each column we will have precisely the same number of resultant columns as in the original Table. If any two of the results of a given Table are alike then their negations are alike, contrary to Table definition. Therefore we have only distinct results in one-to-one correspondence with the original Table. But the Table has every value state of the variable length. Therefore the negated Table must have those same value states. Q.E.D.

Note that the columns in a given Table, $T(2^n)$, may be placed in any desired order. The order we choose is arbitrary, or a convention. If $T(2^n)$ is constructed by columns we can change the rows by changing the columns around, etc. Table $T(2^n)$ forms a useful Group because if we constrain all changes to columns and rows shifting, every column and row remains distinct for any instance. The proof can proceed by induction on the pigeon-hole principle.

REDUCTION OPERATIONS. Also we have

DEFINITION: REDUCTION OPERATORS.
Operators that reduce the number of strings in a formula are called reduction operators.

Logic addition and *multiplication* are reduction operators defined as follows:

Addition	Multiplication
$0 + 0 = 0$	$0 * 0 = 0$
$0 + 1 = 1$	$0 * 1 = 0$
$1 + 0 = 1$	$1 * 0 = 0$
$1 + 1 = 1$	$1 * 1 = 1$

For example, $0011 + 0101 = 0111$, where the operator, +, reduces two strings to one string. Note that 'N' (unary operator synonym to negate a string) is not a reduction operator (nor is positing). It complements the string, but it does not reduce the number of strings. These are Binary Value Strings (BVS).

We can formally define the BVS WFF, as follows:

DEFINITION: BVS WFF (Well-Formed Formula):

For strings of fixed length 2^n where n is non-negative,
1. A string of 0s and/or 1s is a BVS WFF.
2. N followed by a BVS WFF is a BVS WFF.
3. Two BVS WFFs connected by a reduction operator is a BVS WFF.
4. Nothing else is a BVS WFF.

GENERATING TABLE STRINGS. For Table T(4), all sixteen 4-bit value-states can generate from two column base-strings using negation and addition. We do not need any other operators. Once we get a result, we can use it in further operations. Thus, using (N, +):

1.	base	=	0011. Base-string.
2.	N0011	=	1100. 1. Negated.
3.	base	=	0101. Base-string.
4.	N0101	=	1010. 3. Negated.
5.	0011 + 0101	=	0111. 1.+3.
6.	N0111	=	1000. 5. Negated.
7.	1100 + 0011	=	1111. 2.+1.
8.	N1111	=	0000. 7. Negated.
9.	1100 + 1010	=	1110. 2.+4.
10.	N1110	=	0001. 9. Negated.
11.	1100 + 0001	=	1101. 2.+10.
12.	N1101	=	0010. 11. Negated.
13.	1010 + 0001	=	1011. 4.+10.
14.	N1011	=	0100. 13. Negated.
15.	0010 + 0100	=	0110. 12.+14.
16.	N0110	=	1001. 15. Negated.

This completes the task for all sixteen possible value-states.

We could write every line in terms of the two base-strings and all the steps with two operators (N,+). For example, the result 0110 is expressible as:

$$0110 = N(N0011) + N(N0011 + N0101) + \\ N(N0101 + N(N0011 + N0101)).$$

Where in the right side we see only the base strings 0011 and 0101 as operands. When a string is expressed in terms of base-string operands only, this is called its **normal form**. In minimal logics this normal form is unique in its combinations.

> **Note:** Whitehead and Russell used the negation and the or, (N,+), operators bias in their *Principia Mathematica*..

Using a similar method it is also possible to generate the sixteen 4-bit value states using only binary negation and multiplication on the column base-strings.

If we can generate every possible value-state of a base-string length with a given set of rules (operations) on the column base-strings, we say that for that logic system the rules are complete. Thus the operators (N,+) define a complete T(4) logic. This suggests the question: Can this system of logic rules produce $T(2^n)$, where $n > 2$ is arbitrary?

> **DEFINITION:** A VCS logic is complete if the operators of the system can produce the Table, $T(2^n)$, from the n column base-strings.

Every time we extend by one value the number of values we are operating on, we double the column string lengths. This is evident in T(4) from the fact that if we put one more value into the dyadic function row, the 0 must combine with each of the original four pairings and the 1 must combine with each of the original four pairings. This makes eight rows. If we continue the extension and put one more value to each row we must combine *0* with each of the eight values already obtained and then *1* with each of these same eight and together this makes sixteen, etc..

Continuing the argument, by induction we see that the string lengths are always expressible as a power of 2, since each time we put in one more value column in the function table we double the column string length.

For Example: To obtain 8-bit value states without introducing triadic operators, we can take our addition table and put one more column of values onto it in the following way:

$$0 + (0 + 0) = 0 + 0 = 0$$
$$0 + (0 + 1) = 0 + 1 = 1$$
$$0 + (1 + 0) = 0 + 1 = 1$$
$$0 + (1 + 1) = 0 + 1 = 1$$

This adds 0 to each possibility. Now we add 1 to each possibility:

$$1 + (0 + 0) = 1 + 0 = 1$$
$$1 + (0 + 1) = 1 + 1 = 1$$
$$1 + (1 + 0) = 1 + 1 = 1$$
$$1 + (1 + 1) = 1 + 1 = 1$$

Thus, reading top-down on both of these additions, we produce the 8-bit resultant string 01111111.

We can easily determine every possible combination of length-8 by producing every binary numeral from 00000000 to 11111111. The string 11111111 is the largest binary numeral for 8-bit numerals and in decimal notation is equal to 255. Counting from 00000000 this makes 256 8-bit value states. By convention we define operator sets on the three base-strings: 00001111, 00110011, 01010101, that will produce the other 253 permutations of all possible combinations using the binary values $\{0,1\}$. Every possible string is expressible in terms of operations on the base-strings only. For example, if we use the minimal operators, NAND or NOR, there are 256 possible results for either case. If we use only the OR operator we cannot exhaust all the possibilities, we find only 128 results. You can show this tediously by exhaustion. A shorter method is shown below.

Just as $T(4)$ can be represented entirely as ABCD we can represent $T(8)$ entirely as ABCDWXYZ, etc.. Since each letter represents both 0 and 1 we have 256 possibilities represented. Can we produce every 8-bit string using only negation and addition?

Even if it is possible to generate every 8-bit value state using only negation and addition, it should be evident that going through 256 possibilities is a lot of busy work. It is therefore more interesting to discover methods that are general enough to prove the statement without actually doing the detailed work.

I reason this in the following way.

> **PROBLEM:** Show that the operations of negation and addition on the strings 00001111, 00110011, 01010101 generate every possible 8-bit value state.

> PROOF: First look only at the first four value states in a string independently of the last four. We can do all the combinations that we did above to generate in the first four positions the sixteen possible value states for 4-bit strings. We do this using 00110011 and 01010101. Adding 00001111 to all possible results gives us sixteen distinct results, that is, throwing out duplicates, we have 16 different states in the first half of the string and the last half of the string is 1111.

We show this as: ABCDWXYZ + 00001111 = ABCD1111.

Next negate the 00001111 to get 11110000. Adding this to ABCDWXYZ so that
$$ABCDWXYZ + 11110000 = 1111WXYZ.$$

Negating these two forms in turn gets us the forms:

$$N(ABCD1111) = ABCD0000$$
$$N(1111WXYZ) = 0000WXYZ.$$

We can still keep the ABCD and WXYZ because we know they represent all sixteen 4-bit strings and we noted already that when all sixteen 4-bit strings are negated, this still leaves us with all sixteen 4-bit strings, by *complementary equivalence* theorem. But adding these two results together gets us:

$$ABCD0000 + 0000WXYZ = ABCDWXYZ.$$

Since ABCD and WXYZ can vary independently of each other, ABCDWXYZ has 16x16 combinations and permutations. But every possible form gives us 256 forms. By the pigeon-hole principle, PROBLEM: Q.E.F.

From this kind of reasoning we can generalize the problem as follows:

> **COMPLETENESS PROBLEM:** Using only negation and addition operating on $T(2^n)$ column strings, generate all possible strings of the table $T(2^{(n+1)})$.

PROOF: The general argument is by induction on the column string lengths. We have already shown that it is complete for $T(4)$ and furthermore if it is complete for $T(4)$ then it is complete for $T(8)$. Let us take as our induction assumption that it is true for $T(2^k)$ column strings, where integer $k > 2$, and show this implies it is complete for $T(2^{(k+1)})$.

$$T(2^k) = ABCD....$$

Add ...0000 to ABCD... such that it doubles the length of ABCD... into the form ABCD...0000 :
Take the first column base-string length-$2^{(k+1)}$, it is of the form: 0000....1111
        Add this to ABCD....0000
        + <u>0 0 0 0....1111</u>
        ABCD....1111

which doubles the length of the $T(2^k)$ strings.

Negating 0000....1111 we get 1111....0000 and adding this to the front of ABCD... (which for argument ends with WXYZ) gives us the form:
        1111....WXYZ

Negating these last two results and adding them gives us:

$$\begin{array}{r}\text{ABCD}....\ 0\ 0\ 0\ 0\\ +\ \underline{0\ 0\ 0\ 0....\text{WXYZ}}\\ \text{ABCD}....\text{WXYZ}\end{array}$$

where ABCD.... and ....WXYZ are functionally independent of each other.

But the form ABCD....WXYZ, where the two halves are independent, represents every possible form corresponding to the $2^{(k+1)}$ base-string lengths. We built the table $T(2^{(k+1)})$ from the table $T(2^k)$ used for the induction assumption, using only $\{N, +\}$. Therefore, etc.. PROBLEM Q.E.D.

> **Note:** More precision in the representation for the general argument would require using something like $X_1 X_2 X_3 X_4....X_{(k-2)} X_{(k-1)} X_{(k)}\ X_{(k+1)}$. I leave such clutter to the imagination of the active reader. Thus $Z = X_{(k+1)}$, etc.

CONCLUSION: Boolean arithmetics based on n column base-strings are complete under the operations of negation and addition for any normal form string length-2^n.

Negation and addition as used in the *completeness problem* are two operators found in Boolean arithmetic, therefore Boolean arithmetic can produce all value-strings that correspond to the length of whatever base-strings are effective in a given problem. Therefore, etc..

It is just as easy to show that Boolean arithmetics are complete for negation and multiplication on n base-strings. However, it can also be shown from the dual relation between addition and multiplication. Therefore I shall now present the notion of duality as it applies to these Boolean arithmetics.

DUALITY. As long as we are handling single values, the dual of one value is simply the other value. But with value-strings we have two possible oppositions, namely, the opposition of string values and the opposition of string order. To obtain the dual of 0001 we must use two procedures. If we first negate, N0001 = 1110, then reverse this order to 0111, the dual value-string for 0001 is 0111. The order of these two operations is arbitrary.

The reversal of order for any resultant string is easy to obtain. The value order of a conventional base-string is reversed merely by negating the base-strings. The effect of negation on all such base-strings is that the resultant strings are reversed. Therefore, this is a characteristic of conventional base-strings. A conventional base-string is always dual to itself. We can express this as a general theorem:

BASE-STRING DUALITY THEOREM: A base-string is dual to itself.

PROOF: The proof is based on the definition of duality and on the *Base-String Reversal Theorem* that the negation of a base-string reverses it.
1. To reverse the order of any given base-string, negate it, by *Base-String Reversal Theorem*.
2. Negate the resulting string.
This is a double negation which returns it to its original form by *Double Negation Theorem*.
3. But this is also equivalent to reversing the order and then negating the result. This fits the definition of string duality.
4. Therefore, etc.. THEOREM Q.E.D.

COROLLARY 1. Strings in mirror symmetry such as 00111100 are dual simply to their negation. This dual is 11000011.

COROLLARY 2. Base-strings and also their negations are each in themselves complementary symmetric strings, therefore self duals. 11110000 is a complementary symmetry that is dual to itself.

DUALITY AND SYMMETRIES. The general *algorithm* for finding the dual form of a string is re-stated as:

DUALIZATION ALGORITHM: For value-strings, reverse the string order and negate the result, yields the dual value-string.

This is just a re-statement of forming the dual of any string. We can obtain the possible oppositions of string order by negating the base-strings that generate it. This follows from the observation that base-strings, when negated, immediately produce a reverse order in any well-formed BVS. Therefore the resultant string order is found reversed. Therefore, etc., Q.E.F.

DUALITY CORRESPONDENCE THEOREM: If the dual form of F is D, then the dual form of D is F.

That is: if NR(F) = D then NR(D) = F, where F and D are BVS WFFs.

PROOF: The proof is based on the rules for forming the dual resultant. I shall apply these rules to some general formula. Call this formula of value-strings, F, and its dual, D. To obtain D from F we must negate the column base-strings that form F, find the result, and then negate the result. We write this:

$$NR,F = D$$

where the R indicates that the column base-strings within F are each negated, and the N indicates the whole formula is negated. I say that if NR,F = D is true, then its dual form, F = NR,D, is true. Suppose NR,F = D is true. But D is a BVS WFF. Therefore we may form its dual by negating its column base-strings and then negate the resultant formula. If we negate one side of the equality then we must negate the other side as well or the equality is not sustained. We can write this step:

$$(NNR)F = NNR,F = N,D$$

which gives us: R,F = N,D

This reduction follows from the *Double Negation Theorem* and *Base-String Reversal Theorem*. By negating the column base-strings of both sides we get:

$$RR,F = RN,D$$

which gives us: $F = NR,D$

But this shows that the equation $F = NR,D$ follows from the equation $NR,F = D$, and similarly $NR,F = D$ follows from $F = NR,D$; which was to be done. Therefore, etc.. THEOREM: Q.E.D.

> **DEFINITION: A BVS THEOREM** is a comparison between two BVS WFFs.
>
> **PRINCIPLE OF DUALITY THEOREM:** If a Boolean arithmetic comparison is true, then its dual form is also true.

This theorem is also known as the Principle of Duality for Boolean arithmetic. There are three cases to consider.

> **CASE ONE:** Suppose that $BVS_1 = BVS_2$, I say that:
> $NR,BVS_1 = NR,BVS_2$.

Since it is given that $BVS_1 = BVS_2$, then negating the column base-strings on both sides reverses both resultant strings. Therefore from the value-string THEOREM we have
$R,BVS_1 = R,BVS_2$. But we can negate both sides and have
$NR,BVS_1 = NR,BVS_2$,
which was to be done. Case One Q.E.F.

> **CASE TWO:** Suppose that $BVS_1 > BVS_2$, I say that:
>
> $NR,BVS_1 > NR,BVS_2$.

Since it is given that $BVS_1 > BVS_2$, then negating the column base-strings on both sides reverses both resultant strings. Therefore we have $R,BVS_1 < R,BVS_2$. The inequality reverses. But then we negate both sides and this reverses the inequality again and we have $NR,BVS_1 > NR,BVS_2$, which was to be done. Case Two Q.E.F.

> **Note:** Since this comparison must be true for every corresponding pair of bits, this implies that the BVS_1 result is a string of 1s, the BVS_2 result is a string of 0s. Note also this takes care of '<>'.

CASE THREE: Suppose that $BVS_1 \geq BVS_2$, I say that: $NR,BVS_1 \geq NR,BVS_2$.

The reasoning follows CASE ONE and TWO.
  Since $NR,BVS_1 \geq NR,BVS_2$

  is equivalent to

  $(NR,BVS_1 > NR,BVS_2)$ OR $(NR,BVS_1 = NR,BVS_2)$

  in corresponding bit pairs, one pair at a time, therefore, etc..

Therefore, similarly. Case Three Q.E.F.

Therefore, if a Boolean formula comparison is true, its dual formula comparison is also true. Therefore, etc.
THEOREM: Q.E.D.

This completes the foundation outline for VCS reduction operations. We have left to show how BVSs are solved. Each string in a BVS has a set of formulas that resolve into that string. There is an infinity of formulas for each string. Since all formulas are primes or reducible, our concern is how to find all the primes for a given value string. The reducible formulas are made up of primes and require reduction operators. For VCS we conclude that for any argument the longest prime string is 2^n for $T(2^n)$. This should be evident.

SECTION TWO: FTS Logics

FTS FIRST ORDER BINARY-VALUE LOGIC. The first complete Formula-Transformation-System (FTS) logic was worked out by George Boole by 1848. He purposely modeled logic formulas after algebra. Boole worked out two formula logics. They could be used separately or together. These formula logics include both an infinite-valued logic with values in the Real interval [0,1] and, a special case of this interval, binary-valued logic using the values {0,1}.

> **Note:** I already noted that contrary to abstract set theory, the empty set is not a member of this set, {0,1}. There is a sense in which a false statement is an empty set, but if we choose to do this we lose the power to distinguish false propositions from propositional forms, a distinction that logicians still find useful. $\varnothing$ is not a value in binary-value logic arguments.

These formulas use operators on the logic values and value-variables to produce a resultant value or resultant formula.

In binary Boolean form one can explore the task of finding all possible conclusions within a given formula length limit, a task that can easily be defined for finding proofs in a mechanical way for sentential or class logics. In binary logic it is possible to give rules for generating formulas of infinite length. (It can also be shown that there exist an infinity of formulas of infinite length that require rules of infinite length, but these in particular are undecidable formulas.) Gödel's incompleteness theorems were directed at proving the existence of finite formulas that are undecidable. Such formulas require an infinity of possible values and therefore do not apply to logics of finite values.

FTS LOGICS. This is not an exposition of axiomatic FTS logics. What usually pass for axioms in an FTS Logic are derivable theorems from a VCS Logic. Therefore, linking the two logics into one system eliminates the need for posturing axioms. Quine agrees. It also eliminates the circular reasoning of using logic axioms to prove logic is consistent.

The Law of Excluded Middle is a characteristic of crisp logics. This Law is not a characteristic of fuzzy logics. This discussion is limited to crisp logics. A discussion of fuzzy logics is given in Addendum 4.

DEFINITION: FTS WFF (Well-Formed Formula):
1. A monadic operator $\{N, \varnothing\}$ followed by a, b, c, ... , x_1, x_2, etc., is a FTS WFF.
2. A dyadic operator followed by two FTS WFFs is an FTS WFF.
3. Nothing else is a FTS WFF.

GENERAL REMARKS. It is easy to show that not all WFFs (VCS or FTS) of a crisp system can be theorems. This is not a proof concerning completeness or incompleteness nor does it prove that we can find them all. It merely shows that this partition set of WFFs has no empty member.

METATHEOREM: In a system of logic, not all of the WFFs are theorems.

PROOF: This statement follows from the fact that all WFFs contain all theorems and their respective negations. Their respective negations are not theorems. Therefore, not all of the WFFs are theorems, Law of Excluded Middle, etc., Q.E.F.

REDUNDANCY. An observable characteristic of WFFs is redundancy. This term indicates that for a given value-string we have more than one possible formula (VCS or FTS). When we have more than one prime formula for a given value-string they are called **synonyms**. In an FTS logic based on axioms we must have a transformation rule that subsumes the synonyms under the required value-string or they may lie out of reach of the logic system (the logic system is incomplete, it contains undecidable formulas). We can avoid this by using a VCS logic based on operational definitions, i.e., a basis without axioms.

OCKHAM-DeMORGAN CONVERSION RULES. The essence of duality in logic is found as early as the 13th century in the works of William of Ockham (though, earlier, the Pseudo-Scot apparently used these laws without expressly stating them). William of Ockham gave no deliberate development of duality. His rules were not known to DeMorgan (self-educated 19th Century scholar) who independently rediscovered them. Today these conversion rules go under the misnomer of DeMorgan's Rules.

Ockham's Conversion Rules may be translated from Latin as:

> 1) From the negation of a conjunction proposition to the disjunction of the negations of its parts, and conversely.
> 2) From the negation of a disjunction proposition to the conjunction of the negations of its parts, and conversely.

These two Rules are dual to each other. A dual relation for these conversion rules is established with the immanent reversibility of the corresponding dual terms 'disjunction' and 'conjunction'. Reverse the positions of these terms in a thoroughgoing exchange and each Rule converts to the other. These conversion rules were not understood as dual in Ockham's time.

Using Hilbert-Ackermann symbols we express them as:

> 1) $-(P \& Q) = -P \lor -Q$
> 2) $-(P \lor Q) = -P \& -Q$

Note that it is lines 1 and 2 that are dual to each other.

The equality in each case is another kind of relation called a comparison. It is the nature of this comparison that suggests a general principle of duality for comparisons.

-(P & Q)	-(P v Q)
1 0 0 0	1 0 0 0
1 0 0 1	0 0 1 1
1 1 0 0	0 1 1 0
0 1 1 1	0 1 1 1
↑	↑

DUALITY FOR COMPARISONS. In defining comparisons between formulas we want to connect these formulas with a comparison symbol that is not to be considered as an operation. The usual comparisons are: $>$, $<$, $=$, and their negations. It is clear that we could speak of a duality of comparisons, but it is more powerful and convenient to define these comparisons as metalogical and therefore as independent of the duality of formulas that are being compared. This is recommended in order to have an invariant way of comparing formulas during formula transformations. We shall hold '$\equiv$' and its negation as the logical equivalence operators and therefore it is not invariant under dual transformations. This does not preclude that under certain conditions the inequality may reverse.

If we agree to these conventions then we may express the Principle of Duality for equality comparisons as a metatheorem in the following way:

METATHEOREM:
DUALITY COMPARISON FOR EQUALITY. If we have an equality comparison between two formulas, the duals of these formulas are also equal.

PROOF: The proof follows without difficulty. The gist is, since we have an equality between two formulas, they are both true or both false, their resultant strings are thus alike. If we take the dual of both formulas, both result strings change to their dual forms, i.e., their value strings are still alike. Therefore, etc., Q.E.F.

METATHEOREM:
DUALITY COMPARISON FOR INEQUALITY. If we have an inequality comparison between two formulas, the inequality is reversed for the dual forms of these formulas.

PROOF: Since we have an inequality between two formulas, one is always true and the other is always false. For the tautology, its dual is a contradiction. For the contradiction, its dual is a tautology. Thus the comparison reverses. Therefore, etc., Q.E.F.

Proofs for the other comparisons are variations on these.
FORMULA DUALITY. The duality concept for logic formulas can be defined in terms of the value-strings we associate with them. These value-strings are arranged in rectangular matrices called value-tables.

In this exposition of formula duality the values that are determined by formula shall be represented by {F,T} to distinguish them from values calculated by Boolean arithmetic, {0,1}. If you see a table in {F,T} then the values were determined by FTS, if you see a table in {0,1} then the values were determined by VCS. This code artificially reminds us which system is under discussion. '0' is a synonym for 'F'. '1' is a synonym for 'T'. Note that the interpretation need not be restricted to {true, false}.

The simplest truth table in an FTS binary-value logic is for a proposition and its negation. Positing and negating are both monadic operations but it is not the custom to write a sign for positing, therefore, where it is understood to occur is left empty. We contrast the operational pair (∅, -), using the tables for the monadic operators:

```
Ø P   P    and   -P  P
  F   F          T   F
  T   T          F   T
```

DUALITY OF (Ø, -): The monadic operators Ø and - are each a dual correspondence to itself. Thus, P is dual to itself, -P is dual to itself.

PROOF: If we take the value-string of P, namely FT, and negate it and reverse the order, we find ourselves back with the same value-string as before we started, namely FT. If we take the value-string of -P, namely TF, and negate it and reverse the order, we find ourselves back with the same value-string as before we started, namely TF. Therefore, etc., Q.E.F.

This is why we can agree that P is dual to itself and -P is dual to itself as Quine teaches us (though he did not teach us why).

DUALITY OF AND/OR: The dyadic operators AND and OR are dual to each other, i.e., form a dual correspondence, one to the other.

PROOF: This proof proceeds by the Principle of Exhaustion. Using Truth Tables, we can define AND (.) and OR (v) for each possible pair of (P, Q) with the monadic operations (Ø, -):

```
P Q  -P -Q  P.Q  PvQ  P v-Q  P.-Q  -PvQ  -P.Q  -Pv-Q  -P.-Q
F F  T  T   F    F    F      F     T     F     T      T
F T  T  F   F    T    F      F     T     T     T      F
T F  F  T   F    T    T      T     F     F     T      F
T T  F  F   T    T    T      F     T     F     F      F
```

Since the only difference found among these formula pairs associated with dual strings is the dyadic operator, we may conclude that these dyadic operators are dual to each other for these components (P,Q) paired with (Ø,-), the monadic operations.

Therefore we may conclude by the Principle of Exhaustion that AND and OR are operators that are dual to each other. Q.E.F.

The generality of this proof is founded on an induction on the dyadic characteristic of these operators, anywhere you find a dyadic operator, there are only two operands, etc.. P and Q may represent any FTS WFF.

> **DEFINITION: OPPOSED BOUND-VARIATION.** Given pairs of corresponding dual values or corresponding dual terms, for every occurrence of one of the values or terms in one context the dual value or dual term occurs in the other context.

EXAMPLE: We know from the tables that we can write Ockham's first conversion Rule as

$$-(\varnothing P . \varnothing Q) = \varnothing(-P v -Q),$$

from which we note that the two formulas put in this equation represent both the monadic operators ($\varnothing$,-) and the dyadic operators (v,.) in opposed bound-variation. The contexts are on opposite sides of the equal sign. We also note that if 'P' and 'Q' are put in opposed bound-variation in the WFFs that this would have no effect on the value-strings. This non-effect is a characteristic of non-ordered dyadic operators. When we say that order of operands in a dyadic operation is not significant we are saying that changing the order has no effect on the value-strings. Therefore it is equivalent to say that these two formulas also represent (P,Q) in opposed bound-variation. That is to say, that on one side or the other, the places of P and Q can be swapped in bound-variation.

> **DEFINITION: FREE-VARIATION.** For each occurrence of one or the other of a pair of terms in a context, either member may be interchanged at any occurrence of either if the result is a synonym.

For example, if P = Q, they form a free-variation pair of terms for any formula. They have identical truth-strings and may be freely exchanged wherever they occur in any argument. They may occur in free-variation in any given context. If both occur in a statement, one or the other may be regarded as a redundant difference.

OPPOSED BOUND-VARIATION RULES for AND-OR.
Look again at ∅P.∅Q as the dual to ∅Pv∅Q. Note that the monadic operator is dual to itself. Note also that implicit in the value table definition of duality that there is a one-to-one correspondence of all and each element in the dual pairs, e.g.:

∅(∅P.∅Q)
| | | | | | We return to this shortly.
∅(∅Pv∅Q)

The dyadic operators in these formulas occur in opposed bound-variation. To reverse the value-string order we only have to negate the components, P and Q. To exchange the values in bound-variation we only have to negate the result. From this we have both:

∅(∅P. ∅Q) => -(-P.-Q)
and ∅(∅Pv∅Q) => -(-Pv-Q).

Since these relations are duals, and from previous arguments, putting in dual operators forms a dual formula. Thus (∅P. ∅Q) is dual to (∅Pv∅Q), etc.. Therefore, we can write two equations:

∅(∅P. ∅Q) = -(-Pv-Q)
and ∅(∅Pv∅Q) = -(-P.-Q).

In other words, given formulas P.Q and PvQ dual to each other by theorem, to modify one isomorphically to equal the other we must put its monadic operators in opposed bound-variation.

Furthermore, if we negate both sides, reduce double negatives, then we have the Ockham-DeMorgan isomorphic conversion Rules:

1) $-(\emptyset P \cdot \emptyset Q) = \emptyset(-P \vee -Q)$
2) $-(\emptyset P \vee \emptyset Q) = \emptyset(-P \cdot -Q)$

1 and 2 are dual to each other. It is an easy proof to show that given any logic WFF written with the pairs (.,v) and ($\emptyset$,-), that it is equal to that WFF transformed by putting these operators in opposed bound-variation.

THEOREM: Given any FTS WFF written with pairs (.,v) and ($\emptyset$,-), it is equal to that WFF transformed by putting these operators in opposed bound-variation.

PROOF. The proof is based on the fact that ($\emptyset$,-) in opposed bound-variation produces a dual WFF and (.,v) in opposed bound-variation produces a dual WFF and the dual of a dual produces the original value-string. Therefore, etc., Theorem, Q.E.D.

ISOMORPHIC FTS WFFs OPPOSED in (.,v) and ($\emptyset$,-): Two isomorphic formulas in (.,v) and ($\emptyset$,-) with the operators placed in opposed bound-variation are equal to each other.

PROOF: The proof is constructed as follows:
1. Given an FTS WFF in (.,v) and ($\emptyset$,-).
    EXAMPLE: $\emptyset(\emptyset P \cdot \emptyset(-P \vee -Q))$.
    VALUE-STRING: FFFT.
2. Put ($\emptyset$,-) in opposed bound-variation. This reverses the base-string orders which reverse the resultant string order, but we must also switch the monadic operator of the entire formula which negates every component. These two effects, by definition, give the dual string.
    EXAMPLE: $-(-P \cdot -(\emptyset P \vee \emptyset Q))$.
    VALUE-STRING: FTTT.
3. Put (.,v) in opposed bound-variation. This gives the reverse string order and its negation, the dual of step 2.
    EXAMPLE: $-(-P \vee -(\emptyset P \cdot \emptyset Q))$.
    VALUE-STRING: FFFT.
4. Therefore, etc., Q.E.D.

This argument suggests that the Ockham-DeMorgan Rules may each be also described as the **opposed bound-variation rules** for AND/OR.

OPPOSED BOUND VARIATION FOR ORDERED OPERATORS. So far I have established a duality rule for (First Order) logic WFFs that use only the operators (∅,-) and (.,v). The operators (∅,-), (.,v) are non-ordered. To date, no one has suggested an opposed bound-variation for ordered operators. The usual procedure is to translate all ordered operators into a non-ordered operator form and then find the dual. This lack in ordered operators is remedied by defining a dual ordered operator that corresponds to implication.

In Polish notation, the truth table for material implication is represented here as Inclusion (I):

p	q	I pq
0	0	1
0	1	1
1	0	0
1	1	1

Read as an ordered pair "p includes q."

By the value matrix definition of duality we can establish a dual table as material exclusion represented here as Excluded by (E). Thus we have:

p	q	E pq
0	0	0
0	1	1
1	0	0
1	1	0

Read as an ordered pair ""p is excluded by q."

If q is false, it has no effect on p, p may be true or false. If q is true it must exclude p, therefore p must be false, if p is true when q is true then Epq is false.

In the first row of the value matrix for "p is excluded by q" we see that when q is false it can have no effect on p, therefore p is not excluded by q and the statement is false. In row two, when q is true and p is false, p has been excluded, therefore the statement is true. In row three, since q is false it has no effect on p, therefore asserting that it excludes p is false. In row four, our reasoning is similar to row two, q is true, yet since p is true it is not excluded, therefore the exclusion is false. These considerations help when translating statements into formulas.

Again, in these WFFs there is a one-to-one correspondence for each formula element. We can see from the above tables and by definition of duality that:

>   Ipq and Epq are duals, etc..

We see from the above argument that finding corresponding dual operators is a mechanical process that can be done without difficulty.

Ockham-DeMorgan Rules can be extended to include ordered logic corresponding duals:

>   N(Ipq) == ENpNq
>   N(Epq) == INpNq

Forming complete dyadic logics is a mechanical procedure that no longer requires any special creativity.

When we mix ordered logics with non-ordered logics, the number of redundant primes is increased by each system plus their admixture.

FLIP-FLOP LOGICS. Logics can display another characteristic of dual systems. Dual systems propositions are expressible in minimal length formulas, prime formulas.

When we list the prime formulas of a given system and put them in order of length from short to long, the dual list will appear long to short. If we arrange them from easy-to-obtain to more difficult-to-obtain, again the dual systems display opposition. What is easiest to obtain in one is the most difficult to obtain in the other, etc.

The logic systems example for digital computers is the duality between NAND logic (D) and NOR logic (S). The NAND and NOR resultant strings for two variables are expressed as:

$$Dpq - 1110$$
$$Spq - 1000$$

which we note are dual strings.

MINIMAL PRIMES for VALUE STRINGS

3	== 0011 == p		== p
5	== 0101 == q		== q
12	== 1100 == Dpp		== Spp
10	== 1010 == Dqq		== Sqq

The above minimal prime WFFs are each dual to self.

The following are corresponding prime forms with respect to their operators:

14 == 1110	== Dpq		== SSSppSqqSSppSqq
	== **NAND**		
11 == 1011	== DqDpp		== SSpSpqSpSpp
13 == 1101	== DpDpq		== SSqSppSpSpp
15 == 1111	== DpDpp		== SSpSppSpSpp
1 == 0001	== DDpqDpq		== BBppBqq
	== **AND**		
7 == 0111	== DDppDqq		== SSpqSpq
	== **OR**		
0 == 0000	== DDpDppDpDpp		== SpSpp
4 == 0100	== DDqDppDpDpp		== SpSpq
2 == 0010	== DDpDpqDpDpp		== SqSpp
8 == 1000	== DDDppDqqDDppDqq		== Spq
	== **NOR**		

Star Laws: Foundations for Convergence Buoyancy   479

```
6 == 0110    == DDqDppDpDqq        == SSpqSSppSqq
9 == 1001    == DDpqDDppDqq        == SSqSppSpSqq
```

which shows that the shortest dyadic operation in NAND logic (D) is the longest dyadic operation in NOR logic (S) and the longest dyadic operation in D is the shortest dyadic operation in S. This works for all cases except the WFFs dual to themselves or dual to their negations.

It is evident from this display that the dual logics of S and D follow an expectation. What is difficult to evaluate in S is easier to evaluate in D and vice versa. Cases where the formulas are of equal difficulty undergo no change in difficulty, etc. They are dual to their negations.

6 and 9 are equal in difficulty in both logics. This is expected for these expressions of equivalence and its denial, etc..

This shows that when dual logics S and D (NAND and NOR) are compared that the formulas are also in a dual relation according to the number of steps required to evaluate their respective truth-values. Furthermore, this trait is expected between any dual systems. What is difficult in one is easy in the other and vice versa, except for the WFFs that are dual to themselves.

> **DEFINITION: INCLUSIVE FLIP-FLOP SYSTEM.** A pair of systems in dual-symmetry (corresponding formulas occur in free-variation), form an ***Inclusive Flip-Flop System***.

It works for the NAND and NOR binary-value logics. There are additional difficulties in more complex systems generated by the richness of possible interpretations.

FLIP-FLOP SYSTEMS (two systems in dual asymmetry over the same field). When two universes of discourse are mutually incompatible and can only occur in mutual exclusion they are in **dual asymmetry**. This is an extreme form of bound-variation. The mutual exclusion here is between two different texts. They may be freely exchanged, one for the other, but only in their entirety. This may also seem like an extreme form of free-variation, but it is not. We only speak of free-variation within the same context, the same text. The mutual exchange here is between two separate systems, two separate texts. **The Newtonian and anti-Newtonian systems have this dual asymmetry**. Ptolemy and Copernicus systems also have this exclusive dual symmetry.

DEFINITION: EXCLUSIVE FLIP-FLOP SYSTEM. A pair of systems in dual asymmetry (corresponding parts occur in bound variation) form an *Exclusive Flip-Flop System*.

When we mix exclusive flip-flop systems we get contradictions. This mixing of exclusive flip-flop systems is a form of mode mixing. Many errors in reasoning may be traced to mixing modes. Many well-known paradoxes are based on mixing of modes.

CRITIQUE. In some respects the logics we use are like billiard ball physics. The sentences or classes are treated independently of each other and problems of reasoning without crisply independent variables cannot be well represented in binary value logics. Boole recognized this and was the first logician to recognize and implement the possibility of representing partial truths. He discussed this subject in his logic of probabilities, an infinite valued logic with values defined on the closed interval $[0,1]$, more on that in Addendum 4.

KEY WORDS
Value-Computation-System (VCS), Formula-Transformation-System (FTS), Value-States, Complementary Symmetry, Equivalence, Reduction Operations, Duality, Dual Symmetry, Asymmetry, Mirror Symmetry, Complements, Correspondence, isomorphic, Dualization Algorithm, Boolean Variable Formula (BVS), Principle of Duality, Synonym, Redundancy, Prime, Ockham-DeMorgan conversion rules, Formula Duality, Opposed Bound-Variation, Free-Variation, flip-flop logics, systems, subsystems, inclusive/exclusive

END Addendum 3

Addendum 4
DATA-FIELD DUALITY

Introduction. *Data-field duality is a form of dual dereferencing. Reference and dereference are correlatives. The relation is like a map and the territory it maps. This may be in bound-variation or free-variation, according to usage within the constraints. The constraints are defined by the mode for dereferencing. Every mode for dereferencing has a corresponding mode for referencing.*

This does not preclude that various modes of dereferencing may be used for exhibiting the strengths or weaknesses of a given reference-frame. It is weaknesses in dereferencing that may lead to the downfall of a given reference-frame. Another way of saying this is that the rejected reference-frame is not only inadequate to the data-field it references, it is also too far removed for retrofitting. Upgrading systems may require revising, retrofitting, or abandonment for total replacement.

MAPPING. I have remarked that in dual systems there is a one-to-one correspondence. This recalls the notion of 'mapping'. This notion plays a fundamental role in all branches of mathematics. Its use in the production of geographic maps gives us the paradigm.

An example is the Mercator projection where the sphere surface of the Earth is given by rule a one-to-one correspondence to a cylinder. There are many other possible projection rules than the Mercator projection which, in effect, takes a sphere within a rolled cylinder and maps the sphere surface onto the rolled cylinder by a rule, a kind of mirror of shapes. Abstracting this notion we can consider the analytic geometry of Descartes as a kind of mapping, where algebraic relations are mapped onto geometric ones. That is to say, algebra is mapped onto a geometry. For the dual case, Hilbert, in an attempt to prove the consistency of geometry, specifically mapped geometry onto algebra, proving that if one is consistent then so is the other, etc..

Mathematical physics has a kind of mapping of relations and events for physics by mathematical models. Duality may also be regarded as a kind of mapping since a mapping is always possible when we achieve a one-to-one correspondence rule between two systems of thought or between a system of thought and its data-field. This also suggests that duality is only one of many possible modes for mapping. Note here that three fundamental modes of one-to-one mapping are possible;

>system/system,
>system/data-field,
>data-field/data-field.

If our task is to map system A onto system B this implies one procedure order for this relation. If only one procedure order is defined for this task, the dual procedure is defined as a mapping in the reverse direction. This does not imply that system A and system B with a one-to-one correspondence are symmetric to each other. The procedures are dual, that is, mapping A to B and B to A are procedures that are dual to one another. They are in opposition to each other, and they are the only choices available for A—B one-to-one mappings. The procedure of mapping algebra onto geometry (Descartes) is dual to the procedure of mapping geometry onto algebra (Hilbert) but we do not say that algebra and geometry are duals. It is in the order of the mapping procedures that we find the duality.

Two distinct reference-frames in one-to-one correspondence are homomorphic. Homomorphic reference-frames may reference data-fields *heteronymically* or *homonymically*. Homonymic reference-frames use the same name for the corresponding parts and differ only in their dereferencing modes. Heteronymic reference-frames use different names for their corresponding parts and therefore differ in their referencing.

These definitions derive from the respective referencing modes.

> **DATA-FIELD DUALITY:** If reference-frames are homomorphic for two opposed dereference modes over the same data-field spread, then they reference a ***Data-Field Duality***.

When we can reason homomorphically about data-field elements, where the data-field elements are in opposed corresponding pairs, this is data-field duality. It is a difficult notion to make clear. The difficulty arises from trying to reference something that is independent of the reference-frame. How do we reference the data-field without imposing constraints from (implicating it in) the reference-frame (Thesis 3 in Addendum 2)? Two paradigmatic examples of reference-frames that are homomorphic to data-field dualities are given below to exhibit these characteristics paradigmatically. The first example, geometric dereferencing duality, uses two heteronymic reference-frames. The second one, Boolean dereferencing duality, uses one homonymic reference-frame with two instantiation modes. Both cases reference data-field dualities.

GEOMETRIC DEREFERENCING DUALITY

For argument let us regard all possible configurations in space as a data-field for synthetic geometry. The nature of the duality exemplified here is illustrated by another artificial device. Different names are introduced in pairs for proposed dual concepts. The names could just as well have been homonyms. The idea here is that with a given set of names we can apply different content, where the differences are arranged in dual pairs. These are called ***corresponding pairs***, but this time the pairs are not formula pairs, they are pairs taken from the data-field. The point I want to make is that we can use the same formulas (conventional strings of words or symbols) when applied to a context or to a dual context without changing the formulas. Here context implies a mode of selection from a data-field. The data-field may be percepts or concepts. This procedure is a manner of applying the notion of data-field duality.

As long as we have a coherent context from the data-field we can conceive of an anti-context chosen from that same data-field. Thus we can have a fixed statement form or formula, but it can have coherent dual instantiations according to the heteronymic mode.

ARGUMENT. The use of duality in geometry entails a corresponding of geometric objects that are the 'values' of a geometric statement. As in the distribution of logic values, these geometric values must occur in bound-variation between corresponding dual statements. The geometric field in question is an ideal field because it is a conceptual model for geometric elements. The approach for this exposition is in the form of synthetic rather than analytic geometry. For this argument the space is Euclidean.

To see how this duality is conceived requires some preliminary definitions.

1) DEFINITION: MEET. If **a** and **b** are two lines, the point of intersection of **a** and **b** is called the *meet* of **a** and **b**. The lines **a** and **b** are said to be **concurrent** (EXHIBIT 4.1).

EXHIBIT 4.1: meet

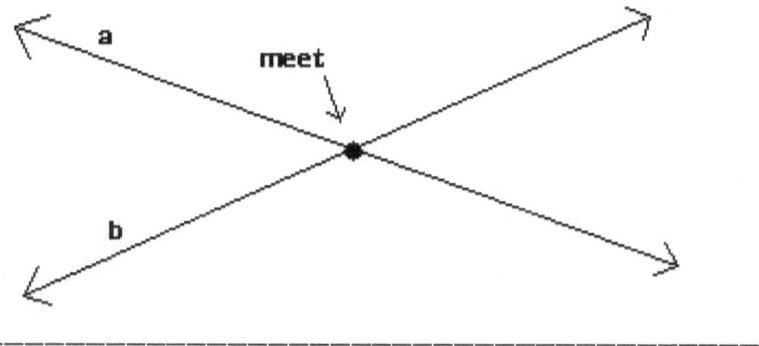

2) DEFINITION: JOIN. If **a** and **b** are two points, the line of intersection of **a** and **b** is called the *join* of **a** and **b**. The points **a** and **b** are said to be **collinear** (see EXHIBIT 4.2).

EXHIBIT 4.2: join

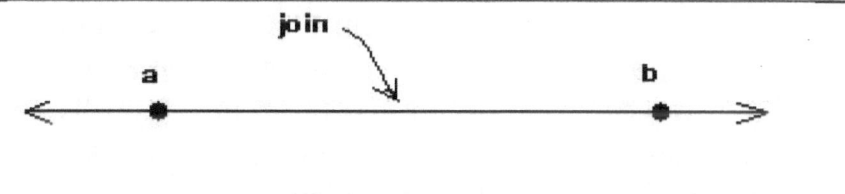

Even though the illustrations are clearly different, the logical relations are homomorphic. The instantiations of the logic are dual in form and we can recognize the following corresponding pairs:

point	line
line	point
join	meet
meet	join

Exchanging the corresponding pairs in bound-variation in one definition will yield the other definition. This is a defining characteristic of a dual relation.

There are two distinct views from which any curve may be regarded, (1) as a *locus* and (2) as an *envelope*. To exhibit this distinction:

> 1) **Locus:** We can define a circle in the following way. On a plane, given a point **c** and any invariant distance from it, the set of points of invariant distance from the point **c** form a circle. We could also conceive that the point revolves around **c** at an invariant distance. There are other ways of proceeding but these two adequately illustrate how to define a circle as a locus.
>
> 2) **Envelope:** We can define a circle in the following way. On a plane, given a point **c** and any invariant distance from it, the set of normal lines of invariant distance from the point **c** form a circle (they appear as tangents to the circle). We could also conceive that a line revolves around **c** normal to an invariant distance. The lines are said to envelope a circle. Also the envelope lines are unlimited in length.

It is apparent that again we may exchange the terms (line, point) in bound-variation and produce the other definition and therefore (locus, envelope) are corresponding dual pairs. These definitions are therefore dual to each other.

From these two points of view a circle can be regarded either as a locus or as an envelope.

DUAL DEFINITIONS FOR GEOMETRY

1) The *degree* of a curve is the number of intersections, real or imaginary, of the curve with any straight line.	1) The *class* of a curve is the number of tangents, real or imaginary, that can be applied to the curve from any given point.
2) A figure formed by three points and their joins is called a *triangle*.	2) A figure formed by three lines and their meets is called a *trilateral*.
3) A curve from which any two, and not more than two points are collinear to a straight line is called a curve of the *second degree*.	3) A curve for which any two, and not more than two tangents are concurrent at a point is called a curve of the *second class*.

Note that a conic section is a curve of the second degree and of the second class.

Here is a partial list of elements in opposition one-to-one correspondence for geometry:

<pre>
        point   line
        line   point
        concurrent   collinear
</pre>

collinear	concurrent
join	meet
meet	join
etc..	
locus	envelope
degree	class
lie on	pass through
triangle	trilateral
etc..	
point of a conic	tangent to a conic
etc..	

We can thus establish a bound reciprocation between the two systems and in general with some practice we may establish the dual theorem at the same time we prove a theorem. This is done merely by showing that it is in the form of a dual equation or some equivalent form. Note that these duals may be formed, one from the other, merely by the bound-variation exchange of the complete set of corresponding terms.

As an example I exhibit a well-known theorem of Pappus (Exhibit 4.3 and 4.4). You can see another presentation of it in Nagel and Newman's book, *Gödel's Proof*, p 64f.

Here I stress that these dual arguments are homomorphic. One could also handle this geometric duality by defining homonyms for the key concepts. Such a homonymic and homomorphic system of reasoning has validity that is independent of which dual model we chose for its instantiation. It is this characteristic that makes this a case of data-field duality.

Star Laws: Foundations for Convergence Buoyancy

Dual Forms for a Pappus Theorem

EXHIBIT 4.3

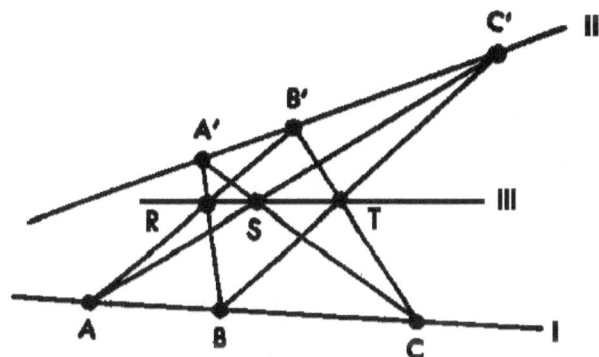

Given A, B, C, three distinct points on line I
        A', B', C', three distinct points on line II
   join A to B', B to A'   joins meet at R
      A to C', A' to C   joins meet at S
      B to C', C to B'   joins meet at T
   I say that the meets, R, S, T, are collinear (on III).

EXHIBIT 4,4

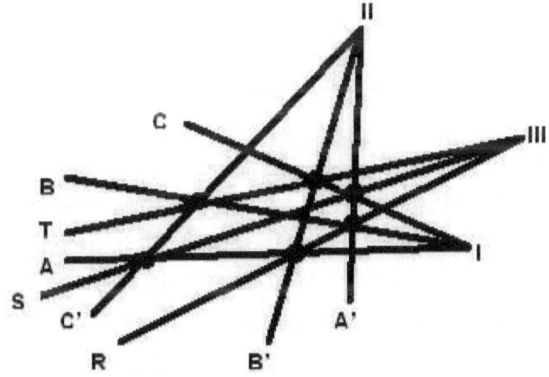

Given A, B, C, three distinct lines from point I
        A', B', C', three distinct lines from point II
   join the meets of AB' and BA'   meets join on R
                    AC' and CA'   meets join on S
                    BC' and CB'   meets join on T
   I say that the joins, R, S, T are concurrent (at III).

In abstracto the two figures are isomorphic, but in their instantiations they are exclusive flip-flop instantiations. They are opposed in a one-to-one correspondence in regard to the geometric values related to and including point and line. Theorems affecting points and lines may thus occur in bound-variation so that knowledge in one system includes the knowledge of the dual system. This suggests that for cases like this, if a given instantiation is coherent with its reference-frame, then the dual instantiation is also coherent with the dual reference-frame.

In order to show this immanent reversibility without confusion, the exhibit shows the well-known artificial distinctions of dual terms for two geometric instantiations that can flip-flop in bound-variation. In particular, I exhibit two models for a remarkable and well-known proof where there is one logical structure for the proof with two completely opposed instantiations. For the dual instantiations, the geometric theorem displays homomorphic logic. It could have been set up with homonyms, but this should be evident.

The instantiations are duals to each other, yet their apodeictic reference-frame could be expressed without any distinctions, one from the other, if we chose to do so.

As mentioned before, a circle may be defined as the locus of points or the envelope of lines. The Principle of Duality springs from the recognition of the fact that a curve may be regarded both as the path of a moving point (or fixed distance from a reference point) and as the envelope of a moving tangent line (or fixed distance of tangents). The first notion was recognized early in the history of geometry. The second idea is less obvious and appears to be due to De-Beaune (1601-1652), a student of Descartes' work. The *Horologium* of Huygens (1629-1693), contains some mention of the properties of the evolute of a parabola and cycloid, i.e., the envelope of the normals (perpendiculars). There are some optical applications to caustics in Tschirnhausen (1631-1708). Leibnitz gave a systematic treatment of envelopes in 1692. Brianchon (1806) showed some advantages in coordinating the two modes of conceiving. The principle of duality was first stated clearly by Gergonne in 1825. He also devised the notion of the class of a curve. Poncelet (1788-1867) gets the credit for the detailed development of the theory. Chasles and Salmon made further applications to metrical properties. The complementary treatment by means of line coordinates is to the credit of Möbius (1790-1868) and Plücker (1829). See *Projective Geometry*, p 108, C. V. Durell.

LOGIC DEREFERENCING DUALITY

BOOLEAN ALGEBRA. Boolean algebra has homonymic dereference duality. Although today it is customary to use the sentential dereference, George Boole preferred to dereference to classes. He was aware of the dereference mode reversibility {class, sentence}, though he did not discuss it as a duality. In effect, George Boole formulated the first homonymic dereferences for logic.

We can start logic with classes and develop toward a sentential logic as Aristotle began and the scholastics continued or we can start with a sentential logic and develop toward a class logic as is currently popular. This choice is arbitrary and the two possibilities are opposed in their orders of presentation. This is also another way of saying that the two presentation modes disagree as to what to regard as fundamental. Treating them as homonymic dereference modes puts this argument in perspective.

CONTINUUM LOGIC. Continuum Logic, as currently defined, is any logic system that dereferences its values to any value in the Real value interval [0,1]. This interval includes two-valued logic systems, finite many-valued logic systems, denumerably infinite-valued logic systems, and any other possible logic system using a value set within the defined interval. Only a logic system that has the capacity to dereference to **any** value in the interval is a Continuum Logic. Continuum Logic divides into two main branches. Historically, the first branch is probability logic, the second branch is fuzzy logic. In this exposition I argue for the distinction of probability logic as a logic in its own right.

Probability logic was first fully worked out by George Boole. On page 262 in his *Laws of Thought* he makes the following statement:

> "...there is another form under which all questions in the theory of probabilities may be viewed; and this form consists in substituting for events the propositions which assert that those events have occurred, or will occur; and viewing the element of numerical probability as having reference to the truth of those propositions, not to the occurrence of events concerning which they make assertion. Thus, instead of considering the numerical fraction p as expressing the probability of the occurrence of an event E, let it be viewed as representing the probability of the truth of the proposition X, which asserts that the event E will occur. Similarly, instead of any probability, q, being considered as referring to some compound event, such as the concurrence of the events E and F, let it represent the probability of the truth of the proposition which asserts that E and F will jointly occur; and in like manner, let the transformation be made from disjunctive and hypothetical combinations of events to disjunctive and conditional propositions."

In effect, by his recognition of this immanently reversible interpretation, Boole defines an infinite-valued logic, a possible Continuum Logic. The values are clearly in the Real interval [0,1], as all probability values must be..

Because Boole's methods for establishing the logic values are algebraic we can say that these values form the domain for a denumerably infinite many-valued logic system. Boole neither states nor denies the use of the transcendental domain. However, we now use this domain to facilitate general proofs. When we use infinitesimal calculus in probability distributions (considered by Gauss and others) we are including values from the transcendental domain. For finite probabilities there is no necessity for this transcendental domain other than allowing for these general proofs.

INFINITE-VALUE LOGIC. Boole's infinite-value logic is the first in the history of logic. It is the first formal recognition of the possibility of partial truths, that truth can be represented as a matter of degree. Furthermore, it can be used for a generalized solution to Aristotle's problem of future contingents, a subject that clearly links probability with logic.

In the Fuzzy Logic literature no mention is made of Boole in this regard. Many of the people involved in fuzzy logic credit Jan Lukasiewicz (1878-1956) with the first infinite-valued logic, but this is a false claim. Lukasiewicz uses the same value interval as George Boole. Lukasiewicz defined negation and the Philonian conditional which can be applied to any system using values restricted to the Real value interval [0,1]. These definitions are:

$$[\sim P] = 1-[P]$$
$$[P] \supset [Q] = 1 \text{ for } [P] \leq [Q]$$
$$[P] \supset [Q] = 1-[P]+[Q] \text{ for } [P] > [Q].$$

If you are experienced in this subject, you cannot help but see the relation with probability calculations. Furthermore, if you only allow the limit values of the Real value interval [0,1], namely 0 and 1, these definitions yield the ordinary two-valued system. Lukasiewicz admits that this set of definitions allows infinitely many degrees of truth "as in the calculus of probabilities." Boole made a direct statement that this conception of partial truths could be used in a logic system. Lukasiewicz prefers this infinite-valued system, but he fails to credit Boole for having done it already.

Today the system of Lukasiewicz is characterized as a denumerably infinite many-valued logic system. This formally leaves out non-denumerable values, such as most transcendentals. This precludes the Lukasiewicz version from the superclass of Continuum Logics. It is a proper subset of the Continuum interval. This explicit limitation allows us to say that Boole's system has wider scope in its applications. Boole does not preclude the possible use of transcendental generalizations in proofs.

Among the fuzzy logicians there are some who claim that Lukasiewicz is the first fuzzy logician (cf Kosko, etc.). Infinite-valued logics can treat truth as a matter of degree, a definatory characteristic attributed to fuzzy logic by some fuzzy logicians. If founding a logic that can treat truth as a matter of degree is evidence of a fuzzy logic, then these fuzzy logicians will have to change their claim for Lukasiewicz and say that George Boole is the first fuzzy logician, the first logician to formalize truth as a matter of degree.

For myself, I wish to deny entirely the title of "Fuzzy Logic" to the subject of the Logic of Probability. The Logic of Probability is well defined and not enhanced by this title. There is a much more powerful and distinct characterization reserved for "Fuzzy Logic" if we do so. Before giving more precision to these distinctions it is useful to review the necessary vagueness in measurement.

MEASUREMENT ERROR. Measurement plays an important role in defining objects in terms of volume, shape, contents, and movement. It is accurate to say that there is a vagueness in all measurements and that this vagueness is summarized by the ***margin of error***. In general, measurements are statistical and we use statistics to calculate the probability of error as well as the margin of error. This margin of error itself is represented as a kind of measure. A theory of measurement has developed in order to cope with the problems of calculating with margins of error. Every practical scientist has to face this problem in handling measurements.

The margin of error is expressed as a spread of variations. It is calculated as a deviation from the presumed measure with '±' some value type. The current types are *mean deviation*, *standard deviation*, or *percent*. An example from error theory for distance and time, analog computers are rarely better than ± 1% accuracy. A 1% deviation going toward Halley's comet could have put any of the 5 probes (1986) off over a million and a half kilometers. Another example, using usual methods, if we tried to find the planetary positions for 60 million years ago, we know these positions may have a significant margin of error. Knowing this has led to research for estimating how far off such calculations might be.

Another type of error for our analysis consists of partition problems in measurement. The king's crown problem given to Archimedes is a good example of a partition problem. How can we tell how much gold is in the crown without damaging the crown?

This crown problem presents us with two distinct notions that will be useful for this analysis. The first is, how do we partition the gold for measurement without damaging the crown. The second is a notion of an entirely different kind.

If the crown is only partly gold then can we say that the crown is gold and only be partly wrong, etc.. The statement, "The crown is gold," may be only partly true. This implies also that it may be partly false. If we could partition the crown in some way and determine that it is 90% gold, this puts the statement of its truth at 0.90, it is 0.10 false. If these numbers were exact, then they do not represent vagueness. If the numbers are inexact, then we introduce a margin of error that may overlap the two categories of True/False. We seem to lose the crispness of the excluded middle.

This, however, need not be the case. If we find a way to measure the gold, we know the rest of the crown has to be the not-gold part, Excluded Middle, even though this truth is expressed as a probability, a probability because of the margin of error. The apparent overlap is not a genuine overlap. We might have 0.90 ± 0.01 gold, 0.10 ± 0.01 something else. This expression is not an appropriate representation for using the Excluded Middle. A more appropriate representation might be that once we determine gold as 0.90 ± 0.01, then the rest is expressible as $1 - (0.90 \pm 0.01)$. This allows crisp use of the Excluded Middle.

We must also decide how we are going to handle adjustments to the example statement, adjustments such as, "The crown is pure gold," or "The crown is only gold," or "The crown is partly gold." These and other forms may be classed as modal forms that we could also consider.

The problems of measurement suggest a number of important considerations in defining vagueness. Given an object, it belongs to a set to a certain degree, we can ask to what degree the object belongs to a set. If the quality defining the set of objects is a measurable quality, then we can seek a way of measuring the degree to which it belongs and does not belong. The crown belongs to gold, 0.90 true or maybe 0.90 ± 0.01 true.

Note that even though Archimedes discovered how to determine the quantity of gold in the crown that the method depended on knowing the weight (specific gravity) of pure gold. Archimedes had no adequate means for determining this purity of gold. It is only fortuitous that gold in its natural state is highly pure, a fact Archimedes presumes but could not have known. He assumes the purity for his solution.

Some fuzzy logicians want to include this vague partitioning in fuzzy sets. For me, this is *not* one of the forms of fuzzy sets. The partition is indefinite, but the Law of Excluded Middle can still work with rigor.

CRISP SETS

In set theory we partition sets of objects and these partitions are also sets. These objects are regarded as clear and distinct. When partitions also are clear and distinct we call the subsequent sets **crisp sets**.

An equivalent description of *crisp set* is: Any set that sustains the power of Excluded Middle. From this we conclude that probability logic is crisp.

VAGUE SETS

Formalizing a system that uses objects that are not clear and distinct has proven to be a formidable task. It has not yet been done to consensus satisfaction. To begin such a defining task we must first agree on the possible types and degrees of vagueness we need to handle. "Vagueness" is not a synonym for fuzziness. I argue that all fuzziness is a form of vagueness, but not all vagueness is fuzzy. Vagueness in this context has no dereference different from ordinary usage. Fuzziness is characterized as formal reasoning without the crisp power of the Excluded Middle.

Vague set partitioning is of two types;

> ONE, we are not sure whether something belongs or not, the definition is vague and fuzzy, the Law of Excluded Middle loses power because there is no distinct either-or separation;

> TWO, takes an object and partitions it such that part of it belongs in one set and part of it belongs in another.

In the second type there are two cases to consider;

> ONE, the partition border is clear and distinct, here the Law of Excluded Middle still holds power;

> TWO, the partition border is vague.

The vague partition border has two cases to consider;

> ONE, the vagueness is like a measurement problem, Excluded Middle can still hold;

> TWO, the vagueness is a transition boundary. Transition boundaries form fuzzy sets, they lose the Excluded Middle.

These considerations exhibit that the main characteristic of Fuzzy Sets is that they lost the power of the Excluded Middle.

FUZZY BOUNDARY PARTITIONING. Given an object, it belongs to a set to a certain degree, we can ask to what degree the object belongs to a set. If the quality defining the set of objects is a measurable quality, then we can seek a way of measuring the degree to which it belongs and does not belong. If the measure is Eudoxian then we say that the measure is not fuzzy.

> **Note:** For this argument we include measurement error margins as *crisp* distinctions. The Law of Excluded Middle is still valid.

There are some well known non-Eudoxian measuring systems. For all of these systems of non-Eudoxian measurement, the magnitudes are vague, but the ordinality remains clear and distinct. For example, the Mohs hardness scale sustains a clear ordinality and this ordinality sustains the Law of Excluded Middle, but we have not yet discovered a Eudoxian measure for hardness. In effect, hardness is a form of surface tension and the Mohs scale indicates only an ordinality for a range of surface tensions. The scale of volcanoes from Hawaiian to Volcano is a vague classification. It too defies Eudoxian measures. It even has overlapping, the scale is fuzzy.

There are at least two forms of fuzziness. The first form concerns the problem of vagueness in certain forms of non-Eudoxian measurement, the second form concerns the partitioning of the object (part of it belongs to the set, part of it does not) in a manner that overlaps. Measure and Partition generate two forms of fuzziness in fuzzy sets.

Fuzzy set partitioning is of two types;

> ONE, we are not sure whether something belongs or not, the definition is too vague for the use of Excluded Middle;
> TWO, it takes an object and partitions it such that part of it belongs in one set and part of it does not. It is a sort of boundary problem.

In the second type there are two cases to consider;

> ONE, the partition boundary is clear and distinct, this is a ***crisp boundary***;
> TWO, the partition boundary is *not* clear and distinct, this is a ***vague boundary***.

The vague boundary has two cases to consider;

> ONE, the vagueness is a measure problem, this is a ***measure boundary***;
> TWO, the vague boundary is a transition problem, this is a ***transition boundary***.
>
> **1.** *Measure Boundary*: the vague boundary is a measurement boundary. There are two types:
> > **a. Eudoxian.** The Eudoxian may be crisp or vague. The Excluded Middle is sustained in both cases.
> > **b. non-Eudoxian.** The non-Eudoxian has vague aspects in matters-of-fact, but the Law of Excluded Middle remains valid for hardness in regard to ordinality.
>
> **2.** *Transition Boundary*: the vague boundary is a transition boundary.
> All transition boundaries are fuzzy boundaries. The Law of Excluded Middle is lost.

SUMMARY of GENERAL BOUNDARY FORMS

> ***Crisp Boundary***: the partition boundary is crisp and allows use of the Law of Excluded Middle.

Vague Boundary: the partition boundary is *not* crisp, which may or not preclude the use of the Law of Excluded Middle.

Fuzzy Boundary: the partition boundary precludes Excluded Middle.

Numbers, Denumerables, and certain other Infinities form clear and distinct sets. Magnitudes may be subdivided into *any* number of finite divisions. This is the in-between category of Galileo which I call ***abfinity*** (going from finite). No instantiation can be infinite yet there is an infinity of these finite instantiations.

EXAMPLE: the counting numbers.

There is an infinity of counting numbers, but no counting number is infinite. Galileo was the first to recognize this distinction. He also understood some of the important implications. Reading his text in the original languages (Italian and Latin) you will never find the contradictory phrase 'infinite number', even though it is wrongly put in every available translation I have ever found (English, French, Spanish, German, Swedish, Japanese, and Russian).

SUMMARY CONCLUSIONS. Geometric Instantiation Duality. Geometric instantiation duality offers a geometric paradigm for exhibiting definatory characteristics of data-field duality. For geometry, its data-field is actually a conceptualization of configurations in space. Whether we can conceive of these spatial configurations as independent of the reference modes is a question I do not address directly. Assuming for this argument that spatial configurations and geometry are independent of each other provides a convenient model for what I am trying to communicate. It is an artificial device, an apodeictic reference-frame that may be used as a paradigm.

We saw two ways of viewing a circle (locus, envelope). These two modes of viewing the circle were said to be dual to one another. This duality was based on the opposition of the points and lines. In one point-of-view the points are fundamental and lines are their joins. In the other point-of-view the lines are fundamental and points are their meets. We have a one-to-one correspondence herein for the oppositions. The opposition in terms of magnitudes is that points have no length and lines have length only. The Pappus theorem shows a remarkable example of a case of homomorphic logic with dual heteronymic exchange, where the fundamental opposition is between points and lines. The argument could be reconstructed in a homonymic exchange.

We could say that the heteronymic representation obtains a systemic duality that maps the data-field duality. It is the fact that the heteronymic representation can be rendered homonymic that we come to terms with data-field duality.

If we have two systems dual one to the other where corresponding terms are distinct we say these duals are heteronymic. Here we can have a system/system duality. If we render the two heteronymic duals into one homonymic system, the duality is no longer expressed by system terms. However, we have remaining the immanent reversibility of the dereference, which here is characterized as data-field duality.

Logic Instantiation Duality. Crisp and Vague are in simple opposition. Included in the Vague are the Probable and the Fuzzy. Crisp and Probable use the Law of Excluded Middle. Fuzzy does not. The oppositions form modes of duality.

Logic instantiation duality offers a paradigm for exhibiting characteristics of data-field duality. For logic, its data-field is actually an opposition between dereferencing to sentences and dereferencing to classes (sets). The particular dualities for this concern are found in these possible dereference modes. The Crisp and the Vague may both refer to sentential logics or to class logics. In this sense they are homonymic dualities. These homonymic dualities are special cases of data-field duality.

KEY WORDS
apodeictic reference-frame, homomorphic reference-frames, homonymic duality, immanent reversibility, mapping modes, sys-tem/system, system/data-field, data-field/data-field, data-field duality, instantiation duality, homonymic, heteronymic, paradigm

END Addendum 4

Addendum 5
REFERENCE-FRAME DUALITY

Introduction. *Science-systems use reference-frame standards in two general modes; reference-frames for intelligibility, reference-frames for measurement. These modes represent a hierarchy because the reference-frame for intelligibility must precede measurement. If a reference-frame for measurement exists it must be a proper subset of a reference-frame for intelligibility. Without a reference-frame for intelligibility there is nothing to measure. Measuring length makes no sense if we do not know what length is.*

The qualitative distinctions in a given intelligibility reference-frame comprise the definition elements for the variables of that reference-frame. Not all variables are for measurement. The measure variables are such that the instantiations can be mapped to Eudoxian or non-Eudoxian orders of magnitude. The non-measure variables dereference to classes that do not have a necessary dereference to ordinality or magnitudes. For example, though we can associate ordinality (e.g., ontogeny, phylogeny, etc.) or magnitude (e.g., size, etc.) with trees, the term 'tree' may be dereferenced independently of these notions.

A measurement system uses operational definitions for associating standard units to the variable instantiations being measured. The numerical values that correspond to the variable instantiations are their measure. It is the definition of the measure variable that denotes what in the instantiation is to be measured.

REFERENCE-FRAMES; SITUS, LOCUS, QUANTUS.

For us, space and time are conditions for comprehension. Emanuel Swedenborg (1688—1772) was the first to write of this. He inspired Immanuel Kant (1724—1804) who used this as a basis for the *Critique of Pure Reason*. Without space and time, the world as we know it dissolves into oblivion. Because they are conditions for comprehension, they cannot ever be attributes or qualities of forms that we perceive. For us they are the conditions for our understanding of these forms, the conditions for the existence of the attributes as we know them.

A condition for existence cannot be an attribute of something that exists. An attribute is not a condition for existence. Aristotle knew this when he said that Being is not an attribute. Being is a condition for existence, therefore etc..

When we try to discuss the conditions for our comprehension we are condemned to circular reasoning. The only way out of the circle is if our conditions for comprehension are *not* under our personal control. Otherwise, we are stuck in a subjective idealism which itself destroys the meaning of subjective since there is nothing objective for making a comparison, for making a contrast. 'Objective' and 'subjective' become apodeictic fictions.

SPACE. 'Length' names the measure of space. For us, space is a condition for understanding our existence. We can conceive of empty space as uniform. By definition, in uniform space every point in it is equivalent, i.e., there are no points possessing special properties. When in space there are no directions possessing special properties we say that this space is isotropic. This again brings us through full circular reasoning. We have only succeeded in finding a name, 'isotropic space', and we are still left with the problem of dereferencing this name.

When we study phenomena occurring near the Earth's surface we find that directions in space are not equivalent, they are not isotropic. A heavier than air body released from the hand always sinks, otherwise it contradicts saying it is heavier than air. In *Star Laws* Book I, I argue that gas molecules migrate along temperature gradients toward the highest temperature. Thus 'heavier than air' has a different connotation that relates to another kind of buoyancy. We reason that a body sinks vertically following a geodesic line toward the center of the Earth and its sinking may be interrupted by the Earth's surface. If we want such a sinking body to change its direction of movement and go up, we must impart an initial velocity in the desired direction. None is required to make debris sink toward the Earth center. This lack of equivalence of different directions with respect to the surface of the Earth is the source of Newton's notion of gravitational 'attraction'. At a vast distance from the Earth, from other stars, in space 'free' from large bodies we suppose that all directions closely approximate equivalence, closely approximate the isotropy of an empty space.

TIME. 'Duration' (time span) names the measure of time. For us, duration is a condition for understanding our existence. We can conceive of empty duration as uniform. By definition, in uniform duration every point is equivalent, i.e., there are no points possessing special properties distinct from any other point. When in duration there are no points with special properties we say that this duration is uniform. Every point in duration is equivalent. However, duration is regarded as non-isotropic. The conceived directions (forward, backward) in it are not regarded as equivalent. This brings us again through full circular reasoning. We have another name to dereference, 'non-isotropic duration'.

The non-isotropism assumed for duration may be an artifact of our perception. Duration as 'non-isotropic' is an assumption based on our interpretation of experience. Other interpretations are possible. For H. G. Wells, time as a fourth dimension is a necessary condition for existence. Just as a line with no breadth cannot exist, an object with no duration cannot exist. He also claimed through the words of his Time Traveler that

> "There is no difference between Time and any of the three dimensions of Space except that our consciousness moves along it." p4, *The Time Machine*.

This implies a space-time continuum (stated in 1894). For us, duration is divided into intervals by cyclic phenomena. This dependence on cyclic phenomena is at once a convenience and an inconvenience. It is a convenience to have cyclic phenomena as references for dividing duration into time intervals, it is an inconvenience that we must depend on the uniformity of these cyclic phenomena which in every case under changing conditions may lose reliability for the assumption of uniformity.

In local Earth observations on a macroscopic scale we assume that duration is uniform. This means that any event, occurring under the same conditions but at different moments in the duration proceed in exactly the same way. If a ball drops from a certain height within a certain duration, it is presumed to sink with the same results a thousand years ago or a thousand years in the future. In detail, this is false. The Earth is gaining matter. Meteor showers and any other space debris converging to the Earth surface make the Earth gain matter. If gravity is directly proportional to the material content and its compaction in a gravitational body then with time the acceleration due to gravity must change. In Newtonian analysis this increases, 'falling' a fixed distance occurs in a shorter time interval.

SPACE AND TIME REFERENCE. A number of important consequences follow if duration is uniform and non-isotropic, space is uniform and isotropic. For us, we cannot imagine existence for a space without duration. Assuming space is uniform and isotropic, it is impossible to determine the position of a particle with respect to space. Assuming duration is uniform and non-isotropic it is impossible to determine the time interval of a particle movement with respect to duration. Reference-frames are required.

EVENT REFERENCE-FRAMES. A body has a definable closed surface (boundary) that divides it into inside and outside.

> **DEFINITION: A body** is regarded as a material system closed by a boundary.

A body, qua body, is as well-defined as its boundary. It is regarded as well-defined when we can always decide whether any given point is on the body (boundary), inside it, or outside it. As a paradigm, take a sphere. Its surface is well-defined and divides the sphere into an inside and an outside. Physical instances are more difficult to define and we often impose geometric models. This limits adequacy.

The state of a body with respect to a space/time reference-frame is called an event.

DEFINITION: An **event** is a system (of one or more bodies) that can be described partially or fully with respect to a space/time reference-frame.

This definition is independent of whether this event is regarded as static or dynamic. Space and time specifications (reference-frames) are used for describing the event as static, dynamic, or a mixture of both. Carefully chosen reference-frames are required to specify an event, reference-frames that do not implicate the event, reference-frames that are independent of the event, i.e., reference-frames that do not introduce pseudophysics or pseudo-measurements.

EVENT REFERENCE-FRAMES. For a given analysis, the space reference-frame is a body or group of bodies regarded as fixed, with respect to which the positions of other bodies are specified; the time reference-frame is a body or group of bodies regarded as in uniform cyclic movement, with respect to which the time spans are specified.

To describe an event we need at least four coordinates, three for space, one for time. The coordinates are also known as dimensions, therefore every event has at least four dimensions.

The line described by a moving particle in a given reference-frame is called its path or trajectory. The locus (shape) of this path depends upon the chosen reference-frame. Though many reference-frames may do for this problem, not all reference-frames show equal convenience. In principle, we choose the one we deem most convenient under the given constraints. If we are going to measure the speed of a car, the Earth's surface is a logical reference-frame. For terrestrial navigation, the Earth reference-frame is best. If we are going to study the movement of the solar planets, the

Earth or the Sun may be the useful reference-frame center. However, our access to the Sun reference-frame is indirect. As long as we are Earth-bound we must establish ways to transform from our geocentric reference-frame to a heliocentric reference-frame.

Ideally, the simplest reference-frame is founded on invariance, invariance in its spatial frame, invariance in its cyclic frame. Uniform cyclic phenomena are a necessity for measuring time spans. Fixed lengths are a necessity for measuring coordinates within space. This is why Newtonian analysis requires inertial reference-frames. Inertia is the only principle Newton requires for finding reference-frames that eliminate pseudo-measurements and pseudoforces.

DUAL REFERENCE-FRAMES. In mechanics the majority of physical descriptions are expressed in terms of the concepts of space, time, and matter. In this subject, physicists have tried to reduce all mechanical phenomena to these concepts. I argue not only that a totally adequate model has yet to be achieved, but that it may not be possible under current modes of thought.

In Physics, as yet, there is no designed general inclusive flip-flop conceptualization, such as the logical NAND and NOR, that produce two compatible systems dual to each other and with the same data field spread. For NAND and NOR, when one mode of reasoning gets long formulas we can flip-flop to the other mode and continue the reasoning with shorter formulas. Ockham's Razor becomes useful again in this sort of duality.

In some ways, a conceptual flip-flop seems to be what we do in the presently accepted analysis of wave-particle aspects of electromagnetic phenomena. At present we recognize a kind of duality in this, but we do not have a theory that exhibits exact understanding of it. They should be exclusive or inclusive flip-flop reference-frames, but they seem to be a mixture of the two modes. This is perceived as an opposition that is contradictory.

This problem may be created by the way the experiments are analyzed. For example, suppose that light is a wave phenomena but our means of detecting it uses particle phenomena. It should be evident that something gets lost and something gets gained in the translation. These losses and gains distort the interpretations. Waves can seem like particles, etc.. This consideration puts doubt into one aspect of the Einstein theory concerning the photoelectric effect. They used Planck's concept of energy quantum with a dual interpretation.

The Einsteins assume that the energy quantum is a property of the radiation rather than of the atoms. The energy of the incident radiation is absorbed in discrete quanta rather than in a continuous stream of waves. This consideration allows them to develop their photoelectric equation that is useful in interpreting certain aspects of the photoelectric effect. Their discrete packets of radiation assumption is equivalent to assuming that light is absorbed in atoms by apparent discrete packets of radiation. This does not preclude that continuous waves can be picked up as if they were discrete packets. There is no reason to choose between the radiation quantum and the quantum in the atom. This choice remains in a state of immanent reversibility as we have no critical experiment for deciding. As the theory they present remains useful, it exhibits that apodeictic fictions have a use in creating Physics theory. Correct use of this formula should admit the dual interpretations in free-variation. Neither mode has an effect on the validity of the formula, the selected mode affects the interpretation of one term in the formula only. It is the term ***photon*** that holds the immanent reversibility when used correctly. Thus we have:

$$h\nu = \mathbf{E}_{on} + \mathbf{I}_{max}$$

where ν is frequency of the incident photon, E_{on} is the photoelectric work function, and I_{max} is the maximum kinetic energy of the emitted electron. Keeping the immanent reversibility in {photon, atom} quanta sustains the equation for either mode of interpretation.

The photoelectric effect was recognized clearly as such by Heinrich Hertz in 1887. Philipp Lenard (1862-1947) observed the variation in electron energy with light frequency. His results were qualitative rather than quantitative. He discovered that the maximum electron kinetic energy is determined by the frequency of the light. Mileva Einstein worked with Lenard for six months. There is circumstantial evidence that suggests that it was her insightful abilities in physics and mathematics that allowed the discovery that the energy of individual ejected electrons increases linearly with the frequency of the light. When Albert published the results under his name he promised Mileva that if the results won a Nobel Prize he would give her all the prize money. He kept his promise.

BASIS FOR MEASUREMENT. It would be convenient if we could show that length, duration, and material quantity are Eudoxian magnitudes. If they are Eudoxian then they can be represented to vary uniformly and continuously in quantity and that any two instances of the characteristic magnitude can be expressed in a simple ratio. For this case, we may choose a standard unit for each characteristic magnitude. For length, duration, and material quantity we can use measured instances of each characteristic as a standard unit, such as: meter, second, and kilogram. Thus a ratio of some instance of a characteristic magnitude and its standard unit can be expressed in terms of a number of standard units, including fractional parts. This number is the measure of the characteristic magnitude.

Unfortunately, measuring these magnitudes is more complicated than we would like. This simplistic description for measurement only works in a local approximation. When we attempt to extend these local observations to astronomical scales we can get into trouble.

With measurement we are bound to approximations. As we progress in techniques of exact measurements we discover that the process of measurement starts to interfere with its purpose. We are at a point in history where we can measure so exactly that the act of measurement itself renders a stumbling block to exactitude. Here the effect of the observer on the observed becomes significant. The act of measuring disturbs (changes) what it is that we want to measure. The instruments for measuring heat and temperature, or the electrical quantities are well known for disturbing the quantities we want to measure.

Before considering some of these difficulties, let us suppose that we can identify characteristic magnitudes, but without necessarily knowing if these magnitudes actually follow the Eudoxian requirements. We must then establish conventions for units of the characteristic magnitudes. These conventions are the standard units.

STANDARDS OF REFERENCE. In 1790, the National Academy of France requested the French Academy of Science to "deduce an invariable standard for all measures and all weights". The appointed Commission fulfilled this request with a system that was both simple and useful for science and commerce. This system is called the metric system, from the Greek word 'metron', meaning a 'measure'. The Commission chose measures in powers of 10, a decimal system. The unit of length was defined as one ten-millionth of the distance from the North Pole to the equator along the meridian that passes through Paris. This unit of length was named the meter, also from the Greek word 'metron'. Measures for volume and mass were derived from the unit of length. This related the basic units to each other in a way that unified the system.

The metric unit for the quantity of matter was named the gram, g. It was defined as equivalent in material quantity to one cubic centimeter of water at its maximum concentration. The concentration of water varies slightly with temperature. It is at its maximum concentration at about 3.98° C. One cubic centimeter (cc) is a cube 1/100 of a meter on each side. This standard is now abandoned. For the exactness we require today, water has too many variations in impurities, in isotopes.

KILOGRAM: The kilogram is defined in terms of a cylinder of platinum-iridium manufactured with a 'mass' close to water.

This cylinder is the only kilogram weight standard we still use (2001 A.D.).

The metric unit of fluid measure is the volume of a cube one decimeter on the side, 1/10 of a meter. It was named the liter, 1000 cc. A liter is slightly larger than a quart (1 liter = 1.056 qt).

The metric unit of time is the second. Prior to 1956, the apparent solar day was used to define the second. Because of the variations in orbital speed and in axis inclination there are slight variations in this day. To compensate for these variations an average apparent solar day was calculated over a one year period. The second was defined as 1/86,400 of a mean solar day, that is, a mean solar day was defined as having 86,400 seconds. This also is inexact. There are minor variations in the Earth's orbit and the mean solar day is not exactly the same each year.

Since 1968 an atomic reference is used as standard for the second. The cyclic events in atoms have the greatest precision that we know of. The second is now defined in terms of the frequency of energy transitions of the Cesium-133 atom. 9,192,631,770 periods of hyperfine transition, 3.26 cm wavelength, is one second. This is called the atomic second.

The meter was redefined in 1961 as being equal to 1650763.73 wavelengths of the $2p^{10}$ - $5d^5$ transition of Krypton-86 ($_{86}Kr$) atom. Light of this wavelength is orange in color. With a Krypton lamp and interferometer the standard meter is produced.

The meter was redefined again in 1983 as the distance covered by light in exactly 1/299,792,458 of a second *in vacuo*. However, as the 'speed' of light is measured with particle devices, it is highly likely that these particles devices have delays that lead to underestimates of light speeds. Therefore, we should expect another definition once this fact is taken into consideration.

TEMPERATURE. Within certain ranges, when heat is generated in a steady state, a thermometer can be used with its highest possible precision (to date not ever very precise). When heat is not generated in a steady state, the procedure for measuring temperatures changes the temperature of the zone in a way where error in measurement is difficult to avoid, difficult to compensate for. These difficulties are most present in rapid non-monotonic changes of state such as explosions generate. This problem is considered in Book I Chapter 5 in more detail.

COMPOSITE MAGNITUDES. Some magnitudes, such as area or speed, are analyzed into more than one magnitude, like or unlike. Area is measured with two lengths, speed is measured with length and time. These are two dimensional magnitudes. Galileo presented with rigor one of the earliest quantification analyses of speed.

GALILEO AND DUAL DEFINITIONS FOR SPEED.

Galileo is a pioneer in using one-to-one correspondence arguments for infinite series. He starts his analysis of motion with a one-to-one correspondence definition of uniform motion.

ARGUMENT. When we make a careful analysis of Galileo's statements concerning speed we find that **s = d/t** is never expressly stated anywhere in his *Dialogues Concerning the Two New Sciences*. Even so, from Galileo's axioms about distance and time there follow theorems from which we could derive **s = d/t** (Galileo never used algebra). Galileo never does the geometric equivalent of this derivation. This is because Galileo never expresses ratios of unlike terms.

Today speed is regarded as an intensional (composite) magnitude with two components, time and distance. In ordinary parlance we talk of speed in two opposed ways, either as distance per unit time or as time per unit distance. It is just as informative to say I walk five miles per hour or I walk one hour per five miles. This order reversal makes these modes of representing speed as dual to each other. These ratios are written as d:t and t:d. For Galileo this is two directions for a one-to-one correspondence, it is not two forms of division.

Which mode best represents speed? These representations are distinct, yet in regard to the information conveyed about time and distance they are equivalently sufficient.

In algebra, when the two modes are expressed as division, they are clearly inconsistent with each other. These expressions look like they could be written as d/t and t/d, respectively. These expressions are incompatible. If we had two distinct names, we could eliminate the problem. The decision as to which mode we use is linked to habit or what we wish to emphasize. Today, when we calculate with speeds we think in terms of our algebra training and use d/t. This results in a magnitude that varies directly with speed. Using d/t simulates our expectation for a one-dimensional magnitude.

A general representation for Galileo's concept might be

$$\text{Speed} = d \text{ km AND s/hr}.$$

In this commutative representation we are not constrained by a fixed order relation. In Galileo's mode of reasoning, resolving speed to a number is more difficult

In reading Galileo we have a difficult time keeping our biased training out of the argument. The form that the definition of speed takes is kept tacit by simply representing speed as a magnitude, apparently the same sort of magnitude as time or distance. Yet time and distance somehow meld together into this single concept, the single magnitude of speed. Speeds can be added to, subtracted from, multiplied, divided, etc., i.e., in Galileo speed is treated as if it were a Eudoxian magnitude. This allows Galileo to use Euclid's *Elements*, Book V, for mathematical manipulations.

One argument I pursue here is that Galileo always relates speed in terms of ratios and proportions or as a magnitude in itself, always with one dimension. This eludes and even prevents the conclusion that speed is expressible as the fraction d/t with two distinct dimensions. The dimensions cannot 'reduce' out. I shall begin this argument.

In preparation for his analysis of motion Galileo went to great lengths to describe two concepts used in what today we call one-to-one correspondences (mapping). The first idea is how two infinite series can be associated *ad infinitum*, such as squares to the counting integers. The second idea is how we can subdivide any magnitude according to any assigned number, *ad infinitum*. Here also a one-to-one correspondence is implied between the magnitude and any of its divisions. This is the pristine principle for forming a number line. Note that all one-to-one correspondences are immanently reversible, A-to-B or B-to-A (inclusive flip-flop). These correspondence orders are dual to each other.

Let us try to understand how Galileo talks about it.

It is in the Third Day that we find Galileo concerned with speed. He begins with the definition of uniform motion.

> "DEFINITION: By steady or uniform motion, I mean one in which the distances traversed by the moving particle during any intervals of time, are themselves equal.
> CAUTION: We must add to the old definition (which defined steady motion simply as one in which equal distances are traversed in equal intervals of time) the word "any", meaning by this, all equal intervals of time; for it may happen that the moving body will traverse equal distances during some equal intervals of time and yet the distances traversed during some small portion of these time-intervals may not be equal, even though the time-intervals be equal." (*op.cit.*, 3rd Day, p197)

COMMENT: The use of the word "any" in this definition has remarkable consequences. Galileo specifically states that "any" means here "all equal intervals of time." Here Galileo shows the importance of exact statement. It is easy to show why his caution is necessary and its rigor is easy to display. Let us consider an example:

A driver is driving a car through a town for a distance that requires one hour. Divided in half-hour periods we might have:

TABLE 1.

time:	1/2 hr	1/2 hr
distance	19 km	19 km

but divided in quarter-hour periods we might have:

TABLE 2.

time:	1/4 hr	1/4 hr	1/4 hr	1/4 hr
distance	9 km	10 km	8 km	11 km

which clearly shows that in the first table it would have been possible that the car traveled at a uniform speed but the finer division set in the second table shows that this was not the case. The motion here is not uniform. The addition of "any" is a powerful way to indicate all possible equal time partitions. This is an example of what Galileo called a third intermediate term between finite and infinite.

Galileo says:

> "...I think there is, between finite and infinite quantities, a third intermediate term which corresponds to every assigned number; so that if asked, ..., whether the finite parts of a continuum are finite or infinite, the best reply is that they are neither finite nor infinite but correspond to every assigned number." (*op.cit.*, 1st Day, p145)

He does not name the concept but he shows a clear definition of its meaning and its use. I name this concept **abfinite** (to replace *ad infinitum*). His use of this concept follows in what may be the first practical application in history of one-to-one correspondences between two distinct abfinite series. With the word "any," a one-to-one correspondence between distance and time is implicit in Galileo's definition of uniform motion. How this is so is next to consider.

At first, in his definition, it looks as if we have a relation (function) where the dependent variable is distance and the independent variable is time. It is time that can take on any interval for an equal sequence of intervals, it is distance that follows as a sequence of equal distances. For Galileo, as we shall see, this does not decide any fixed order nor does it decide which ratio might also represent speed, d:t or t:d, and for good reason. Galileo does not ever represent speed as a ratio of mixed magnitudes. I have also come to believe that because Galileo has represented uniform motion as a one-to-one correspondence that he has eliminated the consideration of searching for another way for defining speed. The job is done. The definition for speed was no longer a question for him.

In a later context he applies his definition of uniform motion to uniform acceleration, a non-uniform motion. His definition of uniform motion (speed) forms the basis for his discussion of uniform acceleration (speed augmenting at a constant rate). This application will not be discussed in this exposition.

That Galileo considers the reversibility of corresponding distance and time is implied in the first two axioms that follow next from this definition.

To put some textual evidence in light, take a look at Axioms I and II for uniform motion.

> "AXIOM I: In the case of one and the same uniform motion, the distance traversed during a longer interval of time is greater than the distance traversed during a shorter period of time.
> AXIOM II: In the case of one and the same uniform motion, the time required to traverse a greater distance is longer than the time required for a less distance."
> (*op.cit.*, 3rd Day, p197)

The first axiom is in regard to distance relative to time, the second axiom is in regard to time relative to distance.

> Distance increases with time.
> Time increases with distance.

Both magnitudes increase directly with speed. Furthermore, the exhibited reversibility is consistent with one-to-one correspondences and does not introduce contradictions.

It is implicit that uniform motion is a form of speed. However, the first formal use of the term "speed" (Galileo actually uses a term that looks more like "velocity," but today velocity is only used as a vector quantity, speed is a better translation) is found in Axioms III and IV.

> "AXIOM III: In one and the same interval of time, the distance traversed at a greater speed is larger than the distance traversed at a less speed.
> AXIOM IV: The speed required to traverse a longer distance is greater than that required to traverse a shorter distance during the same time-interval." (*op.cit.*, 3rd Day, p197).

In this way of speaking, they show a clear sense of reversibility, the more distance traversed the more speed used, the more speed used the more distance traversed, but this reversibility is within a fixed time span. Speed and distance can be put into a one-to-one correspondence in a fixed time span.

If we now look at the theorems, we see that he never refers to ratios between magnitudes that are different in kind. Look and see:

"Theorem I, Proposition I
If a moving particle, carried uniformly at a constant speed, traverses two distances the time-intervals required are to each other in the ratio of these distances." (*op.cit.*, 3rd Day, p197)

COMMENT: We can express this as

$$(s_1 : s_2) \rightarrow \frac{d_1}{d_2} = \frac{t_1}{t_2}$$

where s is a uniformly constant speed, d is distance, t is time. From this Theorem we can derive either t/d or d/t, and the derivations are immanently reversible even though the resulting magnitudes are different.

The ratios clearly express that two distances vary directly with the ratio of their times. Varying together in this way with these Eudoxian magnitudes, time and distance, both can be used to give a kind of measure for speeds.

Today we would prefer writing this theorem statement, *ex equali*, as:

$$\frac{d_1}{t_1} = \frac{d_2}{t_2} = s$$

where s is a constant speed defined as d/t and usually expressed in lowest terms, per unit time.

Using this form we specifically commit ourselves to one of the two possible definitions that Galileo does not explicitly consider. Next:

"Theorem II, Proposition II
If a moving particle traverses two distances in equal intervals of time, these distances will bear to each other

the same ratio as the speeds. And conversely if the distances are as the speeds then the times are equal." (*op.cit.*, 3rd Day, p198).

COMMENT: These implications and ratios are expressible as

$$t_1 = t_2 \rightarrow \frac{d_1}{d_2} = \frac{s_1}{s_2} \quad \text{AND} \quad \frac{d_1}{d_2} = \frac{s_1}{s_2} \rightarrow t_1 = t_2$$

where t_1, t_2 are equal time intervals, d_1, d_2 are two distances, s_1, s_2 are the respective speeds.

These two implications reduce to the logical equivalence:

$$(t_1 = t_2) \equiv \frac{d_1}{d_2} = \frac{s_1}{s_2}$$

When times are equal we see that speeds and distances are directly proportional. From this we could say that, in equal times, speed as a magnitude must increase as distance traversed increases. Here we find that the form seems to require that our representation of speed be d/t. We could have:

$$\frac{s_1}{s_2} = \frac{d_1/t_1}{d_2/t_2}$$

from which it is evident that for $t_1 = t_2$ we have:

$$\frac{s_1}{s_2} = \frac{d_1}{d_2}$$

If we had chosen t/d then we would have found $d_1 = d_2$, which is contrary to hypothesis, *reductio ad absurdum*. Look and see:

$$\frac{s_1}{s_2} = \frac{t_1/d_1}{t_2/d_2}$$

from which it is evident that for $t_1 = t_2$ we have:

$$\frac{s_1}{s_2} = \frac{1/d_1}{1/d_2} = \frac{d_2}{d_1} \quad \text{but} \quad \frac{s_1}{s_2} = \frac{d_1}{d_2} \quad \text{from Theorem II.}$$

Things which are equal to the same thing are also equal to each other (Euclid, Book I, Common Notion 1), from which we have:

$$\frac{d_1}{d_2} = \frac{d_2}{d_1}$$

which is true for non-zero terms only if $d_1 = d_2$, contrary to hypothesis, therefore, etc..

It seems from this argument that the question is settled, Galileo is tacitly using s = d/t. I would like now to argue to the contrary and say that the only reason this conclusion seems plausible is that we are habituated to mixed ratios and are not habituated to this older form of geometric representation of unmixed ratios. When ratios are expressed same in kind to same in kind, this leaves the ratio without its dimensions. In fact, in Galileo's ratios, dimensions can be ignored.

I summarize the next three theorems too so that you can see for yourself that there is no direct reference to d:t or t:d in them either.

Here is a summary of the statements and the algebraic expressions for Theorems III through V:

"THEOREM III, PROPOSITION III
In the case of unequal speeds, the time-intervals required to traverse a given space are to each other inversely as the speeds." (*op.cit.*, 3rd Day, p198).

III: For $s_1 \neq s_2$, $\dfrac{t_1}{t_2} = \dfrac{s_2}{s_1}$

"THEOREM IV, PROPOSITION IV
If two particles are carried with uniform motion but each with a different speed, the distance covered by them during unequal intervals of time bear to each other the compound ratio of the speeds and time intervals." (*op.cit.*, 3rd Day, p198).

IV: For $s_1 \neq s_2$, $\dfrac{d_1}{d_2} = \dfrac{(s_1)(t_1)}{(s_2)(t_2)}$

"THEOREM V, PROPOSITION V
If two particles are moved at a uniform rate, but with unequal speeds, through unequal distances, the ratio of the time-intervals occupied will be the product of the ratio of the distances by the inverse ratio of the speeds." (*op.cit.*, 3rd Day, p199).

V: For $s_1 \neq s_2$ AND $d_1 \neq d_2$,

$$\frac{t_1}{t_2} - \frac{d_1}{d_2} \times \frac{s_2}{s_1}$$

The last of the uniform motion theorems is Theorem VI which states:

"THEOREM VI, PROPOSITION IV: If two particles are carried at a uniform rate, the ratio of their speeds will be the product of the ratio of the distances traversed by the inverse ratio of the time-intervals occupied." (*op.cit.*, 3rd Day, p199).

In formula we have then,

$$\frac{s_1}{s_2} = \frac{d_1}{d_2} \times \frac{t_2}{t_1}$$

from which we can derive:

$$s_1 = \frac{kd_1}{t_1},$$ where **k** is a constant of proportionality.

Choosing appropriate units (a liberty we have when defining new magnitudes), we can have k = 1 and have the current standard formula for speed: **s = d/t**

Galileo does not do this. He keeps only the proportions. Therefore he never requires a constant of proportionality. His ratios are always ratios of like kind of magnitudes. It is as if he does not want to mix apples and oranges in a ratio.

In keeping ratios of like magnitudes the dimensions can be ignored, they are lost.

EINSTEIN THEORIES OF RELATIVITY
Introduction. There are two Einstein theories of relativity. The first is the Special Theory of Relativity which dates from 1905. In 1915 the Einsteins published an additional theory called the General Theory of Relativity. This later theory is said to "extend" the Special Theory. The General Theory of Relativity is the Einstein theory of gravitation. It assumes that we cannot point to any region of space as being totally free from external gravitational influence. The Einsteins argued that because of this permanence that gravitation is an intrinsic feature of space and time. They dereferenced gravitation as the geometry of spacetime.

THE MICHELSON-MORLEY EXPERIMENT
Luminiferous aether was postulated as a medium for the propagation of light. Though James Clerk Maxwell (1831-1879) described electromagnetic phenomena independently of an aether, he believed:

> "...there can be no doubt that the interplanetary and interstellar spaces are not empty, but are occupied by a material substance or body, which is certainly the largest, and probably the most uniform body of which we have any knowledge..." (1879).

Attempts were made to determine the absolute velocity through the hypothetical "aether" that was supposed to pervade all space as the "light medium". The most respected of these experiments was that performed by Michelson and Morley in 1887. Using interference patterns their apparatus could detect an effect even 99% smaller than the expected result. The results of their experiments in terms of the theory then current were interpreted as negative. This was unexpected.

All attempts to detect the Earth's motion through the aether failed. The conclusion was that the Earth did not move through the aether. This left three possibilities; the Earth carries the either along with it, or there is no aether, or there is a contraction along the line of movement.

This contraction hypothesis was made concurrently by Lorentz and Fitzgerald, Lorentz in a short article discussing the experiment and Fitzgerald in his lectures. Lorentz accredits the Fitzgerald claim for concurrence, but it was Lorentz who published the hypothesis and with the exact factor required for determining a contraction in the direction of movement.

It was Albert Einstein, however, who presented in 1905 a physical model for giving the contraction a reference-frame for intelligibility. In a letter to his first wife (1903), Mileva Maric, he referred to this theory as "our" theory. This indicates that his first wife, Mileva Maric, a competent physicist and better mathematician than Albert, played a significant role in its development. He gives her this sort of credit only in private references, Her many collaborations with husband Albert were excluded from all public notice during her lifetime.

We should mention that the Lorentz contraction hypothesis was not readily accepted by many of the accredited physicists. It was objected to because its introduction seemed like an invention artificially devised to account for the phenomenon. The contraction seemed a necessary adjustment for the unwanted result but the physical explanation of Lorentz that the molecules of their own accord effect a shortening in the direction of motion, was judged as too *ad hoc*.

It was not until the Einsteins devised the theory of special relativity that the contraction became the logical consequence of a theory rather than an interpretive adjustment made to match the observation. Even so, in some respects the Einstein presentation was also an *ad hoc* response, but more elaborately developed.

Even though the Einstein theory was not readily accepted on its first appearance, many physicists later accepted it as an explanation for this contraction. The Lorentz contraction factor passed from empirical law into an accredited theory.

These arguments persisted independently of the critics of the 1887 experiment. That the aether moves along with the Earth was not a mainline hypothesis. That the Earth might even be a generator of the aether never entered the conversation.

THE SPECIAL THEORY OF RELATIVITY: The Principle of Relativity. For over 200 years the equations of motion enunciated by Newton were believed to describe nature correctly. The Michelson-Morley experiment seemed to expose an error in these equations.

Newton's Laws of motion were stated assuming that mass, m, is a constant for any given body. Einstein Relativity states that the mass of a body increases with velocity. The Einstein correction formula:

$$m = \frac{m_o}{v\sqrt{(1-v^2/c^2)}}$$

where the 'rest mass', m_o, represents the mass of a body that is not moving and c is the speed of light, which is about 29,979,245,800 cm/s *in vacuo*. For now, the problem of how to determine whether an object is moving or not is ignored.

Thus Einstein introduced a correction factor for Newton's Laws of motion. As Richard F. Feynman (1918-1988) says, if all you want to do is solve problems, then this is all you need to know about Einstein's Special Theory of Relativity.

INERTIAL REFERENCE-FRAME RELATIVITY.
Observations of physical events are based on space-time-mass parametrically represented in terms of a reference-frame. These roughly correspond to the fundamental quantities of classical mechanics (physics), namely; space, time, mass. The difference is that in relativity any separation of these fundamental quantities is artificial, in relativity as developed by Mileva and Albert Einstein these three modes of measure are inextricably intertwined. The usual presentation of the Einstein theory emphasizes the space-time continuum as it was mathematized by Minkowski, but this presentation by Minkowski has to be revised in order to include the implied intertwining with mass. It is more coherent to say that relativity as developed by the Einsteins implies a space-time-mass continuum since mass influences the shape of space-time and space-time influences mass.

Thought experiment (EXHIBIT Addendum 5.1): We conceive three reference-frames, K_0, K_1, K_2. We look from K_0 and see that K_1 and K_2 are going in opposite directions.

K_1 is accelerating at rate g.
K_2 is decelerating at rate g.

The occupants have no way of determining, at that moment, if they are accelerating or decelerating. The problem is undecidable for them. They each weigh 90 kg.

We ask the occupants of K_1 and K_2 to report to us at K_0 how much they weigh, each replies 90 kg. As they pass our observation point our transparent view shows them standing on the same ends of the elevator and their stances look alike. They cannot see out, we can see in.

Imagine being in an elevator in deep space and not accelerating. Under current modes of analysis, whether it has velocity or not is not determinable from within. For the argument we shall presume that in this elevator-frame the laws of physics are valid (recognizable). You judge that you are weightless. Next imagine that suddenly you move toward one end of the elevator with acceleration g. You are then asked to decide whether the elevator is in acceleration or in deceleration. It should be clear that we have no means for deciding from inside the elevator, using the Newton model or the Einstein model. We cannot even use the trick of dropping two balls and see if they approach each other or not (as convergence would show). Within reasonable limits you may determine g but you cannot decide what action generates it.

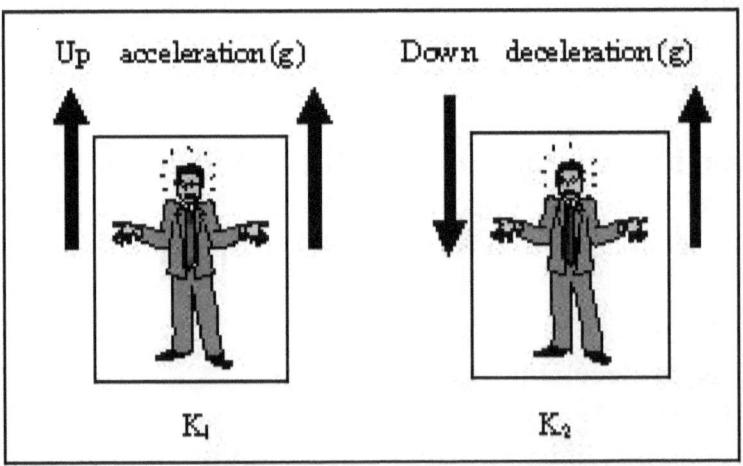

EXHIBIT 5.1: Elevators passing from opposite directions
(passenger image courtesy of Microsoft)

These two cases are dual in their coherence and therefore one is not simpler than the other.

Note: The Einsteins did not consider the dual problem of deceleration. They assumed positive acceleration rather than negative acceleration.

Now you are standing on the Earth with weight W. Is this due to an acceleration or a deceleration (positive or negative acceleration)?

With respect to Newton no special consequence arises from the decision even if judged from another reference-frame.

With respect to the Einstein theory there is a very important consequence which arises once the decision is made. This consequence arises from the necessary conclusion that is produced by their use of the Lorentz transformation formulas. In particular, the reference-frame transformations which obtain the following sequence of reasoning.

The Einsteins presume that weight produced by gravity is equivalent to weight produced by acceleration, and pointedly Albert notes:

> "The rates of the masses of two bodies is defined in mechanics in two ways which differ from each other fundamentally; in the first place, as the reciprocal ratio of the acceleration which the same motive force imparts to them (inert mass), and in the second place, as the ratio of the forces which act upon them in the same gravitational field (gravitational mass). The equality of these two masses, so differently defined, is a fact which is confirmed by experiments of very high accuracy (experiments of Eötvos), and classical mechanics offers no explanation for this equality (*The Meaning of Relativity*, p56, Princeton University Press, 1956)."

They then proceed to show how this relativity accounts for this equivalence.

One of the consequences of not recognizing the qualitative difference between acceleration and deceleration is the following:

Presuming acceleration (positive acceleration)

> "The rate of a clock is accordingly slower the greater is the mass of the ponderable matter in its neighborhood. We therefore conclude that spectral lines which are produced on the Sun's surface will be displaced towards the red, compared to the corresponding lines produced on the Earth, by about 2^{-6} of their wave-lengths." (*op.cit.*)

The observation of this red shift has been claimed from two different sets of experiments. One series is the Pound-Rebka (10% within prediction, 1960), the other is the Pound-Snider (1% within prediction, 1965) This redshift is not to be confused with the shift that astronomers use for estimating stellar distances or use for defending the hypothesis that the universe is expanding. This is a third use of redshift.

Assuming with the Einsteins that gravity and acceleration have an equivalence, we would say that in the case of occupants experiencing g in a decelerating elevator, the gravitational equivalence implies that our weight on the surface of the Earth is derived from a deceleration.

The Earth's surface can resist material debris from approaching the Earth's center, thus the Earth's surface decelerates falling bodies. (Close to the surface of the Earth air increases this deceleration with resistance to movement from friction and from buoyancy.) Weight is thus defined in terms of deceleration. This seems an intuitively acceptable conception, but this is not the first way that physicists analyzed this holding of debris to the Earth surface.

Weight can also be defined in terms of acceleration. Newton defines weight in this way and the Einsteins follow this. Newtonian physics leads to the following equation:

$$W = mg$$

where g is acceleration due to gravity and m is mass. It is assumed here that units are chosen to eliminate the need of a units constant (constant of proportionality). In this equation the deceleration is hidden in the Newton concept of mass. When a force is applied to a moveable mass it will accelerate. If there is no resistance to accelerate, that is, a fixed velocity is instantaneously achieved, then in Newton's system the target mass equals zero. If there is resistance to accelerate, it is the measure of this resistance that allows us to quantify mass, etc.

Cyclic phenomena underpin the measure of time.

QUESTION: If an acceleration slows down cyclic phenomena, does deceleration speed up cyclic phenomena?

RESPONSE: Newton's system does not address itself to this problem, but the Einsteins regard Newton's system as a special case of their Relativity. From the Einsteins we have definite arguments we can present. Consider:

CASE ONE: The elevator accelerates and then stops accelerating. The elevator continues at the last velocity achieved. At this fixed velocity, time is dilated when compared to the elevator starting position.

An elevator is accelerating, it ceases to accelerate and is traveling at velocity v_1. It accelerates again in the same direction and again ceases at velocity v_2. We can say then,

$$v_2 > v_1$$

and furthermore, according to the Einsteins, time t_2, has dilated with respect to time t_1. This physical dilation of time t_2 implies that all cyclic phenomena slow down. This is what the Einsteins believed occurs in a manner directly proportional to the local g.

CASE TWO: An elevator is decelerating and ceases to decelerate at velocity v_1, then it decelerates again and again ceases at v_2. We can then say that

$$v_1 > v_2$$

According to the Einsteins, this implies that time t_2 contracted, the time at velocity v_1 is more dilated then at velocity v_2. This physical contraction of time t_2 implies that all cyclic phenomena speed up.

COMMENT: The physical phenomena are not independent of the Einstein reference-frames. From these two cases we see that they are opposite in principle and it suggests that the choice becomes a critical one. The choice determines the nature of the predictions.

QUESTION: Under acceleration, are cyclic phenomena slowing down or speeding up?
RESPONSE: The answer follows according to wheher the acceleration is positive or negative. If acceleration is positive the cyclic phenomena slow down, if the acceleration is negative (deceleration) then the cyclic phenomena speed up.
QUESTION: In the elevator, can we decide which is the case?
RESPONSE: Nothing in Newton's work or the Einsteins' work shows a way of deciding systemically. The decision must be made through another system of thought or observation. In fact, we observe that the metabolism of as-

tronauts in microgravity does not speed up as the Einstein supposition predicts. The metabolism slows down.

Furthermore, all gravitational bodies generate radiation (Book I, Chapter 3). The Earth gives off radiation on the low energy infrared side, invisible to the eye. If the agitation generating this radiation were speeded up even more, the radiation could include visible light. The Sun, which we reason to have a much stronger gravitation, gives off visible light including very high energy light in the ultraviolet range, x-rays, and gamma rays. This indicates that the agitation and cyclic phenomena that generate radiation on the Sun move on the average at a faster rate than on Earth. There is nothing subtle in this and the little red shift, that the Einsteins hoped to find, has no meaning in this context. We do not need a Theory of Relativity to tell us whether the cyclic phenomena on the Sun are slower then cyclic phenomena on the Earth. The question has no subtlety, and the Einsteins are simply wrong. This also suggests that the Pound-Rebka, Pound-Snider experiments need reinterpretation.

QUESTION: When time dilates, what does temperature do, what does heat do?

RESPONSE: The relativity of temperature seems overlooked by the Einsteins. However, if we state that temperature directly correlates to movement, then when time dilates, the temperature goes down, when time contracts the temperature goes up. This suggests that no matter what temperature is measured in a given reference-frame we can find another reference-frame relative to which it will have a different value. We can calculate this temperature change but its actual measurement is highly problematic. Note that for heat, in material expansion or contraction, there is no change in heat content, though heat per unit volume would change. However, relativistic volume change is not measurable within the same reference-frame in which it occurs. This merits more consideration, but for me it only represents a question of historical interest.

SUMMARY CONCLUSIONS

SUMMARY ON GALILEO SPEED. With dimensions of different kinds, Galileo uses one-to-one correspondences, a relation type that is symmetric and therefore immanently reversible. Galileo never represented speed as d/t. By defining uniform motion in terms of a one-to-one correspondence between distance and time, by persisting in representing all ratios as ratios of like magnitudes, Galileo created two strong obstacles for himself that resisted a d/t representation.

The *Dialogues Concerning the Two New Sciences* was a great light for all who followed in the analysis of motion. This critique hopefully shows another step in coming to terms with Galileo's own mode of thinking. It also suggests that compound magnitudes, such as speed, may allow for more interpretations then presently accredited.

SUMMARY ON EINSTEIN RELATIVITIES. It is to Lorentz's credit that he developed his transform equations that accurately predict a wide scope of results that were measured. The extent of this appreciation for Lorentz is limited by the apparent compensatory nature of the rationale for these transforms. They were designed to compensate for the negative result of the Michelson-Morley experiment, they were deduced in order to sustain the belief in the aether medium for light. Many physicists regarded these transforms as apodeictic fictions (though they did not use this term for them). The Einstein Relativity Theories were developed from new conditions for observation, reflection, and judgment.

Relativity correction factors are claimed significant when particles and objects travel with speeds that are conspicuous fractions of the speed of light. This simple statement has certain complications. Some of these complications come with assigning interpretations to speed and travel (cf Galileo) and how they relate between reference-frames. These notions require reference-frames for judgment. Measurement demands scaled reference-frames. If we are to measure, we must have some scale with respect to which a measurement is made. The fundamental properties in classical mechanics that we measure are length, duration, weight, and inertia. From these properties we derive measurements for space, time, and mass. All these measurements therein were considered absolute and Eudoxian within the practical limits of error margins. All of these measurements have significant reinterpretations in Einstein physics.

Time dilation is not what we observe. What we observe is derived from cyclic phenomena. When we can determine that cyclic phenomena speed up or slow down, then it is possible to hypothesize that time has contracted or dilated, but this is not the only possible conclusion. In fact, I say this conclusion is undecidable by our currently known reference-frames. Einstein theory ignores the deceleration elevator.

Even though Einstein General Relativity implies that space, time, and mass are non-Eudoxian, Relativity measurements are unintelligible without a Eudoxian model for comparison. However, this Eudoxian model becomes an apodeictic construct in the Einstein universe. For the Einstein universe Euclidean space has the status of a fiction.

Note: One of the most mythic figures in 20th Century science, Albert Einstein, completely obliterated the role of his first wife (Mileva) in the formulation of ideas presented only as his own. The catechism physicists who worship Albert Einstein will be reluctant to accredit the genius of Mileva as this would require an adaptation of their current mythic beliefs.

KEY WORDS
Space, Time, duration, event, theory of measurement, standards of measurement, standards of reference, situs, locus, quantus, reference-frames, dual reference-frames, cyclic phenomena, special relativity, inertial reference-frame, isotropic, geocentric, heliocentric, abfinite, temperature

END Addendum 5

Addendum 6
PTOLEMY, COPERNICUS, KEPLER

Introduction. *Ptolemy and Copernicus are mutually exclusive flip-flop reference-frames for the naked-eye Universe. We shall consider how the transition from Ptolemy to Copernicus prepared the way for Kepler to devise one of the first attempts at a celestial mechanics. Ptolemy and Copernicus systems, lacking mechanical principles, stand in sharp contrast to Kepler and Newton.*

PTOLEMY (c100-c178) of Alexandria. Ptolemy's reputation today rests chiefly upon the *Almagest*, once known as *The Mathematical Composition*. But after it had come to be used as the authoritative text in astronomy it was called *The Great Astronomer* to distinguish it from a collection known as *The Little Astronomer*. The Arabs simply called it "The Greatest," prefixing the article *al* to the Greek, μεγιστε, and it is now known as the *Almagest*, including a redundant article. This work of Ptolemy represents a combination and systematization that derives from the astronomical traditions of his Babylonian, Assyrian, and Greek predecessors. The dominant influence is Greek, and the work is imbued with the genius in method and mathematics that is Greek. We accept many of the methods and purposes of Ptolemy. Here are some that we still sustain:

1. We pursue astronomy as a science.
2. We use mathematical models.
3. We explain the unknown through mediate terms starting from principles we accept as known or assumed.
4. We use hypotheses for testing (with some restrictions, cf Newton).

It is true that Ptolemy outlines a picture of the universe which we no longer accept. We do not grant that the Earth is the center of the universe with the Sun circling around it as well as the rest of the universe circling the Earth. We no longer believe in the necessity of perfect and regular movements of the heavens and we do not believe that the heavenly bodies are almost divine. But we cannot deny the great heritage we have from Ptolemy who studied the universe in its vast scale formulating one of the earliest great syntheses to determine appearances.

Recently, Ptolemy has been accused of misrepresentation and manipulation of observations to fit his theory. Though careful studies show discrepancies, there is no direct evidence of intention on the side of Ptolemy who depended on a staff of observers and recorders for his data. Sometimes this included extrapolations and interpolations on observations from previous astronomers. For myself, there is not adequate evidence to merit blatant and dramatic accusations that seem to indicate more interest in the excitement of witch hunting then in historical accuracy. Ptolemy was an astrologer and it was important to him to have an accurate astronomy. This also gives a predisposition to using the Earth as the center.

Placing a time for Ptolemy's life cannot be exact. We have no record of his birth and death dates but we know that the astronomical observations he uses in the Almagest were made from around AD 100 to around AD 178. This includes the reign of the emperors Trajan, Hadrian, Antoninus Pius, and the stoic emperor Marcus Aurelius. We remember that Gibbon's *Decline and Fall of the Roman Empire* begins with Marcus Aurelius. We do not know the relation between the recordings and Ptolemy's life. It is likely that he died in the early years of the decline of the Roman Empire.

Ptolemy worked in Egypt near or in Alexandria. His work in astronomy was a culmination of all that came before him and established him as the greatest authority on the subject until the time of Copernicus (1473-1543) and the period of the 'Copernican Revolution'. We should note, though, that much of this 1,300 year reign was a period in which many kinds of questioning were forbidden by religious forces. This forbidding influence had a general decaying effect on scientific research, though it did not eliminate it.

Ptolemy leaned heavily on secondary sources, including predecessors and contemporary observers. Of his predecessors he likely owes his greatest debt to Hipparchus (c130 BC). Many of the observations that Hipparchus made of the Sun, Moon, planets, and 'fixed' stars were used by Ptolemy. Hipparchus earlier developed the idea that the Earth did not move and it was the center of the universe, i.e., the universe is considered geostatic and geocentric, a point of view useful to Ptolemy as astrologer.

Ptolemy indicates his cosmology in the first eight chapters of Book I. In Chapter One he restates the Greek separation of theoretical and practical knowledge and recalls Aristotle's three genera of theoretical knowledge:

1. The physical (natural—includes biology, etc..)
2. The mathematical
3. The theological

Ptolemy states that he intends to develop the mathematical. The *Almagest* is the first complete treatise on spherical trigonometry. We may note that ancient Greeks regarded astronomy as geometry in motion. The end of Chapter Two states that the Earth is sensibly a sphere and in position lies right in the middle of the heavens. Chapter Three is entitled 'That the heavens move spherically'. Here 'heavens' refers to the so-called 'fixed stars'. They are called 'fixed' because they are assumed to be fixed to an immense sphere which constitutes the 'heavens' Ptolemy refers to. This sphere with the stars fixed on it rotates once a day about a north-south axis. The northern pole is located approximately under the north star, Polaris.

Chapter Four recapitulates and gives arguments for regarding the Earth as sensibly spherical. The arguments advanced here are still valid arguments for establishing the convexity of the Earth. One argument he did not use which we might add is the fact that during lunar eclipses the Earth shadow is always circular which, of course, is only true if the Earth is a sphere. The Pythagoreans (6th century BC), who antedate Ptolemy, used this argument. His chapter is short and may indicate that in Ptolemy's time little argument was required for the Earth's convexity. Eratosthenes (former Alexandria librarian 3rd century BC) gave a very elegant proof of the Earth's curvature, including a measure of the Earth circumference, but Ptolemy does not mention him by name. I emphasize as a curiosity that Ptolemy does not ever mention Eratosthenes anywhere in the *Almagest*.

In Chapter Five Ptolemy gives reasons for considering the Earth as the center of the heavens. His arguments basically say that being off centered would produce an observable eccentricity. His next conclusion is that the sphere of fixed stars is so vast that relative to it the Earth is like a point. In fact, for Ptolemy, no matter from which part of the Earth one regards the stars *there is no parallax*, that is, we can find no naked-eye differences in relative positions.

PARALLAX. With two functional eyes we each experience parallax. Each eye views objects from a slightly different angle. Detectable displacement between these two views is parallax. The displacement of these two images is used for creating three dimensional images. This gives us the sense of depth. Humans can learn to estimate distances using this parallax. This ability is used every time you use your eyes to pick up a glass of water, etc.. However, the resolution of parallax decreases with distance, there is a foreshortening of depth perception as the angle of view narrows with distance. Even a kilometer away flattens our view. When our eye-view appears flat, our eye parallax has passed its limit and is no longer effective. The lines of sight sensed by the eyes are close to parallel. Viewers of the Grand Canyon in America have a vivid experience of this flattening effect.

Astronomers determine the parallax of nearby stars by taking observations from opposite ends of the Earth's orbit. They determine a change of position relative to a background of other objects that are too far away to reveal their own movement. From these observations and a little mathematics they determine the distance to these nearby stars. This is called *heliocentric parallax*. If the equatorial diameter of the Earth is used this is called *horizontal parallax*. The latter may be used for distances within the solar system.

> **Note:** Obtaining a parallax to establish the distance to the Sun had special difficulties because the atmosphere of the Sun displaces the stars behind it and there was difficulty determining an accurate interpretation for this displacement.

To the unaided eye, the horizon seems to cut the sphere of fixed stars precisely in half. This observational evidence backs up Ptolemy's no parallax argument and why Ptolemy could deduce the Earth is like a point in relation to the sphere of fixed stars. He next argues that the Earth is entirely motionless (geostatic). He recognizes that some people say that the Earth rotates and affirms that in some ways it is a simpler conjecture (*Almagest* Chapter Seven).

His argument against a motion of the Earth is a physical one, not a mathematical one. He argues that the Earth might outstrip the air in its movement or that things (including projectiles) in the air might likewise be passed. This argument implies a knowledge of the Earth's size. Under the assumption the Earth rotates, the surface speed for various latitudes could be calculated and shown in many places to be quite large relative to our size, and largest at the equator (over 1600 km/hr).

The Earth's size was first calculated and with amazing accuracy by Eratosthenes. Ptolemy must have known this but again he is curiously silent about crediting the work of Eratosthenes. Eratosthenes is a predecessor of Ptolemy at the great library in Alexandria. Perhaps the name was so well known to the audience of the *Almagest* that Ptolemy regarded it as superfluous to mention it. Yet, he mentions Aristotle.

Today we say that the motion of the Earth does influence atmospheric movements but that in general the atmosphere moves with the Earth. Ptolemy had no belief in any phenomenon that resembled what we call inertia and he gives examples that even disallow the possibility of such a notion in his Universe. Today, that everything on Earth turns with it is ascribed to inertia, a concept associated with Galileo, Descartes, and Newton.

The Greek origin of the word planet is 'wanderer' and the aforementioned bodies are all wanderers relative to the sphere of fixed stars. Thus Ptolemy refers to all of them as planets, including the Sun and the Moon. Only the Earth and the fixed stars do not wander, therefore the Earth is not a planet. The planets are all satellites to the Earth.

In Chapter Eight Ptolemy describes the two different prime movements in the heavens. The first is the daily general motion from East to West. The second is the slower additional movement of the planets, from West to East. This is 29 days for the Moon, one year for the Sun and varies also among the other planets. It is this second and contrary movement which most of the Almagest explains. Note that Ptolemy makes clear that his explanations are hypothetical.

For Ptolemy, astronomy is a science concerned with almost divine beings. The 'heavenly beings' (heavenly bodies) are sensible and they move, they are eternal and impassable (Chapter One). Therefore, except for change in place as they move they are unchangeable. This conception of heavenly bodies as eternal and unchanging except in locomotion is from Aristotle. We read in Aristotle's book entitled, *On the Heavens*, that all Earthly bodies are made up of the four elements; earth, water, air, fire. These elements have simple motions that are regarded as natural to them. Earth and water move downward; air and fire move upward. However, heavenly bodies are not made up of these four elements. The natural movement of heavenly bodies is circular and it therefore follows that they cannot be made from the four terrestrial elements. The conclusion is then that they are made of a different and fifth element. It is from this element we get our word 'quintessence', the fifth essence.

That heavenly bodies suffer no change we read in Aristotle, *On the Heavens*, Great Books, V8: 361 b-c)

> "It is equally reasonable to assume that this body will be ungenerated and indestructible and exempt from increase and alteration, since everything that comes into being comes into being from its contrary... If then this body can have no contrary...nature seems justly to have exempted from contraries the body which was to be ungenerated and indestructible."

Astronomical observations do not directly show that all the heavenly bodies move in perfectly circular paths. Instead; among a number of them we see them changing speed and sometimes changing directions. To rectify this, Ptolemy distinguishes between the **appearance** and the **reality** of these apparent anomalies to perfection. The movements appear to be irregular but really they are regular. His interpretive work is thus guided toward finding the regular movements which the planets 'really' make that would produce apparently irregular movements. It would be convenient if we could decide the relationship between Ptolemy's careful use of the notion of hypothesis and whether he believes his mathematical model accounting for apparent anomalies represents the principles for actual paths of these heavenly bodies. Why this seems to lean toward hypothesis rather than actual paths might be judged or doubted in relation to Book III, Chapter Four, when he begins to deal with the Sun's anomaly and decides how it is to be explained.

Chapter Four, Book III, is entitled: *On the apparent Irregularity or Anomaly of the Sun*. The apparent irregularity consists in the fact that the motion of the Sun is not always uniform, the Sun moves faster in one period of time than in another period of time. We know that in regions with four seasons they are not precisely equal in duration. Ptolemy remarks that the Sun's motion may be explained by either of two hypotheses, that judging from appearances the two hypotheses are equivalent.

The first hypothesis assumes that the Sun involves an epicycle. The epicycle is the device we find throughout Ptolemy's astronomy for explaining apparent anomalies. It is used to explain the apparent motion of all the 'anomalous' heavenly bodies, namely, the Sun, the Moon, the other planets.

Let me briefly outline how an epicycle works in the case of the Sun. To picture this movement to yourself imagine that you are on a Ferris wheel. The Ferris wheel is the deferent circle and your chair moves in the epicycle. The axis of the Ferris wheel is the placement of the Earth and you shall be the Sun turning on your epicycle once around with each turn of the wheel. The locus of your position is then another circle off center from the deferent. With a little imagination you can readily see that by varying the diameters or the rates of turning of the deferent or epicycle that various loci may be generated. This is the first hypothesis. It requires an epicycle. We draw a circle around the Earth as center. This circle is called the deferent. The Sun does not move on the deferent circle. Instead, the deferent turns carrying another circle called the epicycle. Thus the center of the epicycle moves uniformly about the Earth. In this case the epicycle happens to be smaller than the deferent. The Sun is conceived as moving uniformly on the epicycle which itself rotates uniformly on the deferent and thus the Sun moves uniformly about the epicycle center which in turn rotates about the Earth (Exhibit 6.1a).

The center angular speed of the epicycle measured with respect to the Earth, and the angular speed of the Sun measure with respect to the epicycle center, are equal.

The second hypothesis assumes that the Sun moves uniformly in a circle, but that the Earth is not located in the center of this circle. You can plot this new figure from two epicycles on opposite sides of the deferent. The easiest is where the Sun is furthest and nearest from the Earth, connect the two Sun position points, bisect this segment. Using this bisection point as center, construct a circle of radius equal to one half of the originally constructed segment. This construction produces the locus of the Sun on its epicycle (Exhibit 6.1b).

Ptolemy proves this. It is this locus generator that is the second hypothesis. Ptolemy then states that the second hypothesis is simpler because we may assume one circular motion rather than two. He formally proves the equivalence of the two hypotheses in regard to appearances (Bk III; Chap 3).

EXHIBIT 6.1: (a) Epicycles; (b) Eccentric (The Equant).

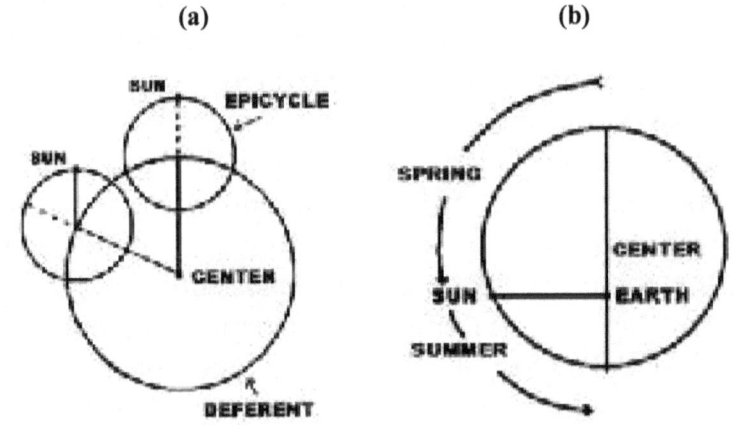

Ptolemy states as cited that the hypothesis of eccentricity is simpler and completely effected by one and not two movements. But one must not presume he accepts the eccentricity as a reality, eccentricities do not represent perfection. In Chapter Three he strongly emphasizes the mathematical equivalence and proves them interchangeable for calculation under certain conditions and therefore use of the eccentricity hypothesis in no way compromises his theme of perfect action. He can always refer to the complete interchangeability of the two hypotheses. Chapter Three proves the equivalence but in Chapter Four for calculation he prefers the simpler hypothesis.

In Chapter Four of Book III, Ptolemy determines the size of the eccentricity of the Sun's locus. Then he considers the Sun's line of apsides. This line connects the Sun's nearest and farthest position from the Earth. The nearest point is named **perigee,** and the farthest point is named **apogee** (Exhibit 6.2a).

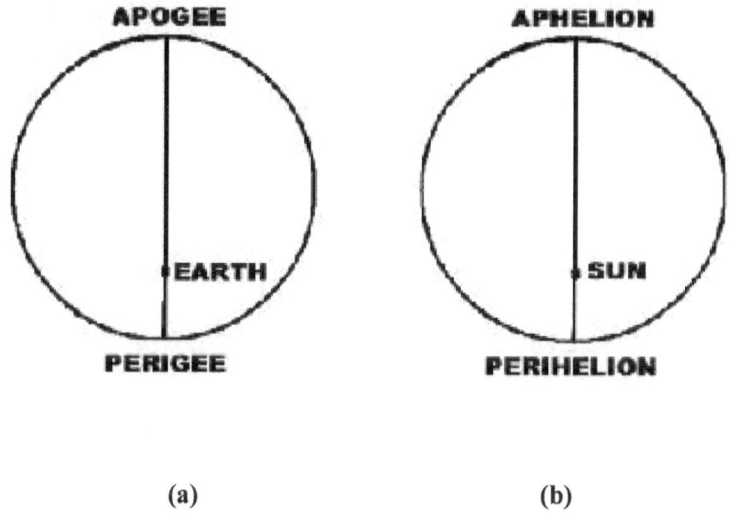

EXHIBIT 6.2: (a) Perigee—Apogee; (b) Perihelion—Aphelion

The Sun's eccentricity turns out to be one twenty-fourth (1/24) of the radius, and the line of apsides is so located that the Sun is at apogee 24° 30' before it reaches the summer tropic (about June 21).

We note that in a dual perspective where the Earth moves around the Sun, a heliocentric theory, that the Earth orbit point nearest the Sun is named **perihelion** and furthest is named **aphelion** (Exhibit 6.2b). With Kepler the line of apsides is the long axis of an ellipse.

PTOLEMY ON GRAVITATION. It is in Book I, Chapter Seven, that Ptolemy speaks of weights falling to the Earth. We read:

> "And so it also seems to me superfluous to look for the causes of that motion to the center when it is once

for all clear from the very appearances that the Earth is in the middle of the World and all weights move towards it. And the easiest and only way to understand this is to see that once the Earth has been proved spherical considered as a whole and in the middle of the Universe as we said, then the tendencies and movements of heavy bodies (I mean their proper movements) are everywhere and always at right angles to the tangent plane drawn through the falling body's point of contact with the Earth's surface. For because of this it is clear that, if they were not stopped by the Earth's surface, they too would go all the way to the center itself, since the straight line drawn to the center of the sphere's surface at the intersection of that line.

... Therefore the solid body of the Earth is reasonably considered as being the largest relative to those moving against it and as remaining unmoved in any direction by the force of the very small weights, and as it were absorbing their fall. And if it had some one common movement, the same as that of the other weights, it would clearly leave them all behind because of its much greater magnitude. And the animals and other weights would be left hanging in the air, and the Earth would very quickly fall out of the heavens. Merely to conceive such things makes them appear absurd." (*Great Books of the Western World*, ed. 1952, vol. 16)

Aristotle divides all local motions into natural and violent local motions. Ptolemy here calls the natural local motions of bodies below the lunar sphere their proper motions. Galileo continues the use of this distinction between natural and violent motions but in Newton it loses all meaning in a general mathematical treatment.

SUMMARY PTOLEMY. Ptolemy admits that a heliocentric system could be used to describe heavenly motions but he argues both from appearances and from physical mechanics that the Earth is the center of the Universe. And as the causes of motion known in his time seem to fit the geocentric theory better, they are useful in defending this reference-frame. In Ptolemy's epoch it is not obvious that objects should 'fall' to all planets, but it is easily shown that objects 'fall' to the Earth. The geocentric theory is the best point of view for Ptolemy as astrologer, though he does not mention astrology in the Almagest. Note also that Ptolemy regards the Earth as the largest solid body in the Universe.

Ptolemy used three devices to improve the geocentric theory; the epicycle, the eccentric, and the equant (Exhibit 6.1). Epicycles and eccentrics are versatile in the possible motions they can describe and it is thus possible to construct a very accurate model of the Universe for predicting planetary positions.

With the epicycles on deferents the change in brightness of the outer planets is explained as well as the retrograde motion in which it occurs. There is no evidence of intention in these results but since the outer planets retrograde when on the inner part of the epicycle it is brightest, which corresponds well with appearances. It is only the inner planets, Venus and the Moon that one anticipates a more exaggerated change in brightness than actually appears. The appearance of the Moon in Ptolemy's movement would have to expand and shrink, sometimes appearing twice the diameter as at other times. This is certainly contrary to observation.

I note here that one feature of Ptolemy's system was that the epicycles of the 'outer' planets all had exactly the same period, precisely one year. Also the positions of the 'outer' planets on their epicycles always match the position of the

Sun relative to the Earth. This correspondence and its difference from the 'inner' planets was a key argument of Copernicus against the coherence or elegance of the Ptolemaic system. The Copernican revolution around the Sun unified all the planetary movements under one principle.

Proposed in the 2nd century AD the Ptolemy model persisted for one-thousand three-hundred years which obligates us to propose some possible reasons for this long acceptance.

1. It was coherent with Aristotle and later with Aquinas who integrated Aristotelian thought with Christian beliefs, thus disbelief could be interpreted as Church heresy.
2. It placed us at the center of the Universe, thus giving us special importance, a perspective consistent with Genesis. The Earth is also the largest solid body in the Universe which is another form of importance.
3. It had common-sense appeal to all; to those who looked at the heavenly movements and to anyone seeing water and earth fall, and feeling air and seeing fire not falling.
4. It explained why no parallax of the fixed stars was visible (naked eye observation) and predicts that it can never be observed.
5. It predicted within 2 degrees the position of the wanderers over long periods of time (within 4 apparent Moon diameters).

Note: On a night of the full Moon hold an ordinary pea at arm's length and you will observe it covers the Moon completely (c 0.65 cm).

Only the last three have any appeal to direct observation.

Why dwell on Ptolemy's system which seems so far from the Newtonian synthesis of celestial and terrestrial mechanics? The *Almagest* is mathematical and gives no physical explanations for causes. When Ptolemy argues against the Earth turning, it is physical in nature but not causal. He does not give a cause for the Earth not turning, he only argues for how we may conclude that it does not turn. Ptolemy's view was that of a number of his predecessors, the natural motion of heavenly bodies is circular. But when we say 'natural' we are presuming that once the motion here is shown to be circular that no further explanation is needed. If we say it is natural for a body to move in a certain way, we are saying that this certain way requires no explanation. Thus for Ptolemy working with the axiom that all heavenly bodies move naturally in circles, there is nothing left to explain except why certain appearances do not show us this regularity. This is what the major part of the *Almagest* is all about; reducing all apparent anomalies to regular circular motion, the natural motion. Copernicus sustains this principle. As an anti-Newtonian model, Convergence Buoyancy reinstates the importance of circular motion as the basis for all other celestial movement forms, though with important differences.

In answer to the question, I say that it is useful for comprehending Newton's synthesis to see how celestial mechanics evolved from systems that had no physical principles for causes. We shall see how the transition from Ptolemy to Copernicus prepared the way for Kepler to devise one of the first attempts at a celestial mechanics. Ptolemy and Copernicus systems, lacking mechanical principles, stand in sharp contrast with Kepler and Newton.

Next we consider how the Ptolemaic system was displaced.

Nicolaus COPERNICUS (1473-1543)
The Copernican Revolution. Copernicus was a Catholic priest, a humanist learned in Greek, mathematics, and as-

tronomy. He was also a jurist and a physician. It was not until 1530 that he provided in the *Commentariolus* a preliminary outline of his heliocentric theory. It immediately attracted great attention. At Rome, Johann Albrecht Widmanstadt lectured upon the new doctrine; Pope Clement VII gave his approval; Cardinal Schönberg entreated the author to make public his full thought upon the subject. In the spring of 1539 Copernicus was visited by young Joachim Rheticus, a protégé of Melanchthon, at the age of 25 he was professor of mathematics at the University of Wittenberg. Rheticus studied the details of the system and in 1540 with the approval of Copernicus published a general account of it entitled *Narratio Prima*. Finally Copernicus was persuaded to allow Rheticus to publish the *De revolutionibus orbium coelestium*. On 24 May 1543 an advance copy was presented to Copernicus on his death bed and he died the same day.

The word 'revolutionibus' which referred only to the motion of planets became a paradigm for any drama of change in beliefs including civil upheaval. Today when we speak of the Copernican Revolution it is often the upheaval of beliefs that is called to mind. The entire World picture was changed by simply exchanging the roles of the Earth and the Sun. This exchange is a dual relationship between the two World systems. Where the Earth had been immobile in the central position, it now was pictured to have both a daily rotation and a yearly revolution, etc., it was just another planet among several. The Sun is fixed and no longer regarded as a planet.

This Copernican shift in view-point is a dethroning, the dethroning of the human privileged position in the universe. This dethroning only follows if you believe that the Copernicus 'revolutionibus' is correct. By the time of Kepler and Galileo the only accredited claim for the heliocentric system was its mathematical equivalence. As for choosing between the two systems, Copernicus had on his side the argument about the brightness of Venus (this argument is off somewhat with the discovery of the phases of Venus, even so Venus clearly varies in apparent brightness, closer to Copernican prediction), he eliminated several epicycles and based the motions of all the planets (including the Earth and excluding the Sun) on one set of principles. Copernicus says in his Preface and Dedication for *De revolutionibus orbium coelestium*:

> "Then in setting up the solar and lunar movements and those of the other five wandering stars, they do not employ the same principles, assumptions, or demonstrations for the revolutions and apparent movements. ... Moreover, they have not been able to discover or to infer the chief point of all, i.e., the form of the world and the certain commensurability of its parts. But they are in exactly the same fix as someone taking from different places hands, feet, head, and the other limbs--shaped very beautifully but not with reference to one body and without correspondence to one another--so that such parts made up a monster rather than a man." (*Great Books of the Western World*, ed. 1952, vol. 16)

He notes in this Preface that other philosophers had proposed that the Earth moved. Cicero mentions Nicetas. Plutarch mentions Philolaus the Pythagorean. Copernicus further states:

> "... I finally discovered by the help of long and numerous observations that if the movements of the other wandering stars are correlated with the circular movement of the Earth, and if the movements are computed in accordance with the revolution of each planet, not only do all their phenomena follow from that but also this

correlation binds together so closely the order and magnitudes of all the planets and of their spheres or orbital circles and the heavens themselves that nothing can be shifted around in any part of them without disrupting the remaining parts and the universe as a whole." (*op.cit.*)

This was the main argument, in his view, that Copernicus offers in his own favor. Note also that we are no longer referring to mere appearances. We are now being introduced unambiguously to the notion of orbits. There is a concern to do more than just account for appearances.

So we see that the Copernican system makes the Earth a planet with a yearly revolution around the Sun. Only the Moon remains as a satellite to the Earth. A second step is required, the daily motion associated with night and day is assigned to the Earth. The heavens no longer rotate so the heavenly sphere itself becomes fixed. With the addition of declination changes the Earth now has three motions:

1) the daily geocentric rotation,
2) the yearly heliocentric revolution,
3) the annual declination changes.

We still appear to be in the center of things and we still say the Sun rises and the Sun sets, but now we are to interpret this as illusory. That we are in the center of the universe is an illusion. In Ptolemy, the Earth is like a point in relation to the size of the universe. In Copernicus, the entire orbit of the Earth becomes like a point and the universe becomes thousands of times larger than it had to be for Ptolemy (divide the value accepted for Earth's distance from the Sun by the Earth's radius). We become indeed much smaller, so much so that the change of scale becomes exceedingly difficult to comprehend. So, not only is mankind removed from the center of the universe, we are made to appear even smaller.

Copernicus and Ptolemy both emphasized the role of hypothesis in astronomy. These hypotheses are formulated to 'save the appearances'. Copernicus claims to have done this with more elegance. Both astronomers maintained that uniform circular motion is the natural motion of heavenly bodies. Copernicus shares Ptolemy's belief in the godlike and perfect nature of the heavenly bodies. His arguments, like Ptolemy's, show influence of the Aristotelian tradition.

Early Objections to Copernican System:

    1. Scripture is violated by what we learn from Copernicus. In the Battle of Jericho, God ordered the Sun to stand still. He would not have done this for a Sun that does not move, etc..

    2. "Common sense", the Sun is clearly seen to rise and move across the sky, at night we can observe that the stars are in fixed positions as if on a sphere that rotates about the Earth, etc.. Common sense helps the difference argument: the Earth is not a planet, it does not wander, it is different from the moving objects we see in the heavens, it is unique.

    3. Physics--The speeds required for a moving Earth are enormous and deemed unbelievable for an inhabited Earth. Ptolemy had presented thought experiments against the Earth moving at such speeds. His arguments were accepted and used against Copernicus. These movement problems were much later regarded resolved as phenomena of inertial relativity.

Tycho BRAHE (1546-1601)

Brahe's contribution to World systems was a compromise between Ptolemy and Copernicus. He maintains the importance of Earth as center of the Universe and has the Sun and

Moon going around the Earth and all the other planets as satellites to the Sun.

Tycho Brahe's fastidious pursuit for accuracy led to the development of giant observation instruments under his design and management. This extended observations to their possible human limits in pre-telescope astronomy. We should not under-estimate the greatness of this achievement. His accurate tables of astronomical data inspired Kepler to great discoveries. It was Brahe's accurate measurements of planetary movements that helped Kepler to discover three kinematics laws of planetary motion.

Johannes KEPLER (1571-1630)

Kepler's *Epitome* was published in 1618 and marks another turning point in astronomy. He already had several paradigms that show equivalent mathematical models are possible for astronomical description. He accepted the heliocentric thesis from Copernicus and in careful examination of Brahe's observation measurements of Mars he found that an ellipse better fit the observations.

His tenacity in applying different curves to the recorded positions of Mars shows that indeed astronomy is a type of experimental science. Instead of manipulating matters-of-fact in the data-field to test the theory, he manipulated the theories to find a fit for the data-field matters-of-fact, a dual form of experiment.

Kepler showed himself more Platonic than Aristotelian. His predisposition in this regard sometimes carried him into subjects that many today consider fancy, such as the Pythagorean idea of the harmony of the spheres, that planetary movement gives rise to music harmonies. As it turns out, many objects of the solar system have harmonic relations. They are said to be in resonance. Like Ptolemy, he also was a respected astrologer. In the tradition of Ptolemy he wanted a system that encompassed the entire universe, the World.

Kepler is most known today for his three laws of planetary motion which are still currently used, though the third law requires certain corrections beyond the scope of this discussion.

> **FIRST LAW**: The planets move in ellipses around the Sun; one of the foci marks the Sun's position and the other is empty.
> **SECOND LAW**: The time of the planet traversing a segment of elliptical orbit is proportional to the area swept out from the Sun to the planet.
> **THIRD LAW**: In terms of distance and time, the squares of the periodic times of two planets are to each other as the cubes of their mean distances from the Sun.

If the periodic times of two planets are T_1 and T_2 and if their mean distances from the Sun are R_1 and R_2 then:

$$\frac{T_1^2}{T_2^2} = \frac{R_1^3}{R_2^3}$$

which displays the formula for the third law.

Let us illustrate this law relative to Mars and the Earth. The periodic time for the Earth is one year and Mars has the periodic time of 1.881 years. Thus:

$$\frac{(1.881)^2}{1} = 3.538 = \frac{R_1^3}{R_2^3}$$

If we say the Earth's distance from the Sun is unity then;

$$\text{for Mars} = (3.538)^{(1\backslash 3)} = 1.524 = R_1$$

which tells us that the planet Mars is 1.524 times as far away from the Sun as the Earth.

Kepler systematically rejects the foundation principles of traditional astronomy. He begins by rejecting the hypothesis that movements must be expressible as uniform and circular. He rejects these traditional arguments as false or inconclusive. We shall here review these arguments and his rejections since this represents a major departure from traditional beliefs.

Kepler asks,
> "But by what arguments did the ancients establish their opinion which is the opposite of yours?
> By four arguments in especial
> 1) From the nature of movable bodies.
> 2) From the nature of the motor virtue.
> 3) From the nature of the place in which the movement occurs.
> 4) From the perfection of the circle." (*Great Books of the Western World*, ed. 1952, vol. 28)

Kepler states that heavenly bodies are subject to laws of motion and first suggests that there are attractions and repulsions analogous to lodestones whose force varies with distance. He does not state the mathematical function by which we can calculate the power of these attractions and repulsions. The notion of forces between heavenly bodies is something new to astronomy and we have here in Kepler an anticipation of Newton's application of universal gravitation. That is, by analogy, Kepler applied a terrestrial phenomenon, magnetism, to a celestial phenomenon. Applying terrestrial phenomena to celestial phenomena, even by analogy, was completely foreign to previous astronomers and an inspiration for Newton.

FIRST ARGUMENT: The ancients said that heavenly bodies were made of a special element. Remember the fifth element which is the 'quintessence' of which the heavenly bodies are made. They suffer no change or irregularity.

KEPLER answers: That he too regards the heavenly bodies as having a special nature but he maintains they are bodies. This already is sufficient for Kepler to account for certain irregularities.

SECOND ARGUMENT: The causes of planetary movement are minds (Gr. νους) which are attached to the planets.

> "And accordingly even the figures of the movements, on account of the very nature of the minds, are most perfect circles." (*op.cit.*)

KEPLER answers: That he denies that minds or gods move the heavenly bodies. He considers it to be a truer philosophy to place the natural cause of motion in the power of the bodies. The possible combination of a celestial body's own natural power with the attraction or repulsion of another results in less perfect movements. *He suggests that the interval between the Sun and planet varies with attraction and repulsion in a sequence.*

THIRD ARGUMENT: The influence of place, this argument is based on the old physics where elements go according to their nature, light objects go up, and heavy objects go down. Thus every body has a motion proper to it.

KEPLER answers: That he grants that heavy bodies move down and light objects move up. He further grants that heavenly bodies traverse closed paths but he remarks that this is no argument for circular paths, merely that the paths must be closed, but also the ellipse is of this kind.

FOURTH ARGUMENT: The perfect circle

KEPLER answers: That he grants that the circle is the most perfect movement figure. But he states that planets are bodies and are not moved by minds (he notes: Not speaking of the Creator's Mind) therefore circles are not appropriate to planetary movement.

KEPLER SUMMARY: In some ways Kepler's work is more revolutionary in attitude and thought than that of Copernicus and yet it is rare to speak of a 'Keplerian Revolution'. Copernicus greatly changed the perspective of astronomy in its possible conception but brought no significant changes in methods. Kepler made the following important changes:

1) Tries to explain celestial motions in physical terms of attraction and repulsion. (The planets are bodies that can be acted upon.)
2) Anticipates Newton in applying terrestrial observations to celestial phenomena, e.g., the magnetism analogy (he knew the work of Gilbert).
3) Makes no distinction between real and apparent; orbits are real, appearances are real; he derives orbits from their appearances.
4) Rejects the old uniform circular movement and replaces it with the more accurate ellipse and his three laws of planetary motion.
5) Denies that heavenly bodies are unchangeable.

This is a formidable list of changes which when put beside the contemporary work of Galileo, places us on the threshold of Newton's great synthesis.

KEY WORDS
geocentric, geostatic, heliocentric, heliostatic, kinematics, dynamics, trajectory, parallax, celestial mechanics, revolution, perigee, apogee, perihelion, aphelion, anomaly, deferent, epicycle, equant

END Addendum 6

Addendum 7
Fermat's Last Theorem (FLT)

* I dedicate my general FLT proof to:
William Darkey and **Donald Cook**
tutors at Saint John's College, MD and NM, USA.

Abstract: A geometric proof for FLT, the original intent is to adhere only to mathematical procedures that could be known to Fermat. It is also accessible to students early in their mathematical studies.

Introduction. Euclid, Diophantus, Fermat, and Descartes did not use negative numbers though they admitted the subtraction operator. In fact, none of these mathematicians nor any of their contemporaries used signed numbers. Fermat and Descartes were contemporaries and corresponded through letters.

     a. For Descartes there are no quadrants, because three of them would require negative numbers. The general problem Descartes solved in his *Geometry* was how to translate equations into geometric figures. His coordinates (sometimes forming a forward or backward 'L') were made with number lines measured by unsigned numbers. The unsigned numbers do not allow applying the concept of quadrant to his use of coordinates.

b. Fermat also developed independently an analytic geometry for expressing formulas as geometric figures using unsigned numbers (found in *varia opera arithmetica* of Fermat, published 1679 by his son, Clement Samuel Fermat). That Fermat was translating formulas into geometric figures suggests a rationale for translating $\mathbf{d^n = a^n + b^n}$ into a geometric representation though today, graphed in quadrants, we get two quite different pictures for even and for odd **n>2**.

c. These early mathematicians used numbers as representing magnitudes or quantities in the sense that Euclid used them in books V and VII of his *Elements*. Today we would interpret such unsigned numbers as absolute values, though this misconstrues the historical facts. These early mathematicians had no conception of unsigned numbers since they did not have 'signed' numbers to compare them to. Thus they had no conception of an absolute value.

The strategy below starts with proofs restricted to unsigned numbers as the first constraint Fermat worked under. This is followed by a treatment extended to signed numbers showing that every configuration of signed numbers is convertible to a form that entails the unsigned number proof or generates an absurdity. Therefore, etc.

Fermat's Last Theorem.

FLT: For unsigned rational numbers **a, b, d,** and integer **n>2**, it is impossible to partition $\mathbf{d^n}$ into $\mathbf{a^n}$ and $\mathbf{b^n}$ such that $\mathbf{d^n = a^n + b^n}$.

We proceed by ***reductio ad absurdum.***
We start with the FLT special case for unsigned *integers*:

FLT for unsigned integers: For unsigned integers **a, b, d, n** where integer **n>2**, it is impossible to partition d^n into a^n and b^n such that $d^n = a^n + b^n$.

For this case I use a geometric construct and the identities it produces, which allows a general proof for FLT based on these identities. The numbers that measure this construct are all from the domain of unsigned numbers.

We start with the case **{a, b, d}** as coprimes, where **a, b,** and **d** are given as unsigned non-zero integers.

Given $a^2 + b^2 = c^2$, then c^2 is an integer but we do not know yet if **c** is rational. Triangle **abc** is a right triangle (Euclid I:48).

For a general proof using geometry we use the constructed triangle **abd** from $a^n + b^n = d^n$ where **n>2**

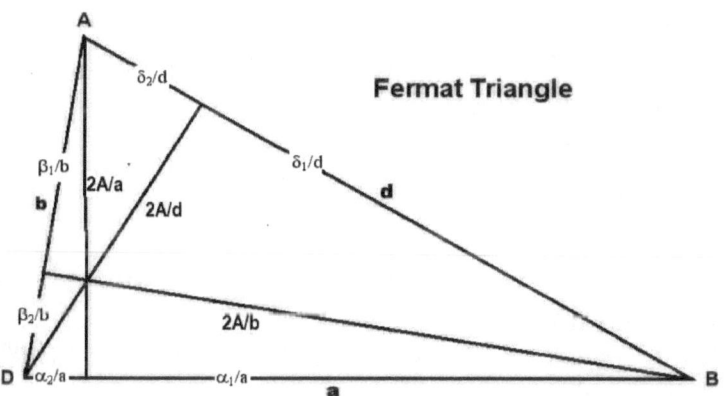

The rationale for this triangle is as follows:

Lemma 1: For **n>1**, d^n cannot be partitioned into two equal parts such that $d^n = 2a^n$,
Proof: First we show that any **nth** root of 2 is irrational for any integer **n>1**.
This proof is by *reductio ad absurdum*.
Suppose that the **nth** root of two, where **n>1** is expressible as p/q where **p** and **q** are relatively prime integers, such that $2^{(1/n)} = p/q$.
Take both sides to the **nth** power:
$2 = (p/q)^n = p^n/q^n$,
and multiply both sides by q^n.
Therefore $2q^n = p^n$.
Therefore p^n is even and therefore **p** is even,
say $p = 2r$.
Then $p^n = 2^n r^n$, then $2q^n = 2^n r^n$,
therefore $q^n = 2^{(n-1)} r^n$.
Since **n>1** we have that **q** is even, making **p** and **q** not coprime, contrary to hypothesis.
Therefore any **nth** root of 2 where **n>1** is irrational.
If $d^n = 2a^n$, we have that, $d^n/a^n = 2$, or $d/a = 2^{1/n}$, a rational number equal to an irrational number, a contradiction.
Therefore, for **n>1**, d^n cannot be partitioned into 2 two equal parts such that $d^n = 2a^n$.
Therefore. etc., Lemma 1 Q.E.F.

Therefore from Lemma 1 we conclude that $a \neq b$.
By symmetry it is arbitrary which is the greater, for argument let $a > b$.
From $a^n + b^n = d^n$ where **n>2**, we then have
$d > a > b > 0$.

Lemma 2: If **n>2** then $a^2 + b^2 = d^2 + R_2$ and R_2 is an unsigned integer.
Proof: Given $a^n + b^n = d^n$ with **n>2**, divide through by $d^{(n-2)}$:
$a^n/d^{(n-2)} + b^n/d^{(n-2)} = d^n/d^{(n-2)} = d^2$
but since $d > a > b > 0$ we have

$a^n/a^{(n-2)} > a^n/d^{(n-2)}$
$b^n/b^{(n-2)} > b^n/d^{(n-2)}$

and adding these we get
$(a^n/a^{(n-2)} + b^n/b^{(n-2)}) > (a^n/d^{(n-2)} + b^n/d^{(n-2)}) = d^n/d^{(n-2)}$

or
$a^2 + b^2 > d^2$

To render an equality and keep integers we add unsigned integer R_2 to the right side such that
$a^2 + b^2 = d^2 + R_2$

Therefore, etc. Lemma 2 Q.E.F.

If $a^n + b^n = d^n$ and $n>2$ we conclude from Lemma 2 that
$a^2 + b^2 = d^2 + R_2$ where R_2 is an unsigned integer.

Therefore $d^2 + R_2 = c^2$
Therefore $d^2 < c^2$ and so $d < c$.

c subtends a right angle (since $a^2 + b^2 = c^2$) and **a** and **b** remain constant. That **d<c** implies **d** subtends less than a right angle (Euclid I:25). However, from **d > a > b** we have **d** is the longest side of **Δabd** and therefore subtends the largest angle of **Δabd**.

Therefore each angle of **Δabd** is less than a right angle, therefore, by definition, we have an acute angled triangle.

Elementary geometry teaches us that the point of concurrency for the altitudes of an acute angled triangle is inside the triangle (see Theorem Addenda below for proof).

Thus from elementary geometry we see that this is an acute angled triangle, **abd**, where **d > a > b > 0**, and the altitudes are concurrent within this triangle.

The respective altitudes may be represented as:
2A/a to side **a** ; **2A/b** to side **b** ; **2A/d** to side **d**

This follows from the fact that the area of a triangle is equal to the base times height divided by 2 (from Euclid I:34). Each side and its corresponding altitude yields **A**, e.g., height **2A/a** times the base **a** and divided by **2** equals **A** where **A** is the area of the $\triangle$**abd**, therefore etc.

In an acute angled triangle the altitudes are interior and therefore partition each of the three sides into two non-zero parts (see Theorem Addenda Lemma Porism).

We can next represent the altitude partitions of the sides as follows:

$$\mathbf{a} = \alpha_1/\mathbf{a} + \alpha_2/\mathbf{a}$$
$$\mathbf{b} = \beta_1/\mathbf{b} + \beta_2/\mathbf{b}$$
$$\mathbf{d} = \delta_1/\mathbf{d} + \delta_2/\mathbf{d}$$

This representation of the partitions is a semi-arbitrary device that renders the proof more concise. The Greek variables represent unsigned numbers.

At this point in the argument we do not know if they are integers or fractions. However, we know that when divided by the indicated denominator, they sum to that same denominator.

And multiplying through by the denominators:

$$\mathbf{a}^2 = \alpha_1 + \alpha_2$$
$$\mathbf{b}^2 = \beta_1 + \beta_2$$
$$\mathbf{d}^2 = \delta_1 + \delta_2$$

From the construction and the Pythagorean theorem (Euclid I:47) we have:

$$\alpha_2^2/\mathbf{a}^2 + 4\mathbf{A}^2/\mathbf{a}^2 = \mathbf{b}^2$$

$$\beta_2^2/b^2 + 4A^2/b^2 = a^2$$

$$\delta_1^2/d^2 + 4A^2/d^2 = a^2$$
$$\alpha_1^2/a^2 + 4A^2/a^2 = d^2$$

$$\beta_1^2/b^2 + 4A^2/b^2 = d^2$$
$$\delta_2^2/d^2 + 4A^2/d^2 = b^2$$

and multiplying through by denominators and applying Euclidian Common Notions, etc.:

$$\alpha_2^2 + 4A^2 = a^2b^2$$
$$\beta_2^2 + 4A^2 = a^2b^2$$
$$\text{therefore } \alpha_2 = \beta_2$$

$$\delta_1^2 + 4A^2 = a^2d^2$$
$$\alpha_1^2 + 4A^2 = a^2d^2$$
$$\text{therefore } \delta_1 = \alpha_1$$

$$\beta_1^2 + 4A^2 = b^2d^2$$
$$\delta_2^2 + 4A^2 = b^2d^2$$
$$\text{therefore } \beta_1 = \delta_2$$

and

Lemma 3: Under the special case FLT constraints, in $a^2 + b^2 = d^2 + R_2$ the unsigned integer R_2 must always be even.

> **Proof:** Checking parity, **a** and **b** cannot both be even, because from $d^n = a^n + b^n$, this would require that **d** is even and therefore **a, b, d** would not be coprimes, contrary to hypothesis.
>
> If **a** and **b** are both odd then **d** is even, therefore, in $a^2 + b^2 = d^2 + R_2$, unsigned integer R_2 must be even in order to keep both the right and left sides even.
>
> If the parities of **a** and **b** differ, their sum is odd and therefore **d** is odd,

therefore, in $a^2 + b^2 = d^2 + R_2$, unsigned integer R_2 must be even in order to keep both the right and left sides odd.
Therefore, unsigned integer R_2 is always even under the special case FLT constraints for *unsigned integers*.
Therefore, etc. Lemma 3 Q.E.F.

From Lemma 2 we have $a^2 + b^2 = d^2 + R_2$

and since $\alpha_1 + \alpha_2 = a^2$; $\beta_1 + \beta_2 = b^2$; $\delta_1 + \delta_2 = d^2$
we have $(\alpha_1 + \alpha_2) + (\beta_1 + \beta_2) = (\delta_1 + \delta_2) + R_2$

From this and the identities: $\alpha_2 = \beta_2$; $\delta_1 = \alpha_1$; $\beta_1 = \delta_2$
we derive: $R_2 = 2\alpha_2 = 2\beta_2$

From lemma 3 we have that the unsigned integer R_2 is even.
Therefore α_2 and β_2 are each unsigned integers.
But $\alpha_1 + \alpha_2 = a^2$ and $\beta_1 + \beta_2 = b^2$
therefore α_1 and β_1 are also unsigned integers.
But again from our identities: $\alpha_1 = \delta_1$; $\beta_1 = \delta_2$
we have $\alpha_1, \alpha_2, \beta_1, \beta_2, \delta_1, \delta_2$ are all unsigned integers.

These preparations allow us to complete this Fermat general power partition theorem in the following way:

From our geometric construction we have a number of similar triangles interior to the Fermat triangle.
First I use that triangle [δ_2/d][2A/d][b] is similar to triangle [β_1/b][2A/b][d].
Both have a right angle and both share angle **BAD**, therefore their third angles are equal (Euclid I:32 and they are similar triangles by Euclid Definition VI:1).
Setting up proportional sides (subtending equal angles) we have that
[δ_2/d]: [2A/d] = [β_1/b]:[2A/b] (Euclid VI:4)

or $[\delta_2/d]/[2A/d] = [\beta_1/b]/[2A/b]$, which simplified gives
$[\delta_2(2A)/d^2] = [\beta_1(2A)/b^2]$

from which we have $[\delta_2(2A)/d^2] = [\beta_1(2A)/b^2]$
Dividing out **2A** from both sides: $[\delta_2/d^2] = [\beta_1/b^2]$.
Multiplying both sides by $(d^2)(b^2)$ we get $[\delta_2 b^2] = [\beta_1 d^2]$.

Since by hypothesis **b** and **d** are coprime we have that no factors of d^2 can divide b^2, and no factors of b^2 can divide d^2. I have also shown that $\{\delta_2, \beta_1\}$ are integers, and therefore from $[\delta_2 b^2] = [\beta_1 d^2]$, I say that d^2 divides $[\delta_2 b^2]$ but cannot divide b^2, therefore can only divide the integer δ_2.

But $\delta_1 + \delta_2 = d^2$ and if $\delta_1 = 0$ then its partition $\delta_1/d = 0$, which nullifies the **d** partition, contrary to hypothesis (the construction constraints), therefore $\delta_1 > 0$.

But since d^2 divides the integer δ_2, and $\delta_1 > 0$, it follows that since $\delta_1 + \delta_2 = d^2$, d^2 must also divide the integer δ_1, therefore d^2 divides each of the two unsigned integers δ_1 and δ_2.

Therefore d^2 is equal to a number greater than itself, contrary to hypothesis, a relation which cannot subsist for unsigned integers.

> **Note:** Through appropriate choices of similar triangles, put in similar ratios, this contradiction can be shown relative to any major side $\{a, b, d\}$, though any one of them is sufficient to complete a proof.

Therefore I say that partitioning an unsigned integer to an integer power, **n>2**, into two unsigned integers to that same power cannot be expressed in lowest terms, contrary to hypothesis.

Therefore, For unsigned integers **a, b, d, n** where integer **n>2**, it is impossible to partition d^n into a^n and b^n such that $d^n = a^n + b^n$. Q.E.F.

Furthermore, this implies that neither non-coprimes nor fractions can be so partitioned, since for non-coprime integers we can divide out the GCD and again apply the above argument. If expressed as fractions, multiplying through by the denominators would produce an unsigned integer to a given power partitioned into two unsigned integers to that same power. If they are coprime, or have a GCD that divided out reduce them to coprime, we can again apply the above argument. Therefore, for *unsigned rational numbers* **a, b, d,** and integer **n>2**, it is impossible to partition d^n into a^n and b^n such that $d^n = a^n + b^n$.

Finally, I say that by hypothesis this theorem is rendered functionally independent of the degenerate partitions of the form $a^n + 0^n = d^n$, which shows **a = d**, contrary to hypothesis.

Therefore, etc., **Q.E.D.**

for Fermat's Last Theorem (under Fermat's constraints)

==

Note: This proof includes the case of **n=4** as well as every other compounded case of integer **n>2**.

==

Porism Extending FLT to Signed Rational Numbers:
The acceptance of signed numbers for FLT in more recent times has changed the theorem statement to include them. This is done in ignorance of Fermat's original conception. Even so, we can easily convert all configurations of signed

Star Laws: Foundations for Convergence Buoyancy 579

rational numbers such that each configuration entails the unsigned number proof or generates an absurdity, as follows:

To render the transformations more concise I keep **a, b, d** as variables of unsigned numbers and assign the signs to the variables. Thus **d** is an unsigned variable and **+d** assigns a plus sign to the unsigned numbers and **-d** assigns a minus sign to the unsigned numbers. **+d** may also be written **(+1)d** and **-d** may be written **(-1)d**. In the universe of signed numbers the unsigned numbers may be regarded as absolute values which are treatable as positive numbers in the universe of signed numbers.

Case One: To show that $(-d)^n = (-a)^n + (-b)^n$, entails $d^n = a^n + b^n$. Assume **a, b, d** are variables for unsigned numbers. Factor out $(-1)^n$, such that we remove **(-1)** that assigns a negative sign to each unsigned variable **a, b, d**. This gives $(-1)^n[d^n = a^n + b^n]$, which entails the unsigned number proof.

Case Two: To show that $(-d)^n = (+a)^n + (+b)^n$, entails $d^n = a^n + b^n$. For odd **n>2** we have a negative number equal to a positive number, which does not subsist for Real numbers. And if **n>2** is even then $(-d)^n = (+d)^n$, and substituting $(+d)^n$ for $(-d)^n$, factor out $(+1)^n$ such that we have $(+1)^n[d^n = a^n + b^n]$, which entails the unsigned number proof. Therefore etc. Furthermore, by multiplying $(-1)^n$ times $(+d)^n = (-a)^n + (-b)^n$ we convert it to the case
$(-d)^n = (+a)^n + (+b)^n$, which is done.

Case Three: To show that $(-d)^n = (-a)^n + (+b)^n$, entails $d^n = a^n + b^n$, when **n>2** is even. Factor out $(-1)^n$ and we have $(-1)^n[(+d)^n = (+a)^n + (-b)^n]$, if **n>2** is even then $(-b)^n = (+b)^n$, Substitute $(-b)^n$ with $(+b)^n$ we have $(-1)^n[(+d)^n =$

$(+a)^n + (+b)^n]$, factor out $(+1)^n$ and we have $(-1)^n(+1)^n[d^n = a^n + b^n]$, which entails the unsigned number proof.

If $n>2$ is odd, given $[(-d)^n = (-a)^n + (+b)^n]$, multiply through by $(-1)^n$, which gives $[(+d)^n = (+a)^n + (-b)^n]$, then note by the rules of operators on signs, when n is odd, adding $(-b)^n$ is equivalent to subtracting $(+b)^n$. thus we can have $[(+d)^n = (+a)^n - (+b)^n]$, or $[(+a)^n = (+d)^n + (+b)^n]$, and factoring out $(+1)^n$ we get $(+1)^n[a^n=d^n+b^n]$, which in form entails an unsigned number proof, etc., but this form also shows a lesser equal to a greater, contrary to the construction constraints, i.e., contrary to hypothesis.

All the combinations and permutations of $\{+,-\}$ are hereby applied to FLT and entail an unsigned number proof or generate an absurdity.

Therefore, FLT holds for signed rational numbers,

Q.E.D. Fermat's Last Theorem with Porism extension to signed rational numbers.

THEOREM ADDENDA
Geometric Lemma for FLT

Geometric Lemma: The point of concurrency for altitude lines of an Acute angled triangle is inside the triangle, for an Obtuse angled triangle is outside the triangle, for a Right angled triangle is on the triangle.

Two Porisms are required for the Geometric Lemma; though Euclid did not show them they associate with Book IV:5 (to circumscribe triangles).

Porism 1: The perpendiculars to the medians of the triangle sides are concurrent.

1. Any triangle can be circumscribed by a circle. Euclid IV:5.
2. Since by definition the vertices of these triangles touch the circumscribing circle, the edges of the triangle connecting these vertices are therefore secants to the circumscribing circle, by Definition of Secant.
3. Since by Euclid III:1 Porism, the perpendicular bisectors of secants are collinear with the diameters of their circle.
4. Therefore from the definition of diameter they must all pass through the center of that circle.
4. Therefore the perpendicular bisectors of these secants are concurrent at the center of the circle circumscribing the triangle.
5. Therefore perpendiculars to the medians of the triangle sides are concurrent. Q.E.F.

Porism 2: The altitude lines of a triangle are concurrent.
1. We take any triangle, ABD (see exhibit below) and construct through each angle vertex a line parallel to the side subtending that vertex angle. Euclid I:31. That is construct: B'D' ∥ DB, A'B' ∥ BA, A'D' ∥ DA.
2. The sides of triangle ABD are transversals to these constructed lines and by Euclid Postulate 5 these lines meet on the side where the transversal angles sum to less than two right angles forming the larger triangle, A'B'D'.
3. Triangle A'B'D' has within it the original triangle, ABD, and three other triangles and an elementary set of observations leads to conclude that all four are congruent one to the other, as follows:
4. Using lines cutting parallels, each of the smaller triangles have corresponding angles that are all equal to each other (note matching numbers).

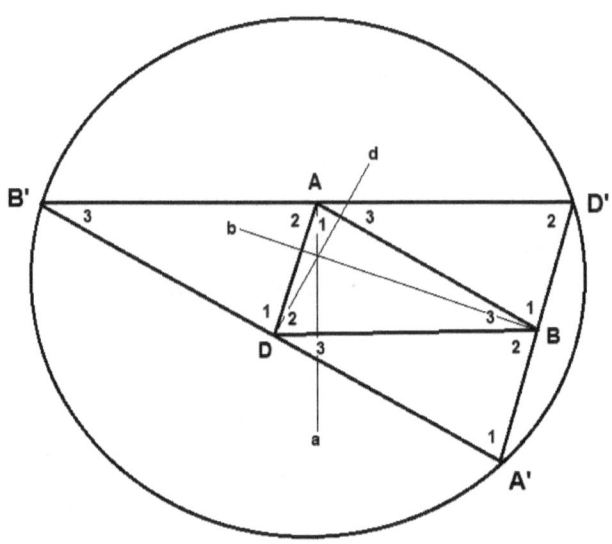

5. Furthermore, the parallelograms show that each of the four triangles has at least one corresponding equal side with an adjacent triangle (Euclid Book I: Common Notion 1). Therefore ASA (Euclid I:26) they are all congruent one to the other.
6. Circumscribing triangle A'B'D' (Euclid IV:5) we know from Porism 1 that the perpendiculars to its medians are concurrent at the center of this circle.
7. Furthermore, by definition, the sides of the larger triangle, A'B'D', are secants of this circle and by the congruences these secants are made up of a pair of equal corresponding sides from the smaller triangles, which show us that the vertices of triangle ABD are the median points of these secants (definition of median).
8. Because these median perpendiculars {Aa, Bb, Dd} pass through the vertices of triangle ABD and from parallel transversals (Euclid I:29) they are perpendicular to the subtended sides of triangle ABD, the median perpendi-

culars of triangle A'B'D' are also the altitude lines of triangle ABD.
9. Therefore, the altitude lines {Aa, Bb, Dd} of the triangle ABD are also concurrent at the center of the circumscribing circle of triangle A'B'D'.
10. Therefore, the altitude lines of triangles are concurrent.
Q.E.F.

Note: Though the diagram is biased toward the triangle used for FLT, this proof remains general because it does not depend on the triangle type used. Any of the three types can be used. In Euclid IV:5 he uses all three types, etc.

Geometric Lemma: The point of concurrency for altitude lines of an Acute angled triangle is inside the triangle, for an Obtuse angled triangle is outside the triangle, for a Right angled triangle is on the triangle.

Euclid did not consider this problem in his *Elements* nor in any of his extant works though it is provable using his *Elements*. Books I, III, and IV. The following proof shows one of many possible strategies

We have three general types of Euclidian triangles distinguished by their array of interior angles; one has one Obtuse angle, one has one Right angle, and one has all Acute angles.
   1. For an obtuse angled triangle there are at least two acute angles else the sum of its interior angles would exceed two right angles, but the sum of the interior angles is equal to two right angles (Euclid I:32). Therefore, there are no right angles in an Obtuse angled triangle. Therefore, none of the sides can be an altitude (by Definition of altitude), therefore the altitudes dropped from the vertices of the acute angles must fall inside or outside of the triangle. If they fall inside the triangle each altitude forms two interior

triangles and the dropped altitude subtends the obtuse angle and therefore neither of the remaining angles can be a right angle or they violate Euclid I:32 and therefore the so-called altitude cannot be perpendicular to the side it was dropped to. Therefore the altitudes from the acute angles both fall outside the Obtuse angled triangle, therefore the concurrent altitude lines (Porism 2) are also outside of this triangle type, etc. It is also true that a triangle is an obtuse angled triangle if and only if it has at least two altitude lines falling outside the triangle. For if one of its perpendiculars was a side of the triangle, by Definition of perpendiculars, a right angle would require another side perpendicular to it, therefore it becomes by Definition of right triangle, a right triangle. And if only one of its perpendiculars fell outside the triangle it would have two perpendiculars that are within the triangle. But this implies that one of its altitudes from an acute angle falls inside the Obtuse angled triangle, which has already been shown impossible. therefore, etc.

2. For a right angled triangle, two sides form the right angle and both are therefore altitudes to the triangle (by Definition of altitudes). Therefore, from the vertex of the right angle we can construct a third perpendicular to the subtending side (Euclid I:12). Thus the point of concurrency (Porism 2) of the altitudes is at the vertex of the right angle and therefore on the triangle. It is also true that if and only if there are two sides that are altitudes that we have a right angled triangle, For if not, we would have a right triangle with no sides perpendicular to one another, which nullifies its definition as a right triangle.

3. Therefore I say that in acute angled triangles all of the altitudes are within the triangle and the point of concurrency (Porism 2) also falls within this form of triangle. For if not it would require having an obtuse angle or a right angle, which it does not have, therefore etc.,

Lemma Porism: In an acute angled triangle the altitudes cut the subtending sides into two non-zero parts. For if not, the altitude would cut at an end of the subtending side or outside the subtending side.

It cannot cut outside the subtending side because this only occurs in an obtuse angled triangle. It cannot cut at the end of the subtending side because this only occurs in a right angled triangle, therefore, etc., Q.E.F.

END of FLT General Proof Addenda.

Commentary: In Fermat's time, signed numbers were not in his Universe of Discourse. Doing the subtraction (3 - 9), you obtain that you still have 6 left to subtract and this does not necessarily require the existence of negative numerals. The minus operator is distinct from the notion of negative numerals, etc., as Whitehead believed should be the case.

A hundred years later (18th Century), Euler was one of the most creative mathematicians to promote the use of negative numbers and he used debt as the model for obtaining a practical application for negative numbers. With the use of negative numbers, infinite descent becomes a possibility. His interest in generalization made him also seek to accredit the use of imaginary numbers, a number form that generated from even roots, **n>1**, of negative numbers. The questions and arguments for the meaningfulness of negative numbers continued on into the 19th Century despite Euler's arguments in their favor.

Euler algebraic Proofs for FLT n=3 and n=4. Finite descent that sustains homomorphic algebraic forms is represented in Euler's *Elements of Algebra*, Petersburg 1770, in his first proof for **n=3** using imaginary numbers (corrected by Lagrange for the failure of unique factorization in imaginary numbers) and **n=4** (using Rational numbers), Euler avoids the possible defective usage of infinite descent by initially establishing a lower boundary of cases that use small numbers, in a manner of speaking, a controlled descent. The Euler proof for **n=4** clearly shows his use of a controlled descent for a *reductio ad absurdum* proof.

> **Note:** Leonhard Euler (1707—1783) was the first to present a near proof that no rational volume can be partitioned into two rational volumes (a defective proof repaired by Lagrange (1736—1813)). He also found a second proof based on Real numbers. This is the first special case of Fermat's Last Theorem [Pierre de Fermat (1601-1665)]. Euler was also successful with the fourth power.

Euler's second proof for **n=3** is based on Rational numbers. If we have a set of positive Rational numbers $\{x,y,z\}$ such that $x^3 + y^3 = z^3$

we can multiply through by whatever denominators there may be and divide out their GCD leaving integers expressible as $a^3 + b^3 = d^3$. Thus it is sufficient to prove the case for integers to cover the Rational numbers.

Others discovered algebraic proofs for **n=5** and **n=7**, etc. None of these efforts led to a generalized algebraic proof.

Kummer later invented classes of numbers that he designated as 'ideals', which included Rational numbers and certain complex numbers. These 'ideals' classes sustain unique factorization. With this realm of ideals he is credited with showing the validity of Fermat's Last Theorem for all odd prime power cases for **n < 100** as well as many other special cases.

Certainly the Andrew Wiles and R. Taylor proof (1994), using higher mathematics, is not expressible in terms familiar to Fermat. In its first presentation by Wiles a fatal error was discovered. Wiles joined with Taylor to correct this error and in its present form, requiring 129 pages, no one has yet discovered another fatal error. As long as none is found it remains the first presentation that still holds a claim to a general proof. Mathematicians still seek *algebraic* proofs that do not use such artificial constructs (e.g., 'ideals', etc.) that Fermat would not have used.

END Addendum 7

APPENDIX A: Number References for Calculations

Earth radius = 6,407 km.

Moon radius = 1,740 km.

Earth-Moon distance = 384,403 km.

The Earth volume is 50 x the volume of the Moon.

The ratio of Earth bulk to Moon bulk is 81/1.

Earth-bulk/Moon-bulk/Earth-volume: Moon-volume is 81:1/50:1.

81/50 = 1.62.

The g for the Moon is 1.62 m/s^2.

The g on the Earth surface at sea level is:

    at the North Pole   f(g) = 9.83210 m/s^2
    at latitude 45 degrees f(g) = 9.80620 m/s^2
    at the equator   f(g) = 9.78038 m/s^2

For approximate calculations we use: f(g) = 9.81 m/s^2.

BIBLIOGRAPHY (Selected and Group-indexed by major use of subject matter). All references have dates and editions stated immediately after the author(s) names.

Contents by Subject Headings:
----
1. Background Literature Review
2. General Systems
3. Physical Cosmology
4. General Relativity
5. History of Science
6. General References
----

1. BACKGROUND LITERATURE REVIEW:

Archimedes. c200, *The Works of Archimedes including the Method*. Translated by Sir Thomas L. Heath, reprint by Encyclopedia Britannica, Inc., Great Books of the Western World, ed. 1952, vol. 11, ed. 1952.

Avduyevsky, V. S., editor, 1985, *Manufacturing in Space: Processing Problems and Advances*, MIR Publishers, Moscow.

Bakouline, Kononovitch, Moroz, 1981, *astronomie generale*, 3rd edition, editions MIR, Moscow. (Traduit du russe par V. Polonski.)

Bate, Mueller, & White. 1971, *Fundamentals of Astrodynamics*. Dover Publications, Inc., New York, reprint.

Berkson, William. 1974, *Fields of Force*. Routledge & Kegan Paul, London.

Boltzmann, Ludwig. 1896 (Part I), 1898 (Part II), Dover Publications, Inc., New York, reprint.

V.B. Braginsky and V.I. Panov, Zh. Eksp. and Teor. Fiz. 61, 873 (1971) [transl. in Sov. Phys. JETP 34, 463 (1972)].

Burke, John G. 1986, *Cosmic Debris*. University of California Press.

Cadogan, Peter. 1981, *The Moon--Our Sister Planet*. Cambridge University Press.

Cagniard, L. 1960, *Introduction a la physique du globe*. Editions Technip, Paris.

Chapman & Morrison. 1989, *Cosmic Catastrophes*. Plenum Press, New York.

Cohen, I. Bernard. 1960, *The Birth of a New Physics*. Doubleday & Company, Inc., New York.

Cooper, Lane. (1972, c1935), *Aristotle, Galileo, and the Tower of Pisa*, Kennikat Press, NY.

R. H. Dicke. 1960, "Eötvos Experiment and the Gravitational Red Shift," 28 (4), 344-347.

----------------. 1963, "Cosmology, Mach's Principle and Relativity," 31 (7), 500- 509.

----------------. 1967, "Gravitation and Cosmic Physics," 35 (7), 559-566.

----------------. 1981, "Interaction-free quantum measurements. A paradox?" 49 (10), 925-930.

Drake, Stilman. 1970, *Galileo Studies*, The University of Michigan Press, Ann Arbor.
------------------. 1989, *History of Free Fall*: Aristotle to Galileo. Wall & Thompson, Toronto.
Doligez, Marcel. 1965, *Gravitation*. Albert Blanchard, Paris.
Emiliani, Cesare. 1992, *Planet Earth*. Cambridge University Press.
Faraday, Michael. 1839-55, *Experimental Researches in Electricity*, Encyclopedia Brittanica, Inc., Great Books of the Western World, vol. 45, ed. 1952.
Feynman, Richard P. 1964, with Leighton, Robert B. & Sands, Matthew, *The Feynman Lectures on Physics*. 3 vols., Addison-Wesley, Reading, MA.
------------------. 1965, *The Character of Physical Law*. The MIT Press, 1992 reprint.
------------------. 1985, *QED*. Princeton University Press, New Jersey.
Galileo. 1638, *Dialogues Concerning Two New Sciences*. Translated by Henry Crew & Alfonso de Salvio, Encyclopedia Britannica, Inc., Great Books of the Western World, vol. 28, and ed. 1952.
Geymonat, Ludovico. 1957, *Galilee*. Editions Complexe, Paris.
Gilbert, William. 1600, *On the Loadstone and Magnetic Systems*. translated by P. Fleury Mottelay, reprint by Encyclopedia Britannica, Inc., Great Books of the Western World, vol. 28, ed. 1952.
Hapgood, Charles. 1962, *Les mouvements de l'ecorce terrestre*. Payot, Paris, Foreword by Albert Einstein.
Heath, Sir Thomas L. 1932, *Greek Astronomy*. Dover Publications, Inc., New York.
Hey, Max H. 1966, *Catalogue of Meteorites*. Trustees of the British Museum, London.
Hilbert, David, & Ackermann, A. 1937, *Principles of Mathematical Logic*. Chelsea Publishing Co., New York.
Hoyle, Fred. 1950, *The Nature of the Universe*. Harper & Brothers, New York.
Huygens, Christian. 1690, *Treatise on Light*. Trans. by Silvanus P. Thompson, reprinted by Encyclopedia Britannica, Inc., Great Books of the Western World, vol. 34, ed. 1952.
Kitalgorodsky, A. 1981, 2nd ed., *Introduction to Physics*. MIR Publishers, Moscow.

Kronk, Gary W. 1984, *Comets, a Descriptive Catalog*. Enslow Publishers, New Jersey.

Lagrange, J.-L. 1788, *Mecanique analytique*. Albert Blanchard, Paris, 1965, two volumes, reprint with new notes.

Le Danois, Ed. 1938, *L'atlantique*. Albin Michel, Paris.

Mach, Ernst. 1912, 7th ed., *The Science of Mechanics*. The Open Court Publishing Company, LaSalle, 1960 reprint translation.

March, Robert H. 1978, *Physics for Poets*. Contemporary Books, Inc., Chicago.

Marchis, L. 1898, *Les modifications permanentes du verre et le deplacement du zero des thermometres*. Librairie Scientifique A. Herman, Paris.

Maxwell, James Clerk. 1891, 3rd ed., *A Treatise on Electricity and Magnetism*. Dover Publications, Inc., New York, 1954 reprint in two volumes.

Motz & Weaver. 1989, *The Story of Physics*. Avon Books, New York.

Newton, Sir Isaac. 1713, 2nd ed., *Mathematical Principles of Natural Philosophy*. Translated by Andrew Motte, revised by Florian Cajori, University of California Press, 1934, reprint by Encyclopedia Britannica, Inc., Great Books of the Western World, vol. 34, ed. 1952.

------------------. 1717, 2nd ed., *Optics*. Reprinted by Encyclopedia Britannica, Inc., Great Books of the Western World, vol. 34, ed. 1952.

Poincare, Lucien. 1906, *La physique moderne*, son evolution. Ernest Flammarion, Paris.

Rosen, Stephen. 1979, *Weathering*. M. Evans and Company, Inc., New York.

Strelkov, S. P. 1978, *Mechanics*. MIR Publishers, Moscow.

2. GENERAL SYSTEMS

Barrow, John D. 1991, *Theories of Everything*. Fawcett Columbine, New York.

Bateson, Gregory. 1991, *A Sacred Unity, Further Steps to an Ecology of Mind*. HarperCollinsPublishers, New York.

Bertalanffy, Ludwig von. 1950, "An Outline of General System Theory." British Journal for the Philosophy of Science, No. 1, pp. 134-165.

------------------------. 1950, *Problems of Life*. New York.

----------------------. 1951, "Problems of General System Theory." Human Biology, No. 23, pp. 302-311.
Bertalanffy, Ludwig von, and Rapoport, A., eds. 1956-, *General Systems Yearbook*. Published by ISSS yearly since 1956.
Buchanon, Scott, 1972, *Truth in the Sciences*, University Press of Virginia.
Canguilhem, Georges. 1977, *Ideologie et rationalite*. Vrin, Paris.
Compte, Auguste. 1852, *Catechisme positiviste*. Garnier-Flammarion, Paris, 1966 reprint.
Davies & Gribbon. 1992, *The Matter Myth*. Simon & Schuster.
De May, Marc. 1982, 1992, *The Cognitive Paradigm*. The University of Chicago Press, 1992 with new Introduction.
Ducrot, Oswald. 1968, *Le structuralisme en linguistique*. Editions du Seuil, Paris.
Dyson, Freeman. 1992, *From Eros to Gaia*. Pantheon Books, New York.
Foucault, Michel. 1966, *Les mots et les choses*. Editions Gallimard, Paris.
Fuller, R. Buckminster. 1975, *Synergetics*. MacMillan Publishing Co., Inc., New York.
Geunon, Rene. 1945, *The Reign of Quantity and The Signs of the Times*. Penguin Books, London, 1972 trans. by Lord Northbourne.
Jones, Roger S. 1983, *Physics as Metaphor*. New American Library, New York.
Jung, C. G. 1971, *Psychological Types*. Princeton University Press, New Jersey.
Korzybski, Alfred. 1950, 2nd ed., *Manhood of Humanity*. Institute of General Semantics.
------------------. 1951, *Le role du langage dans les processus perceptuals*. Institute of General Semantics.
------------------. 1964, 2nd ed., *General Semantics Seminar 1937*. Institute of General Semantics.
------------------. 1979, *Time-Binding: The General Theory*. Institute of General Semantics.
------------------. 1980, 4th ed., *Science and Sanity: An Introduction to Non-Aristotelian Systems & General Semantics*. Preface by Russell Meyers, Institute of General Semantics.

Merleau-Ponty, Jacques. 1974, *Lecons sur la genese des theories physiques*. Vrin, Paris.
Merleau-Ponty, M. 1945, *Phenomenologie de la perception*. Editions Gallimard, Paris.
----------------. 1964, *Le visible et l'Invisible*. Editions Gallimard, Paris.
----------------. 1966, *Sens et non-sens*. Nagel, Paris.
Miller & Orgel. 1974, *The Origins of Life on the Earth*. Prentice-Hall, Inc., New Jersey.
Penrose, Roger. 1989, *The Emperor's New Mind*. Penguin Books, New York.
Poincare, Henri. 1906, *La science et l'hypothese*. Ernest Flammarion, Paris.
----------------. 1908, *Science et methode*. Ernest Flammarion, Paris.
----------------. 1920, *La valeur de la science*. Ernest Flammarion, Paris.
Popper, Karl R. 1980, *The Logic of Scientific Discovery*. Hutchinson, London.
Ramanan, K. Venkata. 1966, *Nagarjuna's Philosophy*. Charles E. Tuttle Company, Rutland.
Sapir, Edward. 1921, *Language*. Harcourt Brace & World, New York.
Strahler, Arthur N. 1992, *Understanding Science*. Prometheus Books, Buffalo, New York.
Toffler, Alvin. 1970, *Future Shock*. Pan Books, London.
Whitehead, Alfred North. 1925, *Science and the Modern World*. The Free Press, New York, reprint.
Whorf, Benjamin Lee. 1956, *Language, Thought, & Reality*. The MIT Press.

3. PHYSICAL COSMOLOGY
Lightman & Brawer. 1990, *Origins, the Lives of Modern Cosmologists*. Harvard University Press.
Peebles, P. J. E. 1993, *Principles of Physical Cosmology*. Princeton University Press.

4. GENERAL RELATIVITY
Ciufolini, Ignazio and John Archibald Wheeler. 1995, *Gravitation and Inertia*, Princeton Series in Physics.
Einstein, Albert, et al. 1923, *The Principle of Relativity*. Dover Publications, Inc., New York, 1952 reprint.

Einstein & Infeld. 1938, *The Evolution of Physics*. Cambridge University Press.
Einstein, Albert. 1956, *Investigations on the Theory of the Brownian Movement*. Dover Publications, Inc., New York.
----------------. 1956, *Out of My Later Years*. Bonanza Books, New York.
----------------. 1956, *The Meaning of Relativity*. 5th ed., Princeton University Press.
Lerner, Eric J. 1992, *The Big Bang Never Happened*. Vintage Books, New York.
Peebles, P. J. E. 1993, *op.cit.*
Robinson, Enders A. 1990, *Einstein's Relativity in Metaphor and Mathematics*. Prentice Hall, New Jersey.
Will, Clifford M. 1986, *Was Einstein Right?*. Basic Books Inc., New York.
Will, Clifford M. 1990, "General Relativity at 75: How Right was Einstein?." Science, Vol. 250, No. 4982, November 9, pp. 770-776.

5. HISTORY OF SCIENCE
Cohen, I. Bernard. 1985, *Revolution in Science*. Harvard University Press.
Koyre, Alexandre. 1955, *A Documentary History of the Problem of Fall from Kepler to Newton; de motu gravius naturaliter cadentius in hypothesi terrae motae*, American Philosophical Society, Philadelphia.
Koyre, Alexandre. 1962, *Du monde clos a l'univers infini*. Gallimard, Paris.
Kuhn, Thomas S. 1970, 2nd ed., *The Structure of Scientific Revolutions*. The University of Chicago Press.
--------------. 1977, *The Essential Tension*. The University of Chicago Press.
Kneale, W. & M. 1968, *The Development of Logic*. Oxford University Press.
Schweber, Silvan S. 1994, *QED and the Men Who Made It: Dyson, Feynman, Schwinger, and Tomonaga*. Princeton University Press.

Swimme, Brian and Thomas Berry. 1992, *The Universe Story: from the primordial flaring forth to the ecozoic era--a celebration of the unfolding of the cosmos*. Harper, San Francisco, CA.

6. GENERAL REFERENCES:

De Broglie. 1937, *La physique nouvelle et les quanta*. Ernest Flammarion, Paris.

Durell, C. V. 1955, *Projective Geometry*, MacMillan and Company Ltd, London.

Euler, Leonard. 1748, *Introduction to Analysis of the Infinite*, Springer-Verlag, trans. into English by John D. Blanton, 1988.

--------------. 1770, *Elements of Algebra*, Springer-Verlag, trans. into English by Rev. John Hewlett (1840), reprint of London: Longman, Orme, and Co.

Heisenberg, Werner. 1930, *The Physical Principles of the Quantum Theory*. Dover Publications, Inc., trans. into English by Eckart and Hoyt, reprint of University of Chicago Press edition.

Klimov, A. 1975, *Nuclear Physics and Nuclear Reactors*. MIR Publishers, Moscow.

Vonsovsky, S.V. 1973, *Magnetism of Elementary Particles*. MIR Publishers, Moscow, 1975 from the Russian edition.

Arnold, V.I. 1963, "Small Denominators and Problems of Stability of Motion in Classical and Celestial Mechanics." Russ. Math. Surv., No. 18, p. 85.

Berge, Pomeau, & Vidal. 1984, *Order within Chaos*. John Wiley & Sons, New York, trans. from French by Laurette Tuckerman.

Bohm, David. 1980, *Wholeness and the Implicate Order*. Ark Paperbacks, London.

Bridgman, P.W. 1936, *The Nature of Physical Theory*. Princeton University Press.

--------------. 1959, *The Way Things Are*. The Viking Press, New York, 1961 reprint.

Capra, Fritjof. 1976, *The Tao of Physics*. Bantam Books, New York, 1980 reprint.

Coveney, Peter V. 1988, "The Second Law of Thermodynamics: Entropy, Irreversibility, and Dynamics." Nature, No. 333, pp. 409-415.

Coveney & Highfield. 1990, *The Arrow of Time*. Fawcett and Columbine, New York.

Feigenbaum, Mitchell J. 1978, "Quantitative Universality for a Class of Nonlinear Transformations." Jour. Statistical Phys., No. 19, p. 25.
----------------------. 1980, "Universal Behavior in Nonlinear Systems." Los Alamos Science, No. 1 (Summer), pp. 4-27.
Hayles, N. Katherine. 1990, *Chaos Bound*. Cornell University Press.
Hirsch, J.E., B.A. Huberman, D.J. Scalapino. 1982, "Theory of Intermittency." Phys. Rev., No. 25, p. 519.
Hofstadter, Douglas R. 1979, *Gödel, Escher, Bach: Eternal Golden Braid*. Vintage Books, New York.
Kolmogorov, A.N. 1979, "Preservation of Conditionally Periodic Movements with Small Change in the Hamiltonian Function." Lecture Notes in Phys., No. 93, p. 51.
Lauwerier, Hans. 1991, *Fractals*. Princeton University Press, New Jersey.
Le Mehaute, Alain. 1990, *Fractal Geometries*. CRC Press Inc., Boca Raton.
Lorenz, Edward. 1963, "Deterministic Nonperiodic Flow." Journal of the Atmospheric Sciences, No. 20, pp. 130-141.
Mandelbrot, Benoit B. 1983, *The Fractal Geometry of Nature*. W.H. Freeman and Company, New York.
May, Robert M. 1976, "Simple Mathematical Models with Very Complicated Dynamics." Nature, No. 26, p. 459.
Meeker, Joseph. 1975, *The Spheres of Life: an introduction to world ecology*. Scribner, NY.
Mills, Robert. 1994, *Space, Time, and Quanta*. W.H.Freeman & Co., New York.
Penrose, Roger. 1989, The Emperor's New Mind. Penguin Books, New York.
Poincare, J. Henri. 1890, "Sur le probleme des trois corps et les equations de la dynamique." Acta Mathematica, No. 13, pp. 1-270.
Prigogine & Stengers. 1984, *Order out of Chaos: Man's New Dialogue with Nature*. Bantam Books, New York.
Ruelle, David, and Floris Takens. 1971, "On the Nature of Turbulence." Commun. Math. Phys., No. 20, p. 167.
Schroeder, Manfred. 1991, *Fractals, Chaos, Power Laws*. W.H. Freeman and Company, New York.

Shannon, Claude E. 1948, "A Mathematical Theory of Information." Bell System Technical Journal, No. 27 (July and October), pp. 379-423, 623-656.

------------------. 1951, "Prediction and Entropy of Printed English." Bell System Technical Journal, No. 30, pp. 50-64.

Shannon, Claude E. and Warren Weaver. 1949, *The Mathematical Theory of Communication*. University of Illinois Press, Urbana.

Thompson, D'Arcy Wentworth. 1942, *On Growth and Form*. Dover Publications, Inc., New York, reprint of Cambridge University Press edition.

Tritton, David. 1986, "Chaos in the Swing of a Pendulum." New Scientist, No. 24 (July), pp 37-40.

Walker, Grayson H. and Jason Ford. 1974, "Amplitude Instability and Ergodic Behavior for Conservative Nonlinear Systems." Phys. Rev., No. 188, p. 87.

END of Bibliography

END of Book II

TABLE OF CONTENTS

BOOK I: Convergence Buoyancy Theory

Apology 9
Prologue: Book Outline 13

PART ONE: Science-Systems Analysis

Chapter 1: Convergence and the Illusion of Free-Fall 17

PART TWO: Science-Systems Design

Chapter 2: CB Inversion Spheres and the Illusion of Mass ... 79

Chapter 3: Impetus Weight and the Illusion of Inertia 145

Chapter 4: Radiation and Heat 197

Chapter 5: The Third Perpendicular 235

PART THREE: Science-Systems Implementation

Chapter 6: A Different Interface for Our Universe 291

Chapter 7: Dialogue 347

Epilogue 369

Bibliography (in Book II: *Star Laws Addenda: Duality Theory*)

...The unexamined life is not worth living.
--- Socrates

TABLE OF CONTENTS

BOOK II: Duality Theory 371

    Abstract 375

    Foreword 377

ADDENDA: Science-Systems Analysis

    Addendum 1: Science-Systems 385

    Addendum 2: Status of Logic in Science-Systems 419

    Addendum 3: Formal Duality in Finite Logics 449

    Addendum 4: Data-Field Duality 483

    Addendum 5: Reference-Frame Duality 507

    Addendum 6: Ptolemy, Copernicus, Kepler 543

    Addendum 7: Fermat's Last Theorem 569

Appendix: Number References for Calculations 587
Bibliography 589

    ...everything that comes into being comes into being from its contrary...
                               --- Aristotle, *On the Heavens*

End of *Star Laws: Foundations for Convergence Buoyancy*

www.ingramcontent.com/pod-product-compliance
Lightning Source LLC
Chambersburg PA
CBHW031659230426

43668CB00006B/52